MESO-OPTICS

Foundations and Applications

Cover:

Design of the meso-optical mirror with ring response (a half) which was fabricated in 1986 with diffraction limited resolution.

MESO-OPTICS

Foundations and Applications

L M Soroko

Joint Institute for Nulcear Research, Russia

World Scientific

Singapore • New Jersey • London • Hong Kong

Published by

World Scientific Publishing Co. Pte. Ltd.

P O Box 128, Farrer Road, Singapore 912805

USA office: Suite 1B, 1060 Main Street, River Edge, NJ 07661

UK office: 57 Shelton Street, Covent Garden, London WC2H 9HE

British Library Cataloguing-in-Publication Data
A catalogue record for this book is available from the British Library.

MESO-OPTICS — FOUNDATIONS AND APPLICATIONS

ISBN 981-02-2700-0

Printed in Singapore.

In memory
of my wife,
Valentina M. Soroko

Abstract

"ALMOST EVERYTHING ABOUT CONICAL WAVEFRONTS" is the motto of this book, which can be conceived as an up-to-date encyclopedic review of the theoretical and experimental investigations, performed by many researchers in the field of meso-optics. With many examples it is illustrated that meso-optics is indeed a rapidly developing branch of modern optics, in which the conical wavefronts are used, though sometimes implicitly, to transfer in space the information about the objects or the electromagnetic energy.

By starting from axicons, various meso-optical elements and new scopes of meso-optics is described. Diffraction free light beams and the localized electromagnetic wavefields is discussed. It is shown that gravitational lens can be accepted as a beamsplitter in the interferometer of the cosmic scale. Of particular interest will be inverse meso-optics, which can be used in electron accelerators for transportation of synchrotron radiation and in micromechanics. Some meso-optical microscopes and meso-optical metrological devices will be suggested. The history of meso-optics is followed from the first gravitational lenses in our universe up to the localized electromagnetic fields developed recently. Some integral transforms, useful in meso-optics, is presented in the Appendix.

PREFACE

The aim of this book is to acquaint the reader with up-to-date theoretical, experimental and instrumental contributions of many researchers to the new branch of modern optics which is referred to as "meso-optics".

The **theoretical** contribution follows from the new insight into the traditional optical imaging systems, such as lens, which focuses the light along two orthogonal directions. It has been discovered that imaging systems with axial symmetry, which focus the light only along one coordinate and not along another, orthogonal coordinate, can be used effectively for solving some awkward or even deadlock problems. In the course of more than ten years we have been using the term "meso-optics" to say that conical wavefronts do contribute principally into the systems for transferring in space the information about the object and also for transmitting the electromagnetic energy at long distances.

The **experimental** contribution covers the observations of various new phenomena with conical wavefronts, such as longitudinal interference of the diffraction free wavefronts, the decisive demonstration that information about the position of the microscopic straight-like objects in 3D space can be converted into two point-like spots with high data compression by means of sophisticated meso-optical imaging system, and the evident description of the exotic phenomenon, "darkness" in optics and meso-optics.

The **instrumental** contribution covers traditional and nontraditional techniques in optics, holography and tomography, which seems to be the most adequate to meso-optics. We have now diamond turning technique, computer generated hologram, precise plotter for manufacturing kinoform with prescribed phase relief and the technique for making "focusators". The first kinoform for visible light with diffraction limited resolution was fabricated in 1984. The first precise meso-optical mirror with ring response was made in 1986. Therefore any proposals presented in this book in the field of meso-optics, for visible light and X-ray radiation, are provided with adequate technology basis.

In addition to meso-optics, by means of which the object point is transformed into the line segment or circle, "inverse meso-optics", which enables one to convert into the point of coherent radiation generated by the source in the form of a straight line segment was also discovered. The direct application of inverse meso-optics for transportation of the X-ray synchrotron radiation and for manufacture of micromechanical parts of the devices is well evident.

What else can we benefit from meso-optics? First of all it must be stated, that the problem of the depth of focus in microscopy was solved by means of the meso-optical Fourier transform microscope, in which the information about position of the straight-like microscopic object has been compressed up to two point-like spots of the diffraction limited dimensions. The new device, confocal meso-optical scanning microscope is to be noticed. All meso-optical microscopes and tomographs, suggested in this book, have no counterparts in the conventional optical microscopy and tomography. These systems offer high productivity and can be applied adequately in microscopic robotics vision, for material testing in industry and in high energy particle physics.

This book is organized as follows. After introduction to meso-optics, the properties of conical wavefronts are explained in detail, and some applications of axicons are described. Meso-optical elements with ring response and their adequate application in the meso-optical Fourier transform microscope is also explained. The theory of this device and its features are presented along with description of some other designs of this device and application of the meso-optical Fourier transform microscope in high energy physics.

The diffraction free wavefields, Bessel–Gauss beams and various devices for generation of the diffraction free fields are described further. Following explanation of the exotic phenomenon "darkness" in optics and meso-optics, the longitudinal interference of the diffraction free wavefronts and its application in ophthalmology are presented. The problem of the localized electromagnetic waves is treated along with theoretical and experimental investigations in this rapidly developing part of meso-optics.

Gravitational lens, which is also a meso-optical element, accomplishes nonlinear transformation of the space coordinates and velocity components of the celestial objects. Typical examples of the gravitational lenses discovered recently and the property of gravitational lens as a beam-splitter is described. The fundamental properties of the critical curve (caustics) which separates the space into areas with different number of meso-optical images are considered with emphasis on the parallax gravitational effect in binary and planetary systems. The main topics of gravitational lens applications are presented.

Two chapters of this book are devoted to confocal meso-optical microscope–microtomographs, new stereo-microscope with high depth resolution and to some meso-optical metrological devices. Some potential new applications of meso-optics in science, industry and medicine is suggested throughout this book.

The history of meso-optics, presented in this book, starts from the early stage of our universe, to first gravitational lenses, followed by publications in the last century and finally up-to-date innovations, many of which were proposed and performed with active participation of the author. The Appendix contains some integral transforms, which are useful in meso-optics.

The author believes that after publication of this book the frontiers of meso-optics will be expanded and its development will be accelerated.

L. M. Soroko
Dubna, Russia

Acknowledgments

I wish to thank V. P. Dzhelepov, corresponding member of the USSR Academy of Sciences, and Director of the Laboratory of Nuclear Problems, Joint Institute for Nuclear Research (Dubna), for support of investigations in meso-optics. Gratitude is also extended to Professor S. A. Bunyatov, Chief of the project Neutrino Detector, and to Professor Yu. A. Batusov for their support and discussions. The development of the meso-optical Fourier transform microscope was accomplished as part of the USSR–Hungary scientific collaboration, and I am grateful to Professor D. Kiss (KFKI Budapest), Chief of the Hungarian scientific group. Warm appreciation is expressed to Gy. L. Bencze, A. Kishvaradi, L. Molnar, G. Nitrai and J. Molnar, as well as to V. V. Tereshchenko. I wish to acknowledge A. Ya. Astakhov, V. I. Prichod'ko, V. A. Drozdov, A. L. Men'shikov, A. V. Nikul'nikov from the Laboratory of Computing Techniques and Automation of JINR, V. P. Koronkevich and I. G. Paltchikova from the Institute of Automation and Electrometry in Novosibirsk for their collaboration.

I must thank Professor Ya. A. Smorodinsky from the Laboratory of Theoretical Physics, of JINR for many discussions, in particular on diffraction free waves, symplectic groups and gravitational lens.

Finally I wish to thank L. Gavrilenko and N. Tikhomirova for excellent typing, G. V. Gorbunova for high quality photocopies and V. I. Krivozoubova for her continuous support and patience.

General view of the Meso-Optical Fourier Transform Microscope (JINR, Dubna).

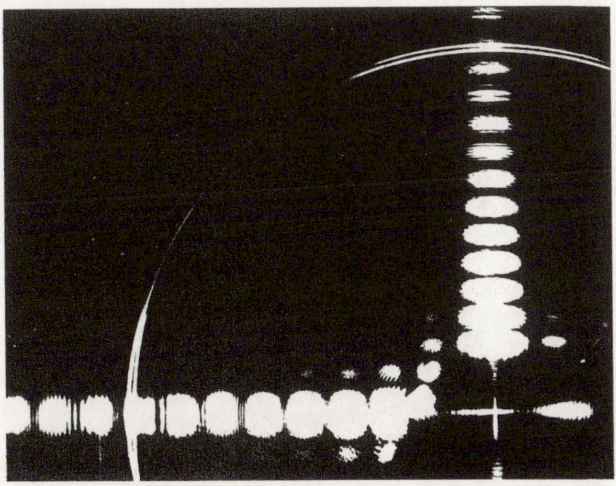

Meso-optical image of the open rectangular with symmetry axes oriented along x and y coordinate axes.

MESOOPTICS $\sqrt{i^i}$ MESOOPTICS

Symbol of Meso-Optics.

CONTENTS

Chapter 1

INTRODUCTION AND SURVEY

MESO–OPTICS is a branch of the modern optics in which the properties of the conical wavefields are treated. The main feature of meso–optics is that by means of the meso–optical elements we can accomplish nontrivial transformations of the space coordinates. For example, the straight line microscopic object can be transformed, with velocity of light, into two point-like images with diffraction limited dimensions. From this example and many others described in this monograph we may conclude that coherent conical wavefields are peculiar eigen-functions of some operator being defined over the class of the straight line objects. In the classical optics any straight line object must be subdivided into a corresponding set of the point-like elements or into a set of spherical waves. It is obvious that such traditional description of the straight line objects is more cumbersome than more productive one offered by the meso–optics.

The compression of the information about the straight line objects arises in meso–optics quite naturally, occurs in the real time and does not demand computer processing systems. In meso–optics we do not use subdivision stage of this information about the straight line object into a set of the point-like elements. The typical meso–optical element does not accomplish the scanning operations which are used in the traditional optics. For example, the estimation of the depth coordinate (z-coordinate) of the straight line segment is performed by the meso–optical element without depth scanning operation. Shortly, we have depth viewing without depth scanning.

Due to these properties, meso–optics offers high productivity, typical to the new meso–optical microscopes for selective observation of the straight line objects, and high selectivity in orientation and in dip angles of these objects.

Meso–optics suits well for precise control of the conical surfaces and also for noncontact measurements of the external surface profile of objects. It is worth mentioning that just in the frame of meso–optics that new stereoscopic microscope with extremely high resolution depth power and with large field of view have been conceived. These new stereo-microscopes suit well for modern microelectronics, micromechanics, microsurgery and microbiology.

Sometimes a fresh nontraditional approach or a technological innovation may seem to be as an unacceptable, strange and even defiant. Its real effectiveness and usefulness can be estimated only in time. However, in an attempt to accomplish such an estimation, we meet various difficulties caused by this new approach and induced by the unpopular way of thinking. Without doubt, MESO–OPTICS may be considered as this category of technological innovations. What is wrong for conventional optics, is proved well for meso–optics: disadvantages in optics are converted into the advantages in meso–optics.

To be convinced of this property inherent in meso–optics, it would be well to compare the principal tendencies of the traditional optics with new avenues offered by young meso–optics. Until now the efforts of opticians were oriented to suppress as much as possible the optical aberrations of the imaging systems. To achieve this aim, there have been published many books and enormous work has been done in the research and technological groups. Meanwhile to solve some deadlock problems in optics, it would be quite reasonable to accept radically opposite line of reasoning.

The problem of depth-of-focus in microscopy presents a typical example. To solve this age long problem in the frame of meso–optics we must introduce in the optical assembly very large, though carefully controlled, aberrations or a significant eccentricity. And it is on the basis of these unusual recommendations, there have been implemented two meso–optical devices, suggested in the last decade. We mean an optical tomographic microscope [1] and a meso–optical Fourier transform microscope [2]. The position of the exit pupil in the first device is extremely eccentric, and the size of the aberrations in the meso–optical mirror in the second one is maximal. In both cases we deal with conical wavefronts rather than with spherical ones. Due to this specific feature, some useful advantages over the traditional devices have been gained. This new approach is a typical one in meso–optics, it drastically differs from the traditional tendencies in optics.

There are some other situations, in which the conical wavefronts define very useful techniques, which lead to upper levels of science and technology. For example, axicon (conical lens) can be used in metrology and in X-ray lasers. Lenses with large longitudinal chromatic aberrations can be incorporated into the optical pick-up transducer for surface profile measurement. Sophisticated axiconic systems and confocal meso–optical microscopes suit well for fast control of the surface profile. Meso–optical autocollimators and multipath interferometers of the Farby–Perot type were suggested for superprecise measurements of the conical wavefronts. Meso–optical element with ring response has been used in the meso–optical Fourier transform microscope for high energy particle physics. The early development of meso–optics is described in the review article [3].

A lucid phenomenological description of the conical wavefronts is presented in Chapter 2. The properties of "wrong" lens, usually rejected as defective, are considered. These lenses were made in conditions with disrupted manufacturing technology, for example, when the optical axis of the lens did not coincide with rotating axis of the cutting or polishing tool. Such "wrong" lens produces conical wavefronts. Two kinds of the manufacturing errors lead to two types of axiconic lens: one with line response and another with ring response. The new term MESO–OPTICS and the general classification of the meso–optical elements are explained.

Next the properties of axicons and polarization states of the output radiation will be considered. There are outlined the main features of some meso–optical elements and systems: conical lens, conical mirror, circular diffracting grating, kinoform-axicons, systems of a lens plus an axicon, systems of two axicons, *W*-axicons and chromatic objective. The experiments with conoid axicons are described. The simple theory of the conical wavefronts is given at the end of the Chapter 2.

Chapter 3 of the book is devoted to the meso–optical elements with ring response (MERR). Various types of MERR are presented, and their properties are treated by ray tracing in the meridional and in the sagittal sections of the MERR. The combined system with point **and** ring response is suggested.

The principle of the meso–optical Fourier transform microscope (MFTM), which is based on the synthesis of Fourier optics and meso–optics, is explained. Some exotic properties of the MFTM are described. A general theory of the MFTM is given and the convolution kernel of the MFTM for straight line objects is derived. A meso–optical analogue of the classical moirè effect, observed in the MFTM, is described. A fast version of the MFTM is suggested. The design of the MFTM with double focusing is presented.

At the end of Chapter 3 the problem of coma aberrations in optics, holography and meso–optics is treated, and two isoplanatic meso–optical systems are suggested.

Diffraction free wavefields and Bessel–Gauss beams are described in Chapter 4. After the vector model of the Bessel beams the properties of the reflaxicon are explained. The phase singularities in the wavefields and "darkness" in optics and meso–optics are treated with some interferometric observation of these new phenomena. A specific interferometric picture referred to as "longitudinal interference of the conical wavefronts" is described. The confocal meso–optical keratometer and profilometers are explained as principally new designs.

Chapter 5 is devoted to localized electromagnetic waves which can be treated as a direct, but nontrivial, generalization of the diffraction free beams. It is stressed that any localized wave consists of Bessel-like components or of properly weighted conical wavefronts. Hence localized waves must be embedded in meso–optics. Three stages of development of the localized waves are explained in detail. A comparison of several types of the localized waves is made. The description of the experimental verifications of the effects predicted by theory is given. The localized waves as exact solutions to the Klein–Gordon equation are explained at the end of the Chapter 5 together with short overview of other approaches such as electromagnetic "missiles" and "bullets".

Gravitational lens as a meso–optical element is treated in Chapter 6. After presentation of the general theory of the point gravitational lens we consider peculiar nonlinear transformations of the space coordinates and of the velocity components which accomplishes gravitational lens and which cannot be observed in the empty space by means of the traditional optical elements. Some typical examples of the gravitational discovered recently in astronomy and in radio-astronomy are described. The property of the gravitational lens as a beam-splitter with time delays ranged from 10^{-12} s to 10^{8} s are explained. Then the gravitational microlensing and the first microlensed events observed in our Galactic are described. The fundamental properties of the critical curves (caustics) which separates the space into areas with different number of the meso–optical are presented with emphasis on the gravitational parallax effects in binary and planetary systems. There are described the applications of the gravitational lens phenomenon to the problem of dark halo objects, giant luminous arcs, large scale inhomogeneities of the universe and for independent estimation of the Hubble's parameter.

The properties of the inverse meso–optics and its potential applications are discussed in the Chapter 7. After the definition of the inverse meso–optics, the necessary conditions for performing of the inverse meso–optical transformations are proved. Two typical applications of the inverse meso–optics are explained in detail: meso–optical system for X-ray transportation, including self-focusing undulator, and meso–optical Cherenkov gas counter of the relativistic charge particles.

Chapter 8 is devoted to up-to-date problems of microscopy and microtomography. It will be shown that meso–optics offers some radical suggestions to solve these problems. After description of several traditional methods of light sectioning for observation of $3D$ objects or $2D$ surfaces in $3D$ space, a confocal scanning microscopes, conventional and meso–optical ones, are described. It is shown that idea of meso–optics can be used to construct principally new stereo-micro- and macro-scopes with very high depth resolution over large field of view.

Some meso–optical metrological devices for testing of the conical surfaces are described in Chapter 9: meso–optical conicometers, meso–optical interferometer, meso–optical profilometers, meso–optical keratometers, metrological system with high order diffraction pattern, meso–optical microscopes for vertical particle tracks.

Very long history of meso–optics, presented in Chapter 10, is started from early stage of our universe, first gravitational lenses, diastoloscope with two conical lenses, cylindrical waves, annular aperture, computer generated optical translator, kinoforms, focusators, meso–optical Fourier transform microscope, confocal meso–optical systems and other innovations in meso–optics. A history of the diffraction free waves and localized electromagnetic waves is analyzed. The stages of the development of the gravitational lens theory and the recent observations of the gravitational microlensing effect will be outlined.

In the Appendix of the book some integral transforms, which are useful in meso–optics, are explained. All presented transforms involve the objects in the form of the straight line segment or aperture boundaries of several straight line segments.

Chapter 2

AXICONS AND CONICAL WAVES

2.1. Conical Waves and MESO–OPTICS

A divergent spherical wave issued from a point-like source is reversed by a perfect lens into a convergent spherical wave which collapses into a point-like image

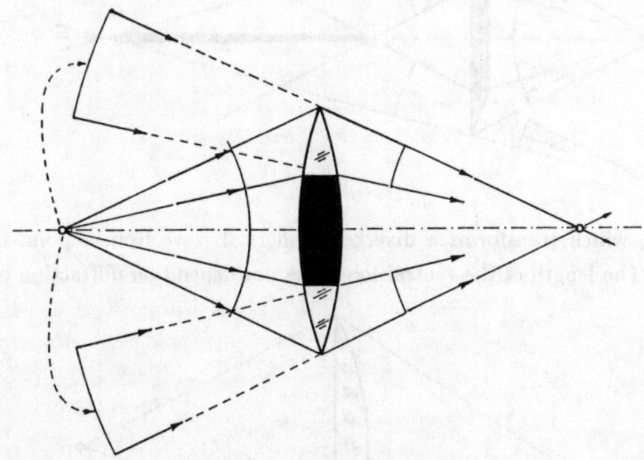

Fig. 2.1. A perfect lens, which transforms a divergent spherical wave from a point-like source into a convergent spherical wave, collapsed into a point-like image.

(Fig. 2.1). In the frame of the geometrical optics approximation this conventional optical imaging element accomplishes an one-to-one transformation

$$(0 - D) \rightleftharpoons (0 - D)\,, \tag{2.1}$$

where $(O - D)$ is zero-dimensional manifold. Any point in the object space goes into a corresponding point in the image space. Here we have a dualism between point and spherical wave. The information about a point of the object is transferred in space by the corresponding spherical wave. Usually any object in $3D$ space is thought as an array of the points. The wavefield produced by the illuminated object can be treated as an equivalent superposition of many spherical waves.

A radically different situation occurs in an axicon shown in Fig. 2.2 in the form of a conical lens. A point-like object O on the axis symmetry is "imaged" by the conical lens into a long straight line segment. The divergent spherical wave is converted by an axicon into a conical wave, which never collapses into a focused point-like real image of the input point. Thus, an axicon is not an imaging element. The transformation, which accomplishes an axicon, is a **homomorphic (irreversible)** one: each point as $0D$ object goes into $1D$ array of points [4]. For conical waves there is no one-to-one correspondence between object space and image space. This feature of the axicon has been regarded as an impenetrable barrier for using axicons as imaging element. Any admixture of conical waves to the spherical ones deteriorates the quality of the image.

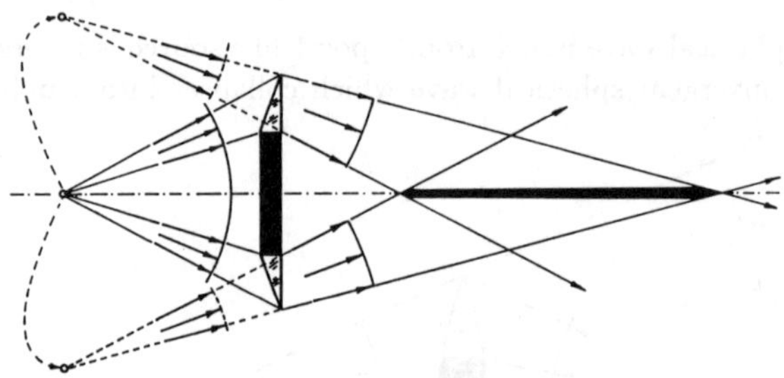

Fig. 2.2. An axicon, which transforms a divergent spherical wave from a point-like source into sphero-conical wave. The length of the central lobe does not depend on diffraction of light.

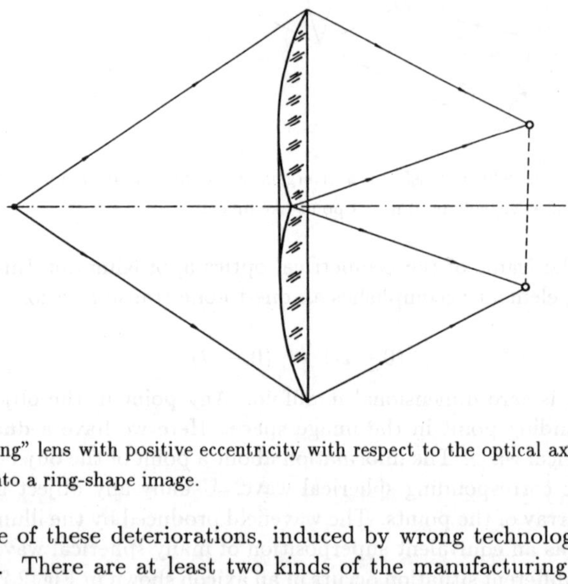

Fig. 2.3. A "wrong" lens with positive eccentricity with respect to the optical axis. A point source is transformed into a ring-shape image.

One source of these deteriorations, induced by wrong technology, will be considered below. There are at least two kinds of the manufacturing errors. In the first one the optical axis of the lens does not coincide with rotating axis of cutting or polishing tool and the eccentricity is positive (Fig. 2.3). The "image" of the point is a circle [5]. The manufacturing error of the second kind is generated in the system with negative eccentricity (Fig. 2.4). The "image" of the point consists of two components: line segment on the axis and circle in the focal plane of the lens, the first component being nearer to the "wrong" lens than the focal plane with circle. Thus, there occur in general case two types of "wrong" lenses, one with ring response and one with line and ring response. These "wrong" lenses are not only of academical interest. They can be used, for instance, in some metrological devices.

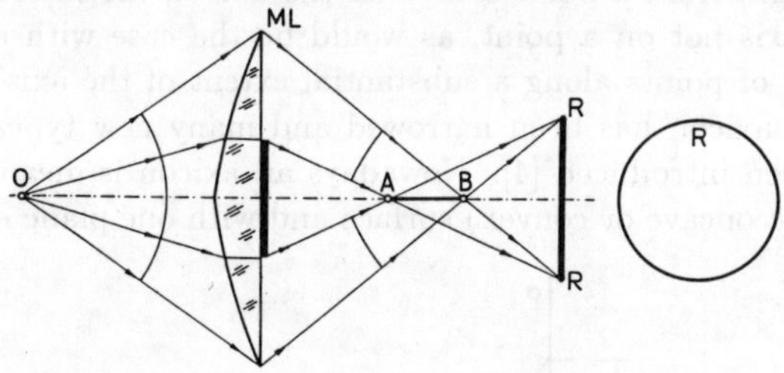

Fig. 2.4. A "wrong" lens with negative eccentricity. A point source is transformed into a line image AB and a ring-shape image RR.

The term MESO–OPTICS has been introduced [4] to describe nontraditional optical elements and devices, in which a point is converted into $1D$ manifold, such as straight line segment, circle or more complicated lines. The term "meso-optics" covers axicons, computer generated elements, kinoforms and circular diffraction grating, which convert the incident wave into a line with specified intensity distribution, as well as various more sophisticated elements and devices.

The geometrical transformations performed by the meso-optical element can be written as

$$\begin{aligned} &(0-D) \to (1-D), \quad (1-D) \to (2-D), \quad (2-D) \to (3-D), \\ &(0-D) \to (2-D), \quad (1-D) \to (3-D), \end{aligned} \tag{2.2}$$

For example, a point goes into a line segment, a line segment goes into a restricted $2D$ surface, and a surface as $2D$ manifold is converted into a restricted $3D$ manifold. In general case, a n-dimensional object goes into a $(n+1)$- or $(n+2)$-dimensional manifold [6].

There are meso-optical elements with linear longitudinal response, with linear transversal response, with linear screw response, with linear spiral response and many others.

There have been suggested meso-optical axial-non-symmetrical elements [7] and meso-optical elements which produce linear images on the lateral surface of a cylinder [8]. The term meso-optics covers nonspherical and nonplane waves: conical, cylindrical, helical [9] and more complex structure.

2.2. Axicons

The simplest axicon [10, 11] is a convergent (positive) conical lens, which converts a point on the symmetry axis into a straight line segment of this axis. In [10] an axicon is defined as "any figure of revolution that by reflection, or refraction, or both will bend light from a point source on the axis of the figure of revolution so as to cross the axis not on a point, as would be the case with a lens, but along a continuous line of points along a substantial extent of the axis". More recently this definition of axicon has been narrowed and many new types of meso-optical elements have been introduced [4]. Nowadays an axicon is meant as a solid cone with one conical (concave or convex) surface and with one plane external surface.

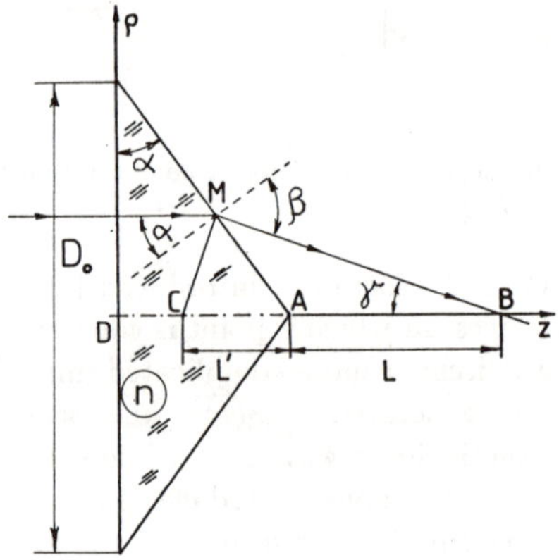

Fig. 2.5. Transformation of light by conical surface with apex in the point A.

Consider the light focusing in a simplest axicon, illuminated by collimated light beam [12]. Some rays are refracted at the conical surface and are bent towards the axis. The focal range of this axicon (Fig. 2.5) is extending from the apex of the cone A up to the point B on the axis. The length of this focal range is equal to

$$\left.\begin{aligned} L &= \tfrac{D_o}{2}\,(\tan\gamma - \tan\alpha)\,, \quad \text{with} \\ \sin\beta &= n \cdot \sin\alpha\,, \\ \gamma + \beta &= 90^\circ + \alpha\,, \end{aligned}\right\} \tag{2.3}$$

where D_o is the diameter of the axicon, α is the wedge angle of the cone and n is the refractive index of the cone material. Another rays are reflected at the same conical surface and produce inside the cone the internal meso-optical image of the length

$$L' = \frac{D_\circ}{2}\left[\tan\alpha - \tan(90^\circ - 2\alpha)\right] . \tag{2.4}$$

For $\alpha = 45°$ this length is equal to $L' = \frac{D_\circ}{2}$. For small wedge angles, $\alpha < 10°$, the following approximate equation is valid:

$$L \cong \frac{D_\circ}{2}/(n-1)\,\alpha, \qquad \alpha < 10^\circ . \tag{2.5}$$

The longitudinal light intensity distribution $I(z)$ on the external part of the meso-optical image is highly non-uniform:

$$I(z) = C_1\left(\frac{z\,D_\circ}{2L}\right)^2 \tag{2.6}$$

where $z = 0$ corresponds to the apex of the cone A, $z_{max} = L$, and $C_1 = \text{const}$. For internal part of the meso-optical image we have

$$I(z') = C_2\left(\frac{z\,D_\circ}{2L'}\right)^2 , \tag{2.7}$$

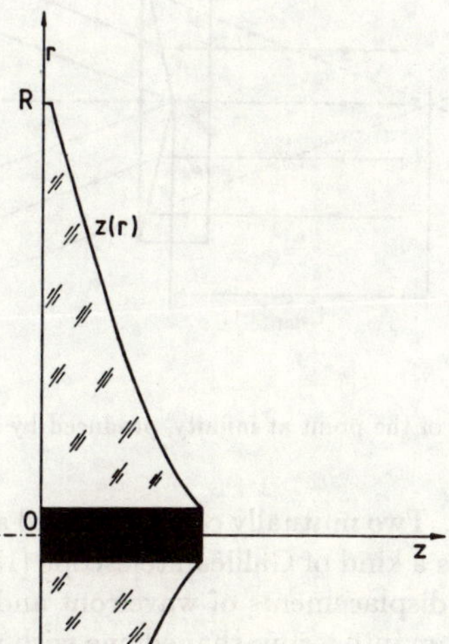

Fig. 2.6. Profile of the typical conoid axicon.

where C_2 is another constant, z'-coordinate is running opposite to z-coordinate and $z' = 0$ on the apex of the cone A. This strong variation of the axial light intensity over the meso-optical image is a principal disadvantage of this traditional axicon. To obtain a perfect system with constant axial light intensity, the generating line of the conical convex surface must be changed. In the small wedge angle α approximation the generating line of this conoid axicon has the form

$$H(r) = \alpha_\circ \frac{D_\circ}{2} \ln\left(\frac{D_\circ}{2r}\right), \tag{2.8}$$

where r is the distance to the optical axis and $\alpha_\circ$ is the constant [13, 14]. A typical conoid axicon is shown in Fig. 2.6.

A convergent axicon produces a real meso-optical image in the image space and a virtual meso-optical image in the object space. The observer looking backward through the axicon sees the virtual meso-optical image of the point-like object in the form of a bright circle. The divergent axicon (Fig. 2.7) produces only a point-like

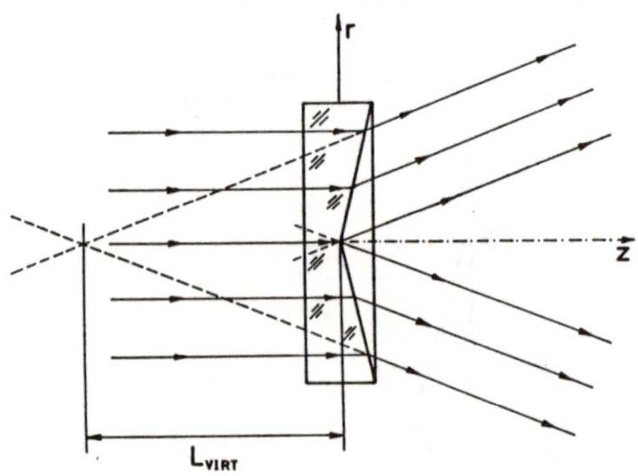

Fig. 2.7. The virtual image of the point at infinity, produced by divergent axicon.

virtual meso-optical image. Two mutually complemented axicons, one divergent and the other convergent, act as a kind of Galilean telescope [15] (Fig. 2.8), which can be used for adjustable radial displacements of wavefront and for converting of a disk-shaped collimated light beam into a ring-shaped one with variable magnification [6]:

$$M \cong 1 + 2\frac{\xi \cdot \alpha}{D}\left(\frac{n-1}{1-\alpha}\right), \quad \alpha < 10^\circ, \tag{2.9}$$

where ξ is the distance between two mutually complimented axicons. This meso-optical telescope serves as a beam expander without energy losses.

The radial reversal of the wavefront is performed by two conical lenses placed vis-a-vis, or by system of a conical lens and a plane mirror [17]. Two conical surfaces are present in a solid rod prism and in a rod prism with concave entrance and exit

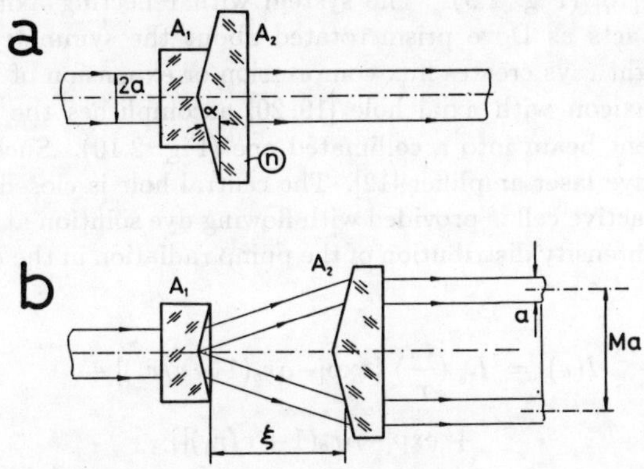

Fig. 2.8. Two mutually complemented axicons, which form a meso-optical Galilean telescope: "a" is the radius of the light beam, A_1 is a divergent conical lens, A_2 is a convergent conical lens, "Ma" is the diameter of the ring-shaped beam, ξ — distance between two axicons and α is the wedge angle.

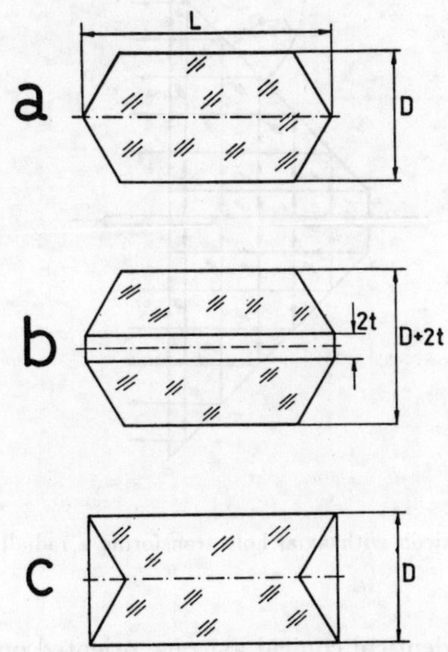

Fig. 2.9. Three types of conical systems, which accomplish the following operations: a) polar reversion with $180°$ rotation of the wavefront (rod prism); b) polar reversion without rotation of the wavefront (rod prism with axial hole); c) cup prism with concave entrance and exit surfaces, D is the diameter of the prism rod, $2t$ is the diameter of the axial hole.

external surfaces [18] (Fig. 2.9). The system with reflecting axial hole or with external cylinder acts as Dove prism rotated about the symmetry axis. Radial inversion of the light rays creates flux compression or expansion of the light beam. The 45° conical axicon with axial hole [19, 20] accomplishes the transformation of radially divergent beam into a collimated one (Fig. 2.10). Such a system has been used in the dye laser amplifier [12]. The central hole is closed with Brewster windows, and the active cell is provided with flowing dye solution at high repetition rates. The radial intensity distribution of the pump radiation in the dye cell is equal to [12]:

$$\begin{aligned} I(r) &= I_{r_\circ} \left(\frac{r_\circ}{r}\right) \{\exp[-\alpha r_\circ(1 - r/r_\circ)] + \\ &\quad + \exp[-\alpha r_\circ(1 + r/r_\circ)]\} \,, \end{aligned} \tag{2.10}$$

where $I_{r_\circ}$ is the light intensity at the surface of the dye solution. The infinite pump intensity on the axis of the cell is a consequence of the ray-optics approximation.

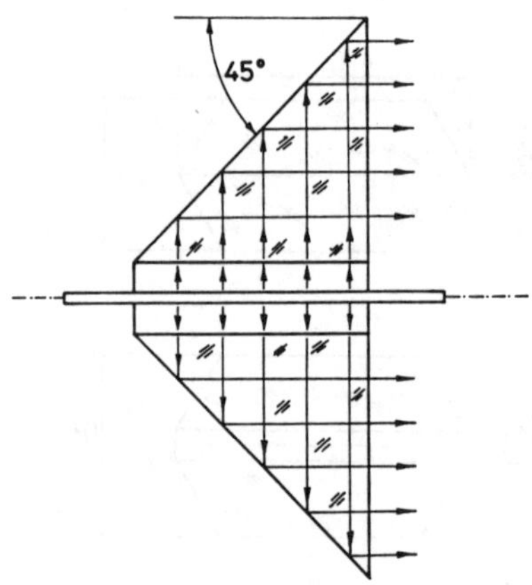

Fig. 2.10. The 45° conical axicon with axial hole transforms a radially divergent beam into a collimated one.

The system with two identical conical axicons, oriented opposite to each other, the apex of the second one being at $z = L$, is shown in Fig. 2.11. The rays of both light beams are parallel to the axis, but the light intensity distribution of the outcome beam is not uniform with maximum on the axis [12, 18]. To compensate this intensity non-uniformity a system of four conical surfaces has been suggested [12]. The first inversion of the intensity distribution is compensated by the second inversion in the second pair of conical surfaces (Fig. 2.12).

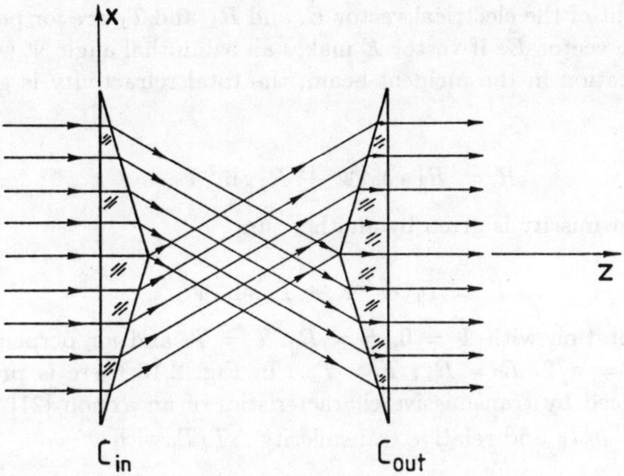

Fig. 2.11. System of two identical axicons accomplishes the inversion of the wavefront.

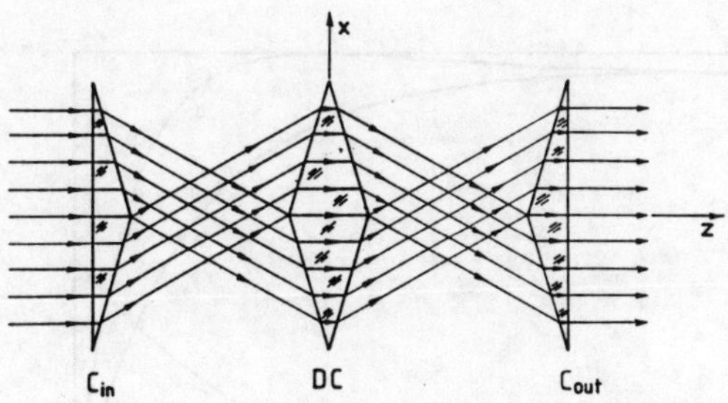

Fig. 2.12. System of four identical axicons accomplishes two mutually complemented inversion of the wavefront.

Of great important are the polarization characteristics of the axicon, illuminated by linear polarized light beam from laser [21, 26, 27, 29, 30]. Recall that reflection or transmission of polarized light beam from or through an inclined plane interface are given by Fresnel equations which define the reflectivity R or transmissity T of this plane interface [21]:

$$\left.\begin{aligned} R_{\|} &= \frac{\tan^2(i-r)}{\tan^2(i+r)}, \qquad R_{\perp} = \frac{\sin^2(i-r)}{\sin^2(i+r)}, \\ T_{\|} &= \frac{\sin 2i \cdot \sin 2r}{\sin^2(i+r)\,\cos^2(i-r)}, \qquad T_{\perp} = \frac{\sin 2i \cdot \sin 2r}{\sin^2(i+r)}, \end{aligned}\right\} \tag{2.11}$$

where i and r are the angles of incidence and refraction, respectively, with $r = \sin^{-1}\frac{\sin i}{n}$, n is the relative refraction index of the plane interface, $R_{\|}$ and $T_{\|}$ are for parallel component of the electrical vector $\vec{E}$, and $R_\perp$ and $T_\perp$ are for perpendicular component of the vector $\vec{E}$. If vector $\vec{E}$ makes an azimuthal angle Ψ with plane of the linear polarization in the incident beam, the total refractivity is given by the sum

$$R = R_{\|} \cos^2 \Psi + R_\perp \sin^2 \Psi \,, \tag{2.12}$$

and the total transmissity is given by another sum

$$T = T_{\|} \cos^2 \Psi + T_\perp \sin^2 \Psi \,. \tag{2.13}$$

For parallel orientation with $\Psi = 0$, $R = R_{\|}$, $T = T_{\|}$ and for perpendicular orientation with $\Psi = \pi/2$, $R = R_\perp$, $T = T_\perp$. In Fig. 2.13 there is presented the polarization induced by transmissive characteristics of an axicon [21] with wedge angle α: $T_\perp$; $T_{\|}$; $T_\perp/T_{\|}$ and relative transmissity $\Delta T/T_{\|}$ with

$$\frac{\Delta T}{T_{\|}} = \sin^2(i - r) \,. \tag{2.14}$$

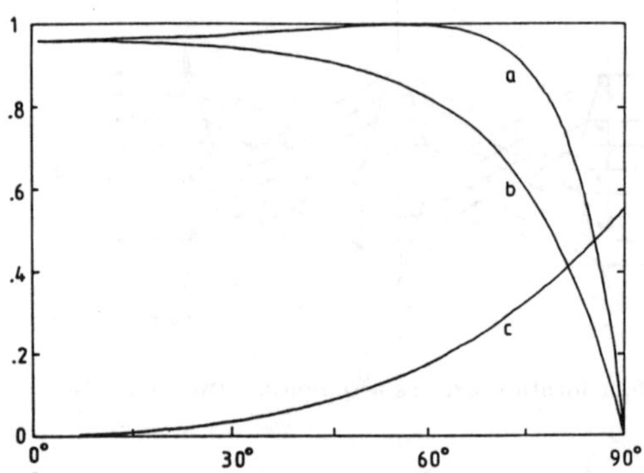

Fig. 2.13. Polarization performance of a conical surface: a) transmissity $T_{\|}$; b) transmissity $T_\perp$ and c) the ratio $\Delta/T_{\|}$, $\Delta T = T_{\|} - T_\perp$, versus of the incidence angle i .

The $T_{\|}$ transmittance is maximal at the Brewster angle. The total transmissity through an axicon versus azimuthal angle Ψ for different axicons is given [22] in Fig. 2.14.

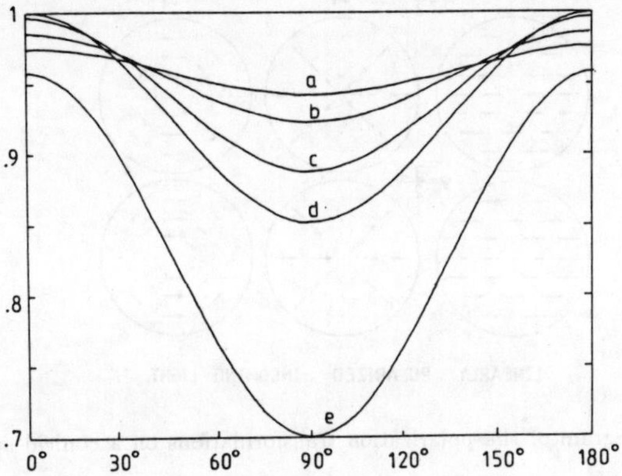

Fig. 2.14. The total transmissity of the linear polarized beam through an axicon versus azimuthal angle Ψ for various wedge angles α: a) $\alpha = 30°$, b) $\alpha = 40°$, c) $\alpha = 50°$, d) $\alpha = 60°$, e) $\alpha = 70°$.

To describe the polarization characteristics of an axicon, the natural components of the polarization vector are introduced: tangential (t) and radial (r) components. An x-linearly polarized light beam in a Cartesian (x, y) coordinate system is defined by their components of the electrical vector $\vec{E}(E_x, E_y, E_z)$:

$$\left.\begin{aligned} E_x &= A \sin(\omega t - kz)\,, \\ E_y &= 0\,, \end{aligned}\right\} \tag{2.15}$$

where $\omega = 2\pi\nu$, ν the frequency of the light wave, $k = 2\pi/\lambda$, λ the wavelength and A the amplitude of the light wave. The equivalent $r-$ and $t-$components of the electrical vector $\vec{E}$ are

$$\left.\begin{aligned} E_r &= A \sin(\omega t - kz) \cos\Psi, \\ E_t &= A \sin(\omega t - kz) \sin\Psi. \end{aligned}\right\} \tag{2.16}$$

Now we consider the passage of the linear polarized light beam through a conical surface. Fig. 2.15 shows vector diagrams for input and output waves in this axicon. The arrow length is proportional to the amplitude of the corresponding components. The case of circularly polarized input beam is presented in Fig. 2.16.

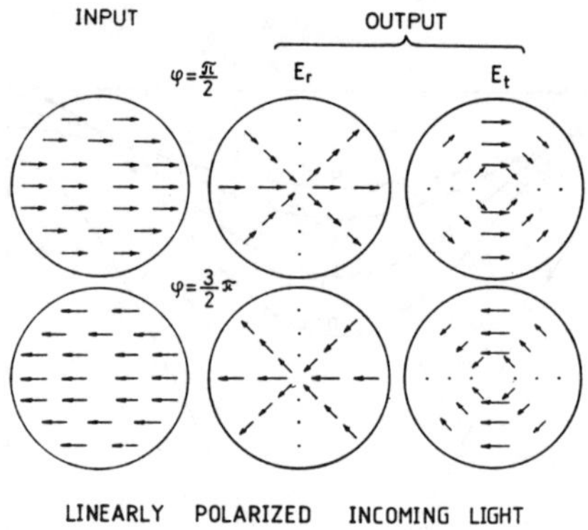

Fig. 2.15. The vector diagram of the polarization transformations on a conical surface for linear polarized beam.

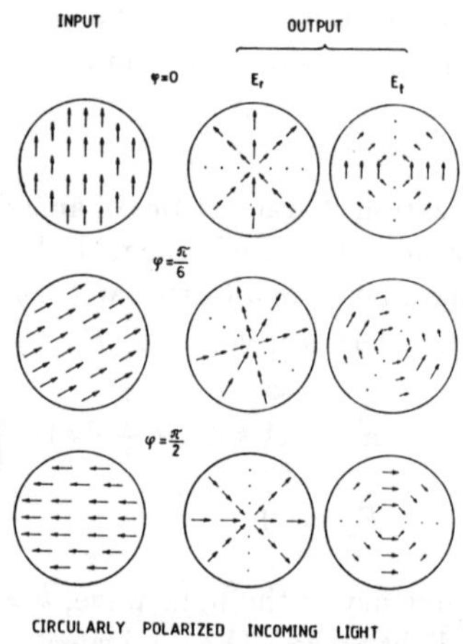

Fig. 2.16. The vector diagram of the polarization transformations on a conical surface for circularly polarized beam.

The reflection losses of the light at conical surfaces induces the polarization of the beam. A cone with Brewster angle $\alpha_B = 34°$ for quartz glass with $n = 1.48$ transmits the $t-$component without loss and the $r-$component of polarization with efficiency $\eta = 0.86$. In the system with four conical surfaces [12] shown in Fig. 2.12, the transmissity of the $t-$component will be equal to $\eta^4 = 0.86^4 = 0.55$.

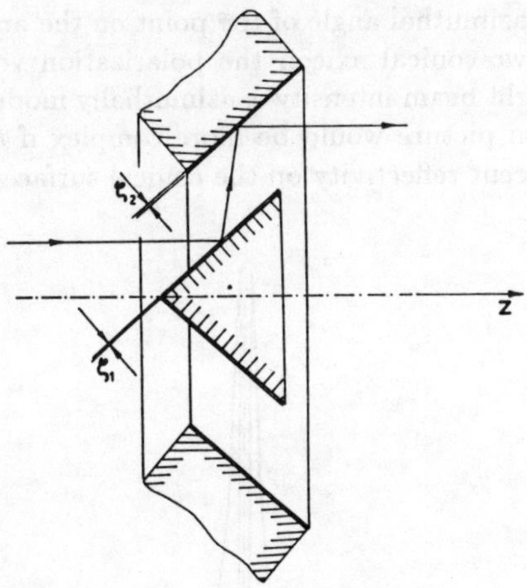

Fig. 2.17. Axiconic system with single-layer coating on two metallic mirror cones with different thickness ζ_1 and ζ_2 of this coating.

However, the dielectric coating on the conical surfaces represents more radical technique for generation of purely $r-$polarized light beam. For axiconic system with single-layer coating and with two metallic mirror cones [23] (Fig. 2.17) this aim can be achieved by equalizing of the reflection coefficients for $t-$ and $r-$components of polarization at two successive reflections. The thickness of the dielectric coatings has been estimated by inverse calculations. The dielectric layer in this design corrects the polarization change, protects the mirror surface and also enhances the polarization independent reflectance for angles above a certain value. The analogous technique of the multilayer dielectric coating has been suggested in [24]. The phase change between $t-$ and $r-$polarization component was chosen equal to $\pi/2$ rad. Such a coating consists of quarter wave pairs of high and low refractive index thin layer. The physical thickness of each layer is given by

$$d = \frac{\lambda_o}{4(n^2 - \sin^2\theta_o)^{1/2}}\,, \tag{2.17}$$

where λ_o is the vacuum wave length and θ_o is the incidence angle. The system with multilayer dielectric coatings works without large loss of reflectivity.

A special attention deserves so called *W*-axicon proposed for use in a laser with annular gain medium [28, 31, 32]. The typical *W*-axicon is shown in Fig. 2.18. Even for perfect dielectric conical surfaces of the *W*-axicon there is observed a peculiar angular transformation of the linear polarization of the input light beam which illuminates the central cone (Fig. 2.19). The double reflected annular-shaped light beam reveals the polarization picture [26] with polarization vector, which is rotating as 2φ, where φ is the azimuthal angle of the point on the annulus. It is interesting to note that in the two-conical axicon the polarization vector is the same over annulus, though the light beam intensity is azimuthally modulated over the annular cone. The polarization picture would be more complex if $t-$ and $r-$polarization components have different reflectivity on the conical surfaces.

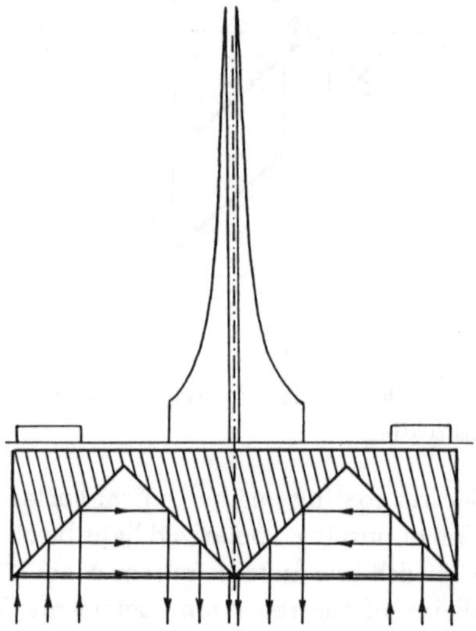

Fig. 2.18. Typical *W*-axicon. The input annular beam with constant intensity distribution is transformed into the circular beam with rising intensity distribution to the optical axis.

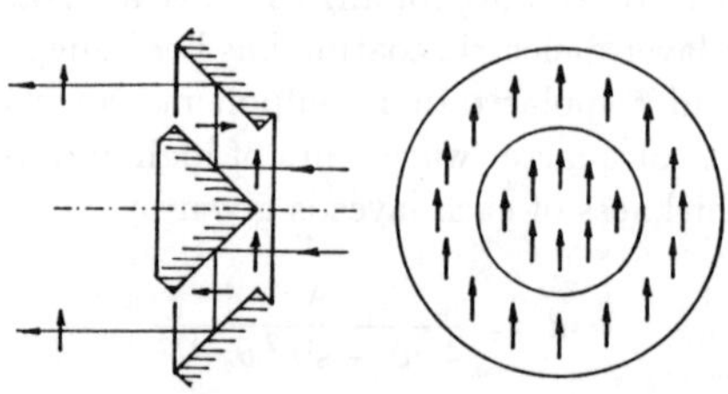

Fig. 2.19 The polarization transformation in *W*-axicon for linear polarized central beam.

The phase shifts after reflection at a perfect dielectric layer are equal to zero for r-component if the incident angle being less than Brewster's angle and equal to π-rad for angles larger than Brewster's angle. For $t-$component these phase shifts are equal to π-rad for all incident angles. For imperfect dielectric coatings or for a metal the phase shifts are varying continuously with incident angle from zero to

π-rad. If the difference between $t-$ and $r-$phase shifts is not equal to zero- or π-rad, the resultant polarization will be elliptic with axes parallel and perpendicular to the linear polarization observed in the case of perfect dielectric surfaces. To preserve the polarization of the light beam in the *W*–axicon, a system with central cone having obtuse or acute angle was suggested [26]. The first reflection occurs at the angle less than Brewster's angle and the second reflection occurs at angle greater than Brewster's angle (Fig. 2.20). In the design [30] a transmissive radially birefringent halfwave retarder was introduced in the ray path of the annular beam and a single-layer coating which equalizes $r-$ and $t-$components of polarization. The same results may be achieved if annular transmissive element is made as a combined system of a linear birefringent element plus linear dichroism element with axes which coincide with the radial and azimuthal directions at a given point of the annular cone.

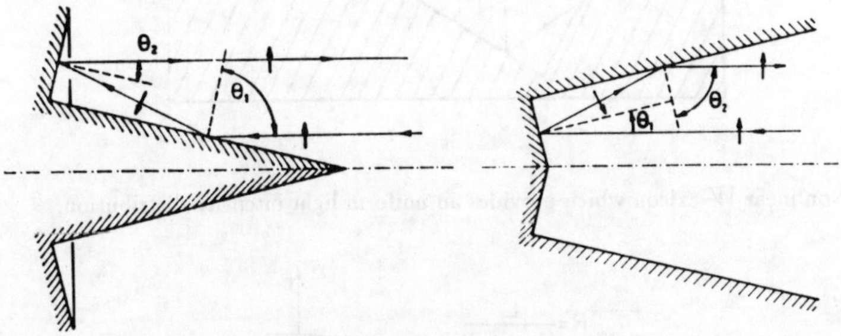

Fig. 2.20. *W*–axicon which preserves polarization of the incoming beam.

In the attempts to use *W*–axicon in the laser resonator we are faced with problem of light intensity non-uniformity in the disk-shaped central beam [33]. Annuli of increasing radii in one cone are reflected onto annuli of decreasing radii in the other. The light beam intensity is rising hyperbolically to infinity at the axis. However, this behaviour is observed only in traditional linear *W*–axicon. The *W*–axicon with nonlinear generating lines of both cones does generate uniform disk-shaped beam. The cross section of such cones are chosen according to the following demands [33]: a) all rays inside and outside of the mirrors must go parallel and must be of equal length between two references planes, and b) the area of corresponding partial annuli on both mirrors must be of constant ratio. Such an approach has proved its worth both for two-part mirror system and for *W*–axicon. Fig. 2.21 presents a cross section of the nonlinear *W*–axicon, which transforms annular light beam into a central disk-shaped beam of uniform light intensity distribution. The technological problem of the manufacturing of a such nonlinear *W*–axicon was resolved by the diamond mashining.

In the parabolic axicon [34] the generating line of the conical mirror has the form of an arc of the parabola, and this generating line is rotating about the axis, which crosses the parabola's axis and which is perpendicular to this axis. The axiconic system (Fig. 2.22) with central 45° half-angle mirror cone and with such a parabolic axicon focuses the collimated light beam into a point-like aberration free image with collecting solid angle up to 1.95 str.

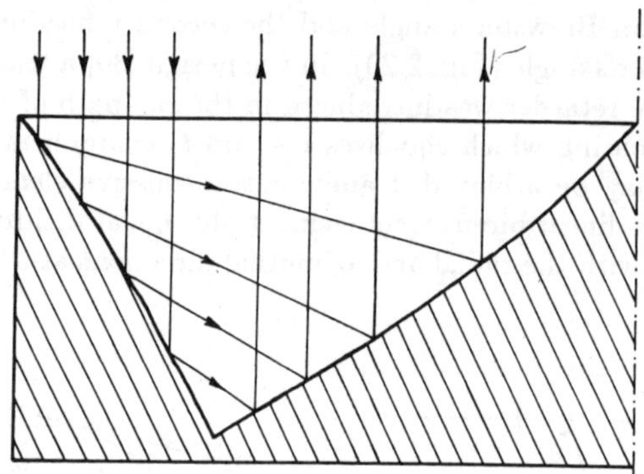

Fig. 2.21. Nonlinear *W*–axicon which provides an uniform light intensity distribution.

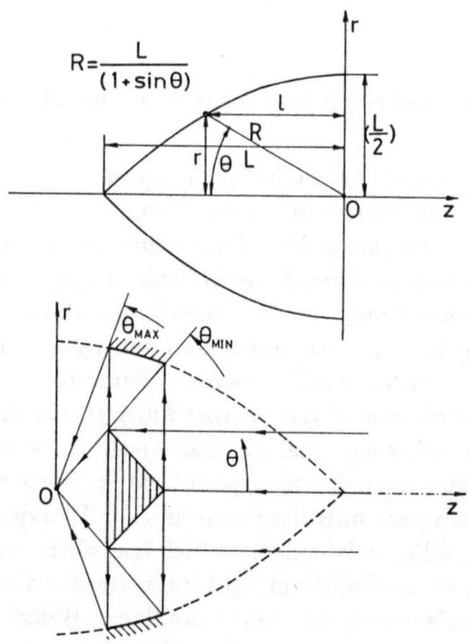

Fig. 2.22. System of 45° mirror cone and parabolic axicon with large collecting solid angle.

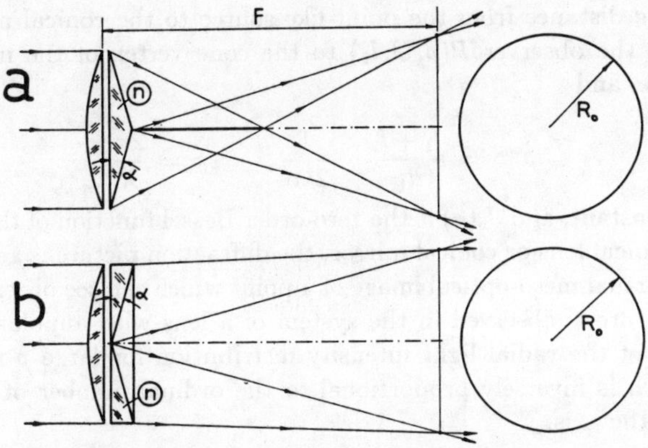

Fig. 2.23. System of an positive lens and axicon: a) with convergent axicon, b) with divergent axicon.

A positive lens combined with one axicon in the form of a conical lens, convergent or divergent, converts a point on the axis into a ring-shaped real meso-optical image [16] (Fig. 2.23). The radius of this ring is equal to

$$R_\circ \approx (n-1)\cdot\alpha\cdot F\,, \qquad \alpha < 10^\circ\,, \tag{2.18}$$

where F is the focal length of the positive lens.

2.3. Side-lobes of the Meso-optical Image

Due to axial symmetry, the diffraction picture of the meso-optical image with inherent side-lobes can be interpreted as an interference of two overlapped conical waves defined by wave vectors $\vec{k}_L$ and $\vec{k}_R$ (Fig. 2.24). The radial spacing of these interference fringes a is equal to

$$a = \frac{(\lambda/2)}{\sin(\theta_{12}/2)}\,, \tag{2.19}$$

where θ_{12} is an angle between two wave vectors $\vec{k}_L$ and $\vec{k}_R$. The intensity of the radial interference fringes is decreasing with distance to the optical axis. More rigorous treatment of the axiconic side-lobes shows [35] that radial light intensity distribution inside the diffraction picture of the conical mirror is defined by equation

$$I(\rho, l_1, l_2, \alpha) = C_\circ\,\frac{1 - 4\,\alpha\, l_1^2/(l_1+l_2)^2}{(l_1+l_2)^2}\, J_\circ^2\,(\rho/\rho^{cm})\,, \tag{2.20}$$

where l_1 is the distance from the point-like source to the conical mirror, l_2 is the distance from the observer $P(\rho, O, l_2)$ to the cone vertex of the mirror, ρ is the radial distance, and

$$\rho^{cm} = \frac{l_1 + l_2}{l_1} \cdot \frac{1}{2k\alpha}, \qquad k = \frac{2\pi}{\lambda}, \tag{2.21}$$

where C_o is constant, and $J_o(\bullet)$ is the zero-order Bessel function of the first kind. In both cases, conical lens or conical mirror, the diffraction picture is generated by the circle-shape virtual meso-optical image of a point which can see observer. The same diffraction picture is observed in the system of a lens with annular aperture [36]. The envelope of the radial light intensity distribution for large ρ is a hyperbolic function, which is inversely proportional to the ordinal number of the maximum starting from the axis.

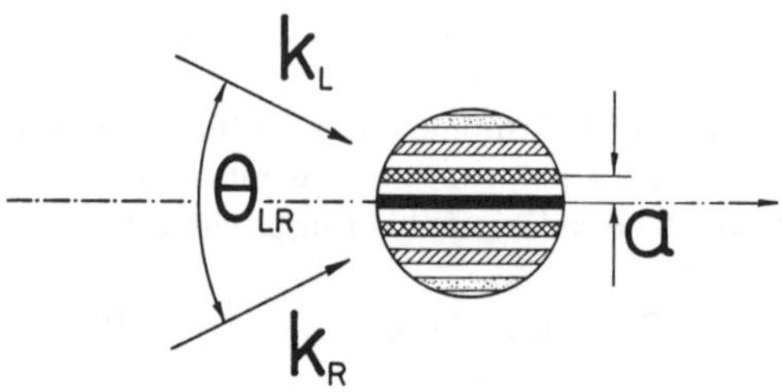

Fig. 2.24. The production of the axial symmetrical diffraction picture in the region of the overlapping of two conical waves with wave vectors k_L and k_R going from two sides.

The radial light intensity distribution observed in the system of a lens with open circular aperture is defined by equation (Airy disk):

$$I_o(\rho) = C \left| \frac{2\, J_1(\rho)}{\rho} \right|^2, \tag{2.22}$$

where $\rho = \pi\, Dr/\lambda f$, $J_1(\bullet)$ is the Bessel function of first order of the first kind, D — diameter of the aperture and f — the focal length of the lens. The envelope of the function $I_o(\rho)$ for large ρ is falling inverse proportional to the square of the ordinal number of the maximum.

The intensity of the diffraction pattern in the vicinity of a circle produced by the meso-optical element with ring response is given by the function [37]

$$I_{RING}(\rho) = \left| 2\, \frac{J_1(\rho_{1/2})}{\rho_{1/2}} \right|^2, \tag{2.23}$$

where

$$\rho_\circ = \frac{\pi D}{\lambda f}(r - R_\circ), \tag{2.24}$$

D is the diameter of the incident light beam and $R_\circ$ is the radius of the circle. The width of the central maximum and the radial spacing of the interference fringes in the vicinity of the meso-optical circle is twice the width of the Airy disk. This is due to the fact that only a half of the total aperture gives the contribution to one *arc* of this circle. The other half of this aperture produces an opposite arc of this circle.

Now we compare the side-lobes structure in axicon with that of circular and spiral diffraction gratings. A circular diffraction grating has circularly concentric grooves of constant radial spacing (pitch) on a plane surface [38]. A point source S on the symmetry axis is transformed by the circular diffraction grating into a system of secondary virtual meso-optical images in the form of circles. Only one circle S_1S_1' of the first diffraction order is shown in Fig. 2.25.

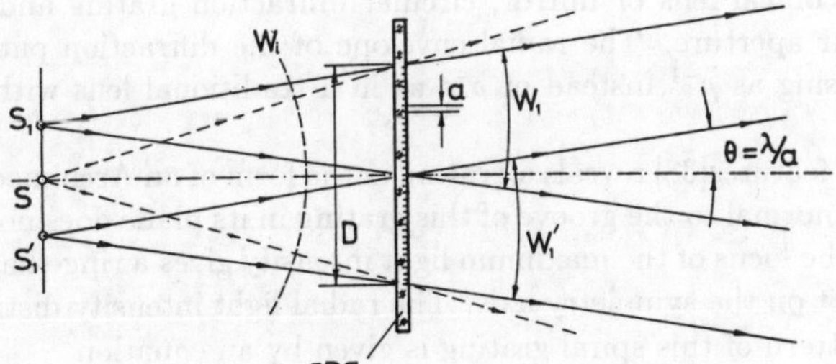

Fig. 2.25. Circular diffraction grating (CDG) and its properties.

This diffraction order produces a real meso-optical image on the symmetry axis. The length of the focal range L is given by equation:

$$\frac{1}{L} = \frac{2\lambda}{D\,a} - \frac{1}{l_1}, \tag{2.25}$$

where l_1 is the distance from the point source S to the circular diffraction grating with external diameter D. For $2\,\lambda\,l_1 > D\,a$ the length L is finite. The overlapped diffraction waves have both spherical and conical components in the meridional cross section. If the diameter d of the finite size source is such that $dD < l_1^2$, then all wavelets are spatially coherent. A contrast diffraction pattern is observed near the symmetry axis over the length L. The radial distribution of the light intensity of this diffraction pattern is

$$I(\rho) = \frac{4\pi^2\lambda l_1 l_2}{a^2(l_1+l_2)^3} J_\circ^2 \left(\frac{\rho}{\rho_\circ^{CDG}}\right), \tag{2.26}$$

where l_2 is the distance from circular diffraction grating to the observer $P(\rho, l_2, \varphi)$ and

$$\rho_\circ^{CDG} = \frac{a(l_1+l_2)}{2\pi l_1}. \tag{2.27}$$

For $l_1 \gg l_2$ we have

$$I(\rho) \approx \frac{4\pi\lambda\, l_2}{a^2\, l_1^2} J_\circ^2 \left(\frac{\rho}{\rho_\circ^{CDG}}\right), \qquad \text{with } \rho_\circ^{CDG} = \frac{a}{2\pi}. \tag{2.28}$$

From Eq. (2.28) follows that in this case when the intensity of the diffraction pattern is proportional to l_2, the scale of the diffraction pattern does not change with distance, and, what is more important, the diffraction pattern, defined by Eqs. (2.26) and (2.28), is achromatic.

Thus, $J_\circ^2$-dependence of the radial diffraction pattern is inherent to many meso-optical systems: conical lens or mirror, circular diffraction grating and lens with an narrow annular aperture. The radial envelope of the diffraction pattern in all systems is decreasing as ρ^{-1} instead of ρ^{-2} as in a traditional lens with full open aperture.

An interesting features[38] reveals a grating in the form of an Archimedean spiral of pitch "a". The normal to the groove of this grating in its plane does not cross the optical axis, and the locus of the maximum light intensity gives a ring-shaped region around a dark spot on the symmetry axis. The radial light intensity distribution of the diffraction pattern of this spiral grating is given by an equation

$$I(\rho) = \frac{4\pi^2\lambda\, l_1\, l_2}{a^2(l_1+l_2)^3} J_1^2\, (\rho/\rho_\circ^{ARCH}), \quad \text{with } \rho_\circ^{ARCH} = \rho_\circ^{CDG}. \tag{2.29}$$

This property of the Archimedean spiral grating can be explained as follows. From Fig. 2.26 we see that in the cross-section of this spiral grating the upper part of the spiral does not coincide with the lower one. These two parts are in antiphase with each other giving rise to the phase difference equal to π rad. The amplitude of the light wave on the symmetry axis will be zero in any cross-section. The phenomenon of "darkness" in optics and meso-optics is treated in Section 4.5.

2.4. General Theory of Axicons

To design the axicon with focus segment of arbitrary length, position and desired on-axis light intensity distribution, several approaches have been developed [39–43]. The approach based on the energy conservation principle and ray-tracing [39] is explained in Fig. 2.27. Let the interface refractive surface S, described by the function $z = f(r)$, is illuminated by the light wave with surface power density

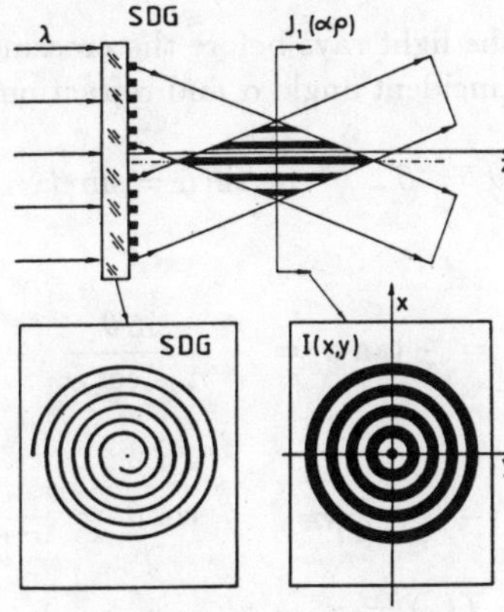

Fig. 2.26. Cross section of the Archimedean spiral grating.

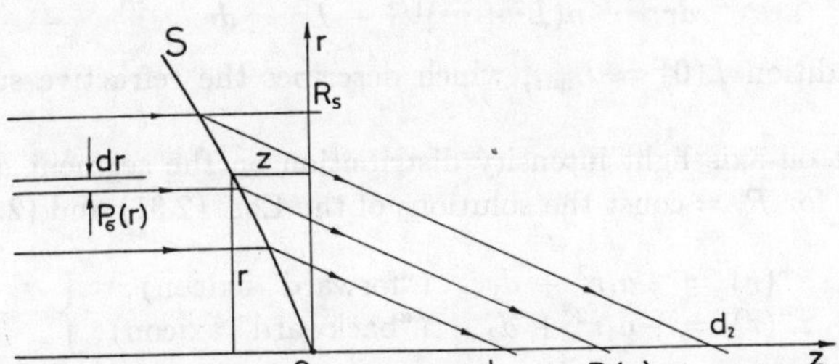

Fig. 2.27. Transformation of the radiation power density $P_\sigma(r)$ into the linear on-axis power density $P_z(z)$ by refractive surface S.

$P_\sigma(r)$. The output on-axis linear power density $P_z(z)$ is given for $d_1 \leq z \leq d_2$. By neglecting the light diffraction, the energy conservation principle yields:

$$2\pi P_\sigma(r) r\,dr = P_z(z)dz \,, \tag{2.30}$$

where dr is the radial increment and dz is the on-axis increment. For one-to-one correspondence $z = z(r)$, the integration of Eq. (2.30) gives

$$2\pi \int_0^r P_\sigma(r')dr' = \int_{d_1}^{z(r)} P_z(z')dz' \,, \tag{2.31}$$

for the case of no ray crossing, or

$$2\pi \int_0^r P_\sigma(r')dr' = \int_{z(r)}^{d_2} P_z(z')dz' , \tag{2.32}$$

for the case of full reverse of the light rays before the crossing the axis.

The refraction law for the incident angle α and refraction angle β gives

$$\alpha + \theta = \beta, \qquad n \cdot \sin\alpha = \sin\beta \tag{2.33}$$

and

$$\frac{dz(r)}{dr} = -\tan\alpha = -\frac{\sin\theta}{n - \cos\beta} . \tag{2.34}$$

As

$$\left.\begin{array}{ll} & \sin\theta = \frac{r}{(r^2+L^2)^{1/2}} , \qquad \cos\theta = \frac{L}{(r^2+L^2)^{1/2}} \\ \text{with} & L = L(z) = z_\circ - z , \end{array}\right\} \tag{2.35}$$

we come to the differential equation

$$\frac{dL}{dr} = \frac{r}{n(L^2+r^2)^{1/2} - L} - \frac{dz}{dr} , \tag{2.36}$$

with initial condition $L(0) = L_{init}$, which describes the refractive surface of the axicon [39].

For constant on-axis light intensity distribution on the segment $d_1 \le z \le d_2$, $P_z = \text{const}$, and for $P_\sigma = \text{const}$ the solutions of the Eqs. (2.31) and (2.32) are:

$$\left.\begin{array}{lll} z^+(r) = & a_1 r^2 + d_1 , & (\text{“forward” axicon}) , \\ z^-(r) = & -a_1 r^2 + d_2 , & (\text{“backward” axicon}) , \end{array}\right\} \tag{2.37}$$

with

$$a_1 = \frac{\pi P_\sigma}{P_z} = \frac{d_2 - d_1}{R^2} . \tag{2.38}$$

From Eqs. (2.37) and (2.36) we derive

$$\frac{dL}{dr} = \frac{r}{n(L^2+r^2)^{1/2} - L} \pm 2a_1 r =$$

$$= \left\{ \frac{1}{[n(1+r^2/L^2)^{1/2} - 1]\,L} \pm 2a_1 \right\} r , \tag{2.39}$$

where plus refers to the “forward” axicon and minus to the “backward” axicon. For new variable $r^2 = \xi$ we have

$$\frac{dL}{d\xi} = a_1 + \frac{1}{2[n(L^2 + \xi)^{1/2} - L]}\,. \tag{2.40}$$

The numerical solution of this equation, which have been found by the iterative method, is given in [39].

For axicon with high angular aperture, the nonparaxial solution [40] is of great utility. The relevant phase transmittance function of the refractive surface

$$t(r) = \exp[i\,k\,\varphi(r)], \quad k = \frac{2\pi}{\lambda}, \quad r = (x^2 + y^2)^{1/2} \tag{2.41}$$

can be found from non-paraxial differential equation [39, 44]:

$$\frac{d\varphi(r)}{dr} = -\sin\theta = -r\,[r^2 + z^2(r)]^{-1/2} \tag{2.42}$$

(Fig.2.28).

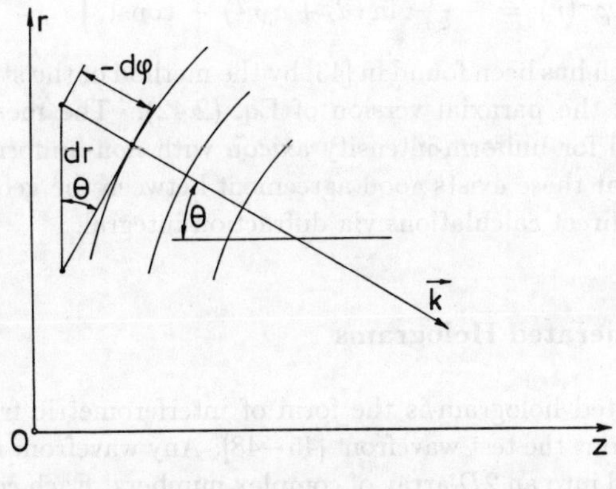

Fig. 2.28. Differential relationship between the retardation phase Ψ, the distance r from the optical axis and orientation angle θ of the wave vector $\vec{k}$.

For axicons with linearly increasing light intensity we have

$$P_\sigma(r) = P_\sigma = \text{const}, \quad P_z(z) = C_1 z, \quad d_1 \le z \le d_2\,. \tag{2.43}$$

By using Eqs. (2.31), (2.32) and (2.43) we arrive at

$$z(r) = (d^2 \pm a_2 r^2)^{1/2} \quad \text{with} \quad a_2 = \frac{2\pi P_\sigma}{C_1} = \frac{d_2^2 - d_1^2}{R^2} \tag{2.44}$$

and after integrating in r we obtain:

$$\left.\begin{aligned} \varphi^{+}(r) &= -\frac{[(1+a_2)r^2+d_1^2]^{1/2}}{(1+a_2)} + \text{const}\,, \\ \varphi^{-}(r) &= -\frac{[(1-a_2)r^2+d_2^2]^{1/2}}{(1-a_2)} + \text{const}\,. \end{aligned}\right\} \tag{2.45}$$

For $d_1 = 0$, the Eq. (2.45) yields the argument of the phase transmittance function of a classical axicon.

For uniform-intensity axicon with $P_\sigma =$ const and $P_z =$ const we have

$$\left.\begin{aligned} z^{+}(r) &= d_1 + a_3 r^2\,, \\ z^{-}(r) &= d_2 - a_3 r^2\,, \\ \text{with} \qquad a_3 &= \tfrac{\pi P_\sigma}{P_z} = \tfrac{d_2-d_1}{R^2}\,. \end{aligned}\right\} \tag{2.46}$$

A general solutions usable for nonparaxial designs have been found in [40]. In the paraxial conditions for $d_1 \gg R$, this solution is converted into

$$\left.\begin{aligned} \varphi^{+}(r) &= -\tfrac{1}{2a_3}\ln\left(d_1 + a_3 r^2\right) + \text{const}, \\ \varphi^{-}(r) &= +\tfrac{1}{2a_3}\ln\left(d_2 + a_3 r^2\right) + \text{const}. \end{aligned}\right\} \tag{2.47}$$

This solution which has been found in [43] by the method of the stationary phase, follows directly from the paraxial version of Eq. (2.42). The method developed in [40] has been used for uniform-intensity axicon with non-uniform illumination. It was proved [41] that these exists good agreement between the geometrical-optics predictions and the direct calculations via diffraction integral.

2.5. Computer-generated Holograms

Computer generated hologram is the form of interferometric fringes carried a phase information about the test wavefront [45–48]. Any wavefront recorded on the hologram is converted into an $2D$ array of complex numbers. Each complex number is given by a pair of real numbers, which define the structure of the interferometric fringes. For a pure spherical wave with phase function

$$\Phi(x,y) = \frac{\pi}{\lambda f_\circ}\left(x^2+y^2\right) + \text{const}\,, \tag{2.48}$$

there is one-to-one correspondence between phase and spatial frequency [47]. In conventional holography a real $3D$ object is to be written in the hologram. Each point of this real object emits a proper spherical wave. The possibilities of the computer generated hologram are much wider than that of the conventional optical hologram. On the computer generated hologram we may write the information about any waves such as that used in meso-optics and to observe this waves on the reconstruction stage.

The fringe configuration on the computer-generated hologram is presented both in Cartesian (x, y) and in polar (r, φ) coordinate systems. The reference wave in the computer-generated hologram can assume any arbitrary form. Due to this we may use radial or circular interferometric fringes [49]. To produce fringes in the azimuthal direction a helical reference wave with amplitude $exp(i\,\varphi/\varphi_\circ)$ is needed. To generate the fringes in the radial direction we use a conical wavefront $exp(i\,r/r_\circ)$. With these phase elements some optical geometrical transformations can be performed [47, 49, 50]. To accomplish geometrical image modifications or map transformations of general type, a phase filters from the space-variant class are to be used. Its local phase variations act on the light going from definite local object element. Hence any local translation, inversion, reflection, stretching of the object can be performed. For example, by means of the computer-generated hologram of a spherical wave [47] a geometrical transformation which holds the geometrical similarity of the object has been performed. In general case the phase filter consists of interference fringes of unusual or even strange structure. To map the concentric circles into ellipses of varying eccentricity the phase filter of the form

$$\Phi(x, z) = C_1 \left[\log(x^2 + y^2) - (x^2 - y^2)\right] \tag{2.49}$$

should be used. For performing more complicated geometrical transformations or combined phase and geometrical transformations a system of two phase filters is to be used [47, 50].

2.6. Kinoforms and Focusators

A kinoform [51] is a blazed diffraction (or holographic) element with optical relief, the thickness of which corresponds to phase difference no more than 2π rad. The depth structure of the kinoform surface may be continuous or stepwise [52, 60]. As opposite to hologram the diffraction efficiency of the kinoform is close to 100%. A kinoform converts an incident radiation to a wavefront of any desired form. For fast manufacturing of kinoform with axial symmetry a computer-controlled laser photoplotter was constructed [43]. To make a kinoform with high numerical aperture a very precise mechanical system is needed. The minimum step [43, 53, 54] was equal to $\lambda/8$ ($\approx 0.079\ \mu$m for He–Ne laser) and the angular discretization of the order of $1''$.

By means of this technology there were fabricated a kinoform-equivalent of axicon plus lens, by means of which a divergent spherical wave has been transformed into a pure conical wave [55], a kinoform-equivalent of a conoid axicon with constant longitudinal light intensity distribution [55, 56], and a kinoform with ring response [56, 58, 59]. The phase reliefs of various kinoforms are shown in Fig. 2.29. Three kinds of technological errors may deteriorate the quality of the kinoform: the errors in position of the ring zones, the elliptical error and the error of the alignment of the array of photopictures used in the system with stepwise profile [57 – 60]. The performance of a typical kinoform-axicon [55] is $L = 50$ m, $\rho_\circ = 35$ μm at $l_1 = 19$ m for $\lambda = 0.63$ μm and $D = 100$ mm.

a b c d

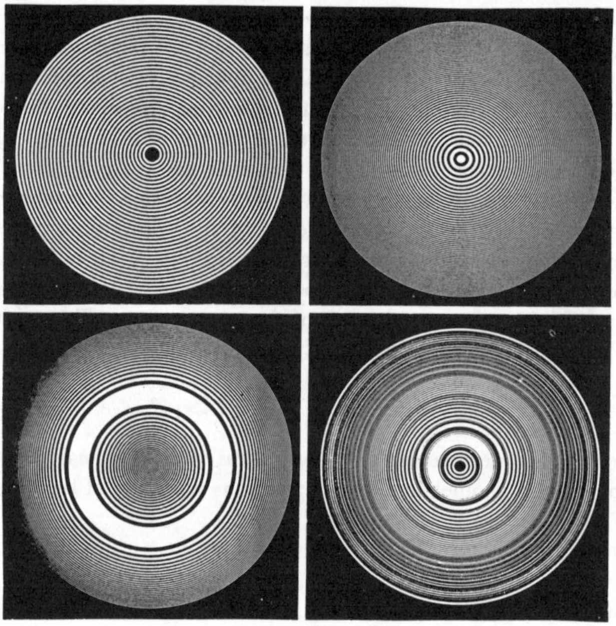

Fig. 2.29. Phase relief of various types of kinoforms: a) axicon, b) lens, c) meso–optical element with ring response, d) meso–optical element with complex ring response.

A diffraction element, called "focusator" [61, 62], generates a focal line of arbitrary form in a one-shot fashion as well as a computer-generated hologram, kinoform or $2D$ holographic translator [63]. On making a focusator we have no real objects. On the first stage there must be calculated a scalar wave function

$$U\,(u,v) \;=\; A\,(u,v)\;\exp\left[i\,k_\circ\,\varphi(u,v)\right], \tag{2.50}$$

in the plane of the focusator, which transforms the illuminated light into the given focal line with prescribed intensity distribution. If amplitude $A(u,v)$ and phase $\varphi(u,v)$ are slowly varying functions over the distance λ, than this inverse problem can be formulated in the frame of the ray optics approximation [64]. At first the amplitude-phase optical elements were searched for [65], but later only the phase relief was studied [66 – 68].

Effective algorithms for solving the inverse problems arisen in the manufacturing of the focusators have been developed [64 – 66]. In rare cases the phase function

$\varphi(u,v)$ can be represented analytically. For example, an axicon-focusator that transforms a spherical wave into a straight line segment on its axis with constant linear intensity density has a phase relief [67]:

$$\varphi(u,v) = \frac{1}{b} \ln\left[2b\sqrt{r^2+(f-br^2)^2} + 2b^2r^2 + 1 - 2fb\right], \tag{2.51}$$

with $r^2 = u^2+v^2$, $b = 4L/D^2$. This phase relief resembles a conoid axicon with logarithmic phase retardation function [13, 14].

2.7. Description of the Conical Waves

Now let us give some insight into the analytical representation of the meso-optical wavefronts. In contrast to traditional optics with its dualism of points and spherical waves, meso-optics offers several another dualisms, such as dualism between straight lines and conical waves. Each conical wave is described by amplitude and phase of the wave produced by coherently radiated line [9]. For luminous straight line of finite length and for wavefront components which are periodic along azimuthal (Φ) and longitudinal (z) directions the wave field, presented by equation

$$\Psi = R(r)\,\exp\left[i\,(2\pi z\cdot\omega_Z + \Phi\omega_\Phi)\right], \tag{2.52}$$

where $R(r)$ is the radial function, must meet the time-independent scalar wave equation in homogeneous medium

$$(\nabla^2 + k^2)\,\Psi = 0 \tag{2.53}$$

where ω_z is the spatial frequency along z-axis, and ω_Φ is the angular frequency along azimuthal angle Φ. The radial function $R(r)$ for propagating waves may be Bessel function $J(\bullet)$, or Hankel functions $H^{(1,2)}_{\omega\Phi}(\bullet) = J_{\omega\Phi}(\bullet) \pm N_{\omega\Phi}(\bullet)$. In general case the complex amplitude of the meso-optical wave is given by equation [9]:

$$u\,(r,\Phi,z,t) = H^{1,2}_{\omega\Phi}\,(2\pi\omega_z z)\cdot\exp[i(2\pi\omega_z\cdot z + \omega_\varphi\cdot\Phi - \omega t)] \tag{2.54}$$

with $\omega_z = \frac{1}{2\pi}\sqrt{k^2-4\pi^2\omega_z^2}$, $\omega = 2\pi\nu = k\cdot c$, where c is the light velocity and ν is its frequency. A unique solution is obtained for integer $\omega_\Phi = n$. At large r a general solution for conical wave has a form:

$$u\,(r,\Phi,z,t) = \frac{1}{\pi\sqrt{\omega_r r}}\cdot\exp\left[i(2\pi\omega_r r + 2\pi\omega_z z + \right.$$

$$\left. + n\cdot\Phi - \frac{\pi}{4} - \frac{n\pi}{2} - \omega t)\right]\cdot\left[1 - O\left(\frac{1}{r}\right)\right], \tag{2.55}$$

where the module of the wave vector is equal to $k = 2\pi\sqrt{\omega_r^2+\omega_z^2}$. The wave vector $\vec{k}$ has two non-zero components:

$$\vec{k}_r = 2\pi\omega_r\,\vec{r}/r, \qquad \vec{k}_z = 2\pi\omega_z\,\vec{z}/z. \tag{2.56}$$

The third component $\vec{k}_\varphi = 0$. The wavefront being perpendicular to the wave vector $\vec{k}$ is inclined to z-axis at the angle $(90° - \theta)$ (Fig. 2.30). The cylindrical wave with $\vec{k}_z = 0$ is defined by $\theta = 90°$. For $\theta = 0°$ the conical wave goes into the plane wave

$$u(z,t) = \text{Const} \cdot \exp(i\,k_z\,z). \tag{2.57}$$

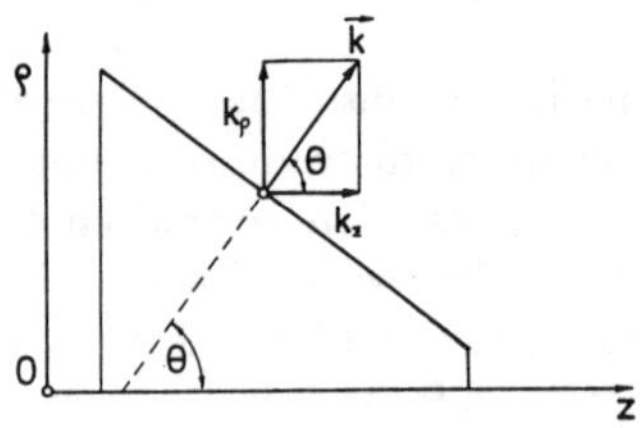

Fig. 2.30. Geometrical explanation of the analytical description of the conical wavefront.

The convergent conical wave with $\theta > 90°$ collapses into a focus line on z-axis and produces circular diffraction pattern with radial light intensity distribution:

$$I(r) = \text{Const} \cdot J_0^2(k_r\,r). \tag{2.58}$$

Any meso-optical wave can be represented by module and by phase of the wavefield

$$u = |u| \cdot \exp(i\Psi). \tag{2.59}$$

To present visually the structure of the phase function, Ψ, a positive definite function

$$t = 1 + \cos\Psi \tag{2.60}$$

is used traditionally.

If the luminous line is parallel to the observation plane, we have $1D$ Fresnel zone amplitude distribution which corresponds to cylindrical wave. The phase function versus transversal coordinate x is described by $\Psi \approx x^2$. If the luminous line is inclined to the observation plane, the loci of equal phase are ellipsis. Almost no phase variation occurs when the observation plane is nearly tangent to the wavefront and the loci of equal phase are parabolas [9].

Helical waves represent propagating waves which are solutions of the Helmholtz equation, Eq. (2.53) with periodicity along azimuthal and longitudinal coordinates.

In addition to the linear phase along z-axis a phase changing proportional to the azimuthal angle Φ occurs. If the observation plane is orthogonal to the coherently luminous line, the azimuthal frequency $\omega_\Phi = n$ denotes the number of spiral arms (branches). The observation surface can be chosen properly, for example, with the aim to receive lucid structure of the function $t = 1 + \cos\Psi$. The change from a luminous line of infinite length adopted in this description model to the line of finite length involves the corresponding diffraction effects.

2.8. Exotic Meso-optical Elements

2.8.1. Right-angle reflaxicon

This right-angle reflective axicon (reflaxicon) (Fig. 2.31) transforms the collimated beam of light into the converging cylindrical wave and provides the propagation invariant fields [69, 70]. In the left part of the Fig. 2.31 the incident ring-shaped beam with constant amplitude for $r_1 \leq r \leq r_\circ$ and zero outside this region is shown. The electric field $\vec{E}^{INC}$ and magnetic field $\vec{H}^{INC}$ of the linearly polarized incident beam are defined by the following components:

$$\left.\begin{aligned} \vec{k}^{INC} &= k\,\vec{e}_z\,, \\ \vec{E}^{INC} &= E_x\,\vec{e}_x\,, \\ \vec{H}^{INC} &= E_x\,\vec{e}_y\,, \end{aligned}\right\} \tag{2.61}$$

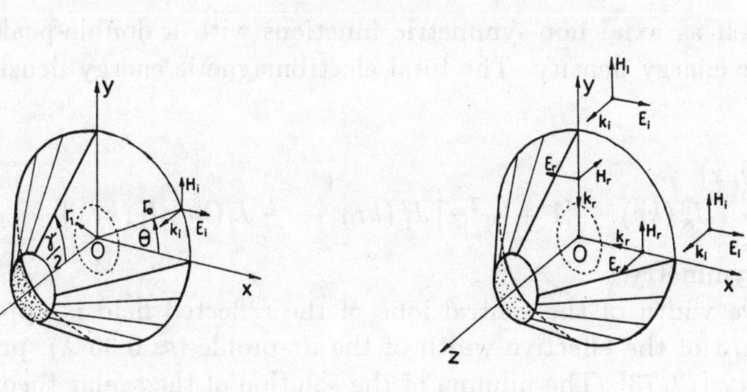

Fig. 2.31. Reflaxicon (axicon with wedge angle 45°): a) the incoming wave with triad $\vec{k}_i$, $\vec{E}_i$, $\vec{H}_i$; b) the reflected wave with triad $\vec{k}_r$, $\vec{E}_r$, $\vec{H}_r$.

with $k = 2\pi/\lambda$, and $\vec{e}_x$, $\vec{e}_y$, $\vec{e}_z$, are unit vectors. In the right part of Fig. 2.31 are shown the reflected waves with vectors $\vec{k}_{REF}$, $\vec{E}_{REF}$, $\vec{H}_{REF}$. The reflected field in the plane $z = 0$ is defined as

$$\left.\begin{aligned}\vec{k}_{REF}(r_\circ,\Phi,0) &= -k\,(\vec{e}_x\cdot\cos\Phi+\vec{e}_y\cdot\sin\Phi),\\ \vec{E}_{REF}(r_\circ,\Phi,0) &= E_x(-\vec{e}_x\cdot\sin^2\Phi+\vec{e}_y\cdot\sin\Phi\cdot\cos\Phi+\vec{e}_z\cdot\cos\Phi),\\ \vec{H}_{REF}(r_\circ,\Phi,0) &= E_x(-\vec{e}_x\cdot\sin\Phi\cdot\cos\Phi+\vec{e}_y\cdot\cos^2\Phi-\vec{e}_z\cdot\sin\Phi),\end{aligned}\right\}\qquad(2.62)$$

All plane-wave components of the reflected field are directed perpendicular to the z-axis and produce a convergent cylindrical wave.

The reflected field in the vicinity of the z-axis is defined by the components [70]:

$$\left.\begin{aligned}\vec{E}(r,\Phi,z) &= 2\pi K E_x\{-\vec{e}_x[\sin^2\Phi\cdot J_\circ(kr)-\tfrac{1}{kr}(\sin^2\Phi-\cos^2\Phi)\,J_1(kr)+\\ &\quad+\vec{e}_y\cdot\sin\Phi\cos\Phi[J_\circ(kr)-\tfrac{2}{kr}J_\circ(kr)]-\\ &\quad-\vec{e}_z\cdot i\cos\Phi\,J_1(kr)\},\\ \vec{H}(r,\Phi,z) &= 2\pi K E_x\{-\vec{e}_x\sin\Phi\cos\Phi\,[J_\circ(kr)-\tfrac{2}{kr}J_1(kr)]+\\ &\quad+\vec{e}_y\cdot[\cos^2\Phi\cdot J_\circ(kr)+\tfrac{1}{kr}(\sin^2\Phi-\cos^2\Phi)\,J_1(kr)]+\\ &\quad+\vec{e}_z\cdot i\sin\Phi\,J_1(kr)\}.\end{aligned}\right\}\qquad(2.63)$$

The time-averaged electric energy density and magnetic energy density are given by:

$$\left.\begin{aligned}w_e(r,\Phi,z) &= \tfrac{\pi K^2}{4}\{\sin^2\Phi\,J_\circ^2(kr)+[\cos^2\Phi+\tfrac{1}{(kr)^2}]\,J_1^2(kr)-\\ &\quad-\tfrac{2}{kr}\sin^2\Phi\,J_\circ(kr)\,J_1(kr)\},\\ w_m(r,\Phi,z) &= \tfrac{\pi K^2}{4}\{\cos^2\Phi\,J_\circ^2(kr)+[\sin^2\Phi+\tfrac{1}{(kr)^2}]\,J_1^2(kr)-\\ &\quad-\tfrac{2}{kr}\cos^2\Phi\,J_\circ(kr)\,J_1(kr)\},\end{aligned}\right\}\qquad(2.64)$$

and exhibit itself as axial non-symmetric functions with a double-peaked central lobe for electric energy density. The total electromagnetic energy density is given by [70, 71]:

$$\begin{aligned}w(r,\Phi,z) &=\\ &= \tfrac{\pi K^2}{4}\{J_\circ^2(kr)+[1+\tfrac{2}{(kr)^2}]\,J_1^2(kr)-\tfrac{2}{kr}J_\circ(kr)\,J_1(kr)\}\end{aligned}\qquad(2.65)$$

and has axial symmetry.

The effective width of the central lobe of the reflected field is approximately $0.99\,\lambda$, about $1/3$ of the effective width of the $J_\circ^2$-profile ($\approx 0.36\,\lambda$), predicted by the scalar theory [72, 73]. The minima of the solution of the scalar theory are well defined, whereas the field in the vector model exhibits no clear minima.

The time averaged Poynting vector

$$\vec{S}(r.\Phi,z) = \vec{e}_z\,\frac{\pi c K^2}{4}\,\frac{2\,J_1(kr)}{(kr)}\,[\frac{1}{kr}\,J_1(kr)-J_\circ(kr)]\,,\qquad(2.66)$$

has only z-component, which oscillates strongly along radial coordinate r.

2.8.2. *Axicon with annular aperture*

Consider a traditional lens shown in Fig. 2.1 anew. If a human eye is placed in the focus of this lens, the observer sees a bright disk with angular opening $\theta = d/f$ where f is the focus of the lens and D is the diameter of the lens aperture. If near a lens a central opaque disk of the diameter $D_a < D$ was put with the aim to increase the depth of field of this lens, the virtual image of the source point assumes the form of a narrow ring. In the limit of high obstruction ratio, $D/(D - D_a) \gg 1$, an observer will see very narrow bright annular picture. We know [36, 74] that the radial diffraction pattern produced by this annular aperture is described by function $J_\circ^2(\bullet)$. In general case a lens aperture can be composed of many alternating opaque and transparent annular zones [75]. To get large depth of focus by modifying the homogeneous aperture of lens, we use very narrow annular aperture and thus decrease the intensity of the transmitted light. The same results would be obtained both for binary and continuous apodization function [76 – 81]. To illustrate this dependence let us estimate the required width t of the annular aperture illuminated by convergent spherical wave of focal length $f_\circ$. The length of the focal range L of this primitive system is equal to

$$\left.\begin{aligned} L &= \frac{f_1 \theta_d}{\alpha} \\ \text{with} \quad \alpha &= D/2\, f_\circ, \qquad \theta_d = \lambda/t \end{aligned}\right\} \tag{2.67}$$

whence it follows that

$$t = \frac{2\,\lambda\, f_\circ}{L\, D}. \tag{2.68}$$

For $L = 10^2$ μm, $\lambda = 0.5$ μm, $D = 10$cm, $f_\circ = 2 \cdot 10^2$ cm the width of the annular aperture of the lens will be equal to $t = 40$ μm and the light efficiency $\eta = 8t/D = 3.2 \cdot 10^{-3}$. It is evident that such a system with low light efficiency has no use in practice.

In contrast to this pure optical design, meso-optical element, such as axicon, posses high light efficiency. All light rays in axicon are propagating practically without diffraction, because an axicon accomplishes focusing of light only in one direction, along the azimuthal angle, and provides no focusing in another direction, along z-axis. Due to this we do not experience diffraction of light when we use axicon. This advantage of axicon results in wide range of applications.

A conventional axicon is described by the linear phase retardation function $\varphi(r)$. A conoid axicon, which exhibits a constant on-axis light intensity distribution over the focal range, is described by the logarithmic phase retardation function [39, 41, 43, 82]. The disadvantage of the conoid axicon is its non-uniform radial resolution [42]. It has been shown that the rays propagating closely to the optical axis are responsible for such lack of the conoid axicon. To improve the performance of the conoid axicon a central opaque disk has to be introduced. The conoid axicon with annular aperture is shown in Fig. 2.32. The principle of this design can be explained, if we express the light intensity at some point $A(r, z)$ by means of the Fresnel diffraction integral [83, 84]

$$I(r,z) = (\frac{2\pi}{\lambda z})^2 \left| \int_o^r \exp\{\frac{2\pi i}{\lambda}[\frac{\rho^2}{2z} + \varphi(\rho)]\} J_o(\frac{2\pi r\rho}{\lambda z})\rho d\rho \right|^2 . \tag{2.69}$$

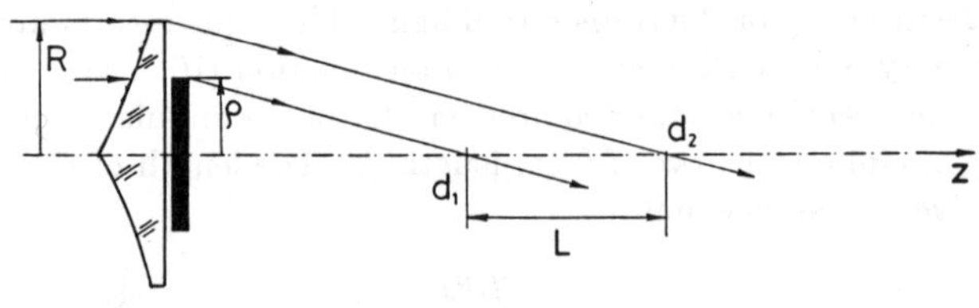

Fig. 2.32. Conoid axicon with annular aperture [42].

For conoid axicon with full aperture the phase retardation function

$$\varphi_F(\rho) = -\frac{R^2}{2(z_2 - z_1)} \ln[z_1 + \frac{z_2 - z_1}{R^2}\rho^2]. \tag{2.70}$$

In the design with annular aperture the logarithmic phase retardation function is extending from $\rho^{min} = R_o$ to $\rho^{max} = R$.

In the axicon with uniform energy flow and constant lateral resolution over the given focal range the phase retardation function is to be chosen as [83]

$$\varphi_A(\rho) = -\frac{R^2 - R_o^2}{2(z_2 - z_1)} \ln[z_1 - \frac{z_2 - z_1}{R^2 - R_o^2}R_o^2 + \frac{z_2 - z_1}{R^2 - R_o^2}\rho^2]. \tag{2.71}$$

The problem of non-diffracting light beam with constant on-axis light intensity is treated in [85].

2.8.3. *Axial non-symmetrical meso-optical elements*

First we consider evident examples of the meso-optical elements which transform a plane wave into a line with desired orientation. The phase relief of the $1D$ zone plate, which produces a bright line parallel to the (x, y)-plane, is given by

$$U_\circ(x,y) = \exp\left[-i\left(\frac{\pi}{\lambda f_\circ}\right)y^2\right], \tag{2.72}$$

where $f_\circ$ is the focal length of this element. To generate a line, inclined at angle γ with respect to (x,y)-plane, another phase relief is chosen

$$U_\gamma(x,y) = \exp\left[-i\,\frac{\pi}{\lambda}\,\frac{y^2}{f_\circ + x\tan\gamma}\right]. \tag{2.73}$$

In general case the phase relief of the meso-optical element has the form [86]:

$$U(x,y) = \exp\left[-i\,\frac{\pi}{\lambda}\,\frac{y^2}{f_\circ + \Delta f(x)}\right]. \tag{2.74}$$

In Fig. 2.33 three typical examples of the $\Delta f(x)$ function are presented [86]: a) $\Delta f(x) = 0$; b) $\Delta f(x) = b \cdot x$ and c) $\Delta f(x) = b \cdot \sin x$, with $b < f_\circ$.

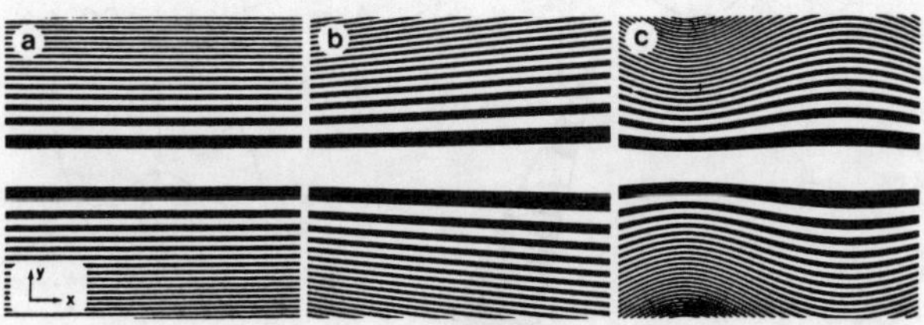

Fig. 2.33. The structure of the $\Delta f(x)$ function in three meso–optical systems: a) without azimuthal modulation, b) linear modulation, c) periodic modulation.

In the polar coordinate system (ρ,θ) the amplitude transmittance of the axial symmetrical Fresnel zone plate is given by

$$U(\rho,\theta) = \exp\left[-i\frac{\pi}{\lambda}\,\frac{\rho^2}{f_\circ}\right]. \tag{2.75}$$

For angular modulated zone plate [7], which generates $3D$ focal curve [8] the intensity transmittance function is defined by equation

$$t(\rho,\theta) = \frac{1}{2}\left\{1 + Re\left[\exp\left(-i\,\frac{\pi}{\lambda}\,\frac{\rho^2}{f(\theta)}\right)\right]\right\} \tag{2.76}$$

with focal parametric function $f(\theta)$ changing along azimuthal angle θ. The angular modulated circular zone plate with this intensity transmittance function, Eq. (2.76), can be decomposed into a sum of angular sectors with different θ and with small angular increment $\delta\theta$ (Fig. 2.34). Each angular sector, defined by focal length $f(\theta)$, produces one point on the z-axis.

In the system with focal parametric function

$$f(\theta) = f_1 \frac{\theta - \pi}{\pi}, \text{ with } f_1 < f_\circ \tag{2.77}$$

a line segment with end points at $z_1 = f_\circ - f_1$ and at $z_2 = f_\circ + f_1$ is generated in the image space.

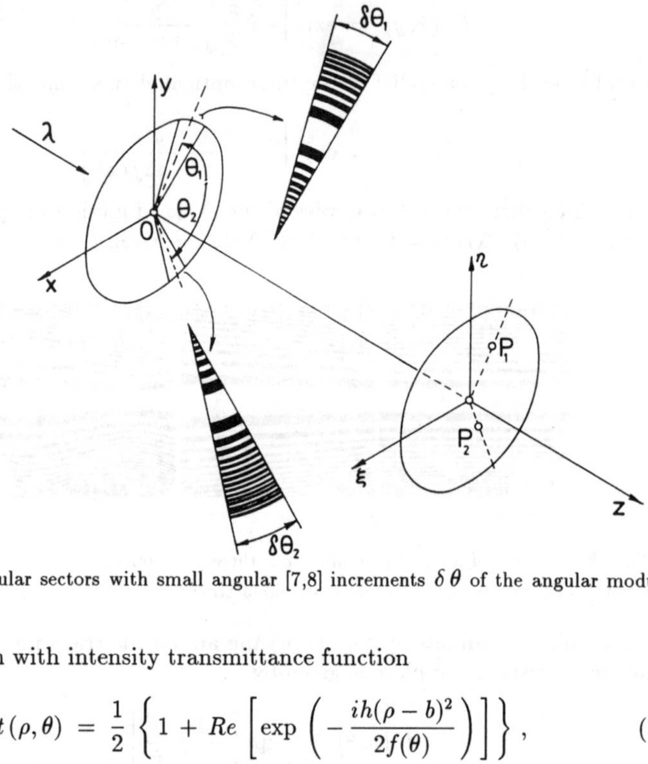

Fig. 2.34. The angular sectors with small angular [7,8] increments $\delta\theta$ of the angular modulated circular zone plate.

In the system with intensity transmittance function

$$t(\rho, \theta) = \frac{1}{2}\left\{1 + Re\left[\exp\left(-\frac{ih(\rho - b)^2}{2f(\theta)}\right)\right]\right\}, \tag{2.78}$$

and with focal parametric function

$$f(\theta) = f_\circ + f_1 \frac{\theta}{2\pi} \tag{2.79}$$

a helical line on the cylindrical surface of the radius "b" is generated.

Another example of the axial non-symmetrical meso-optical element provides on optical microscope with ring-shaped condenser [87]. This microscope is dedicated for observation of the straight line particle tracks oriented at very small angle α to the optical axis of the microscope. The condenser with ring-shaped aperture is provided with exotic circular diffraction grating of ring-shaped grooves. The centers of these grooves are located on the equidistant array of points in the plane of the grating (Fig. 2.35). The distances between adjacent grooves are chosen such that

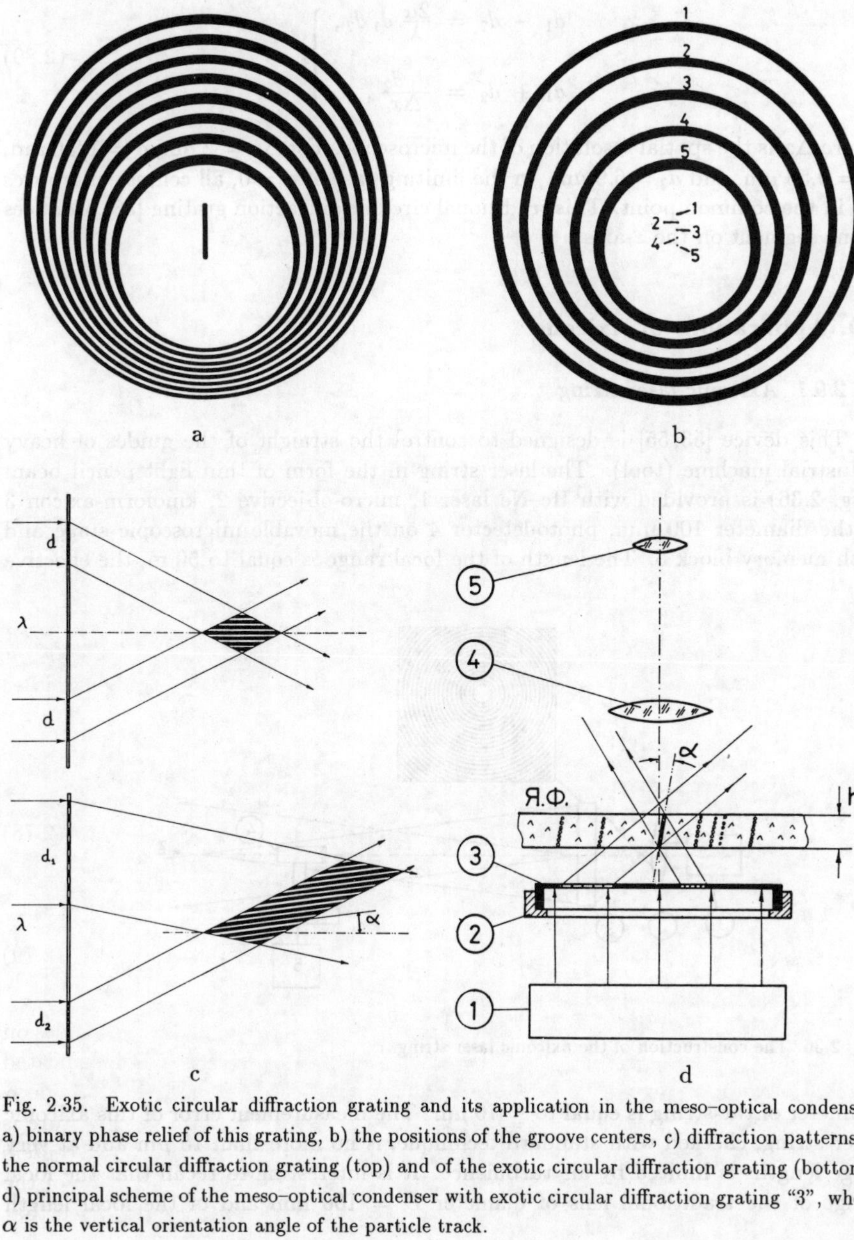

Fig. 2.35. Exotic circular diffraction grating and its application in the meso–optical condenser: a) binary phase relief of this grating, b) the positions of the groove centers, c) diffraction patterns of the normal circular diffraction grating (top) and of the exotic circular diffraction grating (bottom), d) principal scheme of the meso–optical condenser with exotic circular diffraction grating "3", where α is the vertical orientation angle of the particle track.

the maximal d_1 and minimal d_2 distances between adjacent grooves are chosen by the equations

$$\left.\begin{aligned} d_1 - d_2 &= \frac{2\alpha}{\lambda}\, d_1\, d_2\,, \\ d_1 + d_2 &= \frac{d_1\, d_2}{\Delta x}\,, \end{aligned}\right\} \tag{2.80}$$

where Δx is the spatial resolution of the microscope. For $\Delta x = 2\ \mu$m, $\alpha = 1/50$ rad, $d_1 = 9.55\ \mu$m, and $d_2 = 6.9\ \mu$m. In the limiting case of $\alpha = 0$, all centers of grooves are in the common point. This traditional circular diffraction grating [38] produces a line segment on the z-axis.

2.9. Applications of Axicons

2.9.1. Axiconic laser string

This device [53, 55] is designed to control the straight of the guides of heavy industrial machine (tool). The laser string in the form of thin light pencil beam (Fig. 2.36) is provided with He–Ne laser 1, micro-objective 2, kinoform-axicon 3 of the diameter 100 mm, photodetector 4 on the movable microscopic stage and with memory block 5. The length of the focal range is equal to 50 m, the effective

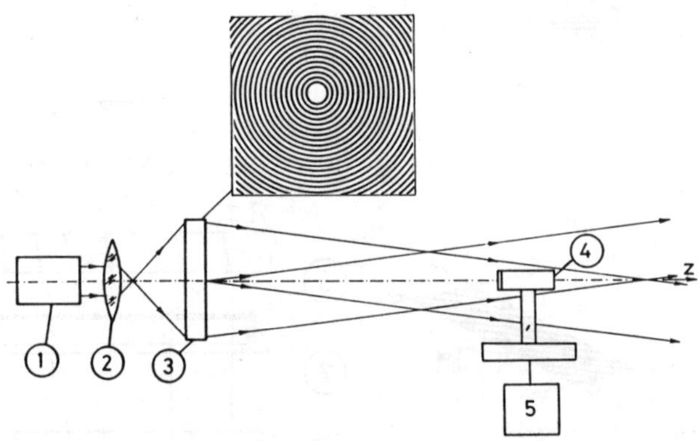

Fig. 2.36. The construction of the axiconic laser string.

diameter of the string is equal to $\leq 70\ \mu$m. The measurement error of this axiconic laser string, checked with standard technique, is no more than 15 μm and at very large lengths is limited by air turbulence. It is interesting to recall that the focal range of the traditional lens of diameter $D = 100$ mm and of the focal length

$f_o = 25$ m is equal to $\Delta z \approx \lambda/\alpha^2 = 50$ cm for $\alpha = d/4f_o$. The factor of merits of this axiconic laser string is equal to 100 : 1.

2.9.2. A pair of convergent and divergent axicons

The focal length of this system (Fig. 2.37) is given by

$$L = D/(n-1)(\alpha - \beta), \tag{2.81}$$

where α and β are the wedge angles of two axicons. This system leaves room for inserting an optical resonator into this system. It was used to focus laser beam of

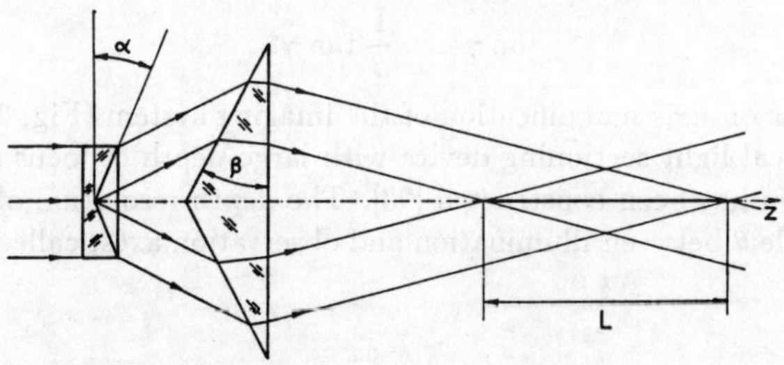

Fig. 2.37. System of a pair of negative and positive axicons with wedge angles α and β.

light in a preionized high pressure gas [88]. A plasma of the length 95 cm and of diameter 3 mm in pure He at pressure of 7 atm has been obtained.

2.9.3. Conoid axicon

The first observation of the optical breakdown in argon at high gas pressure has been accomplished by using an axicon [88]. The length of the spark was equal to 30 mm at pressure 90 atm. These continuous sparks have been produced in pure helium [89] and in air [14, 90]. The energy to the discharge was delivered from sides by means of axicon. The length of the helium plasma was 95 cm and the diameter 3 mm at the gas pressure 7 atm. The conoid with prescribed light intensity distribution was suggested in [82].

Kinoform conoid axicon [56] of diameter $D = 12$ mm provides constant light intensity over the focal range of length $L = 250$ mm. The spot diameter was equal to $d \approx 20$ μm.

2.9.4. Light sectioning with axicon

A tele-axicon which comprise a negative and a positive axicons (Fig.2.37) was used in the illuminator of the object triangulation device [92]. The length of the focal range was equal to 1 m and the spot diameter was no more than 20 μm. The detector system of the triangulation device consists of an imaging lens and of an $1D$ sensor array. To ensure a sharp image of the points on the external surface of the object, the plane of the sensor array was to be tilted at an angle γ with respect to the optical axis. This angle is given by the Scheimpflug condition [94]: the image plane must be inclined at an angle γ' in such a way that object and image planes have common line lying in the plane of the imaging lens. To meet the Scheimpflug condition the angle γ' is chosen according to the equation

$$\tan\gamma' = \frac{1}{\beta}\tan\gamma\,, \tag{2.82}$$

where β is linear on-axis magnification of the imaging system (Fig. 2.38).

A meso-optical light sectioning device with large depth of focus and with high spatial resolution has been constructed [93]. The depth resolution of this device is given by an angle θ between illumination and observation axes, called triangulation angle:

$$\delta z = \frac{\delta x}{\tan\theta}, \tag{2.83}$$

where δx is the spatial resolution of the imaging lens. The device is provided with a kinoform-axicon [52] (Fig. 2.39), which gives a pencil beam of diameter 57 μm, and with rotating scanning mirror. The triangulation angle $\theta = 80°$, the depth resolution $\delta z = 30$ μm for field of view 250 mm, the lateral resolution $\delta x = 170$ μm and the total depth of field $L = z_2 - z_1 = 1700$ mm.

2.9.5. Axicon-scanned Fabry–Perot spectrometer

An axicon-scanned Fabry–Perot spectrometer has been suggested [22, 96, 99] to increase luminosity and resolving power of the traditional spectrometer. The device consists of Fabry–Perot etalon 1, an imaging lens 2, convergent axicon 3 and the detector 4 (Fig. 2.40). The convergent axicon 3 is dedicated to gather the light rays going at an angle Ψ to the axis and to focus these rays into a point on the detector plane at the distance L behind the axicon. This axicon reduces the diameter of interferometric fringes and thus increases the dispersion of the system. Recall that diameter of fringe in the observation plane of the traditional Fabry–Perot etalon is given by

$$D_m = 2f\sqrt{\frac{\lambda}{t}(m-1+\varepsilon)}\,, \tag{2.84}$$

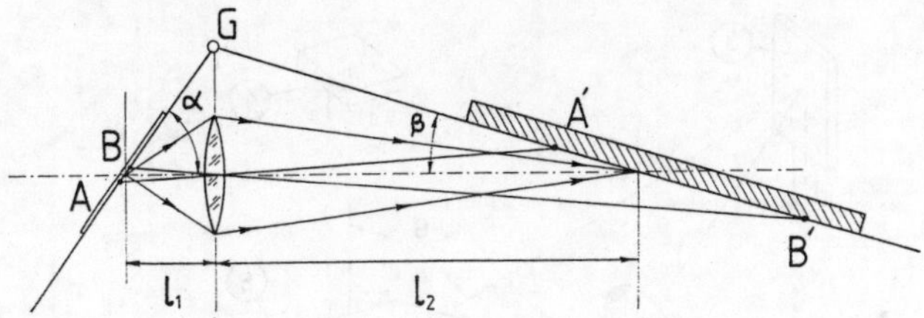

A

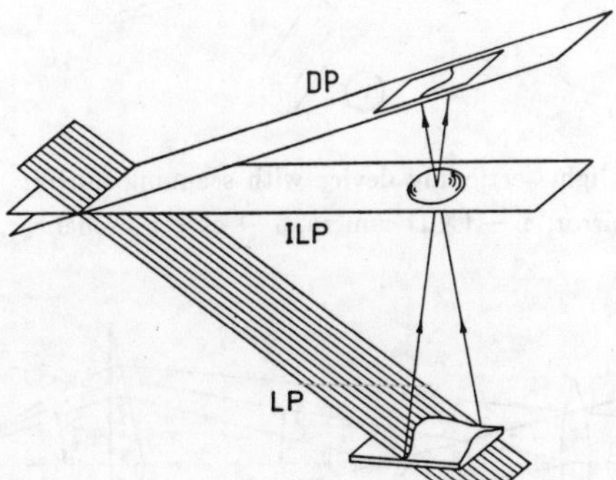

B

Fig. 2.38. A. Explanation of the Scheimpflug condition: the plane of the inclined object and the plane of the inclined image must have common line in the plane of the imaging lens.
B. Mutual arrangement of the detector plane DP, imaging lens plane ILP and the scanning light plane LP in the Scheimpflug condition.

where f is the distance between Fabry–Perot etalon and the observation plane, m is the order of the fringe starting from the center with angle $\Psi = 0$, t is the separation between mirror in the Fabry–Perot etalon, and

$$\varepsilon = D_\circ^2/(d_1^2 - D_\circ^2)\,, \tag{2.85}$$

The dispersion of the axicon-scanned Fabry–Perot spectrometer is by a factor $K_\circ$ more than in the conventional device, where

$$K_\circ = \cot(n-1)\alpha\,. \tag{2.86}$$

However, the insertion of axicon lens with conical surface modifies the light polarization. It has been shown [21, 96] that variance of the transmittance of the axicon-scanned Fabry–Perot spectrometer with azimuthal angle, induced by this

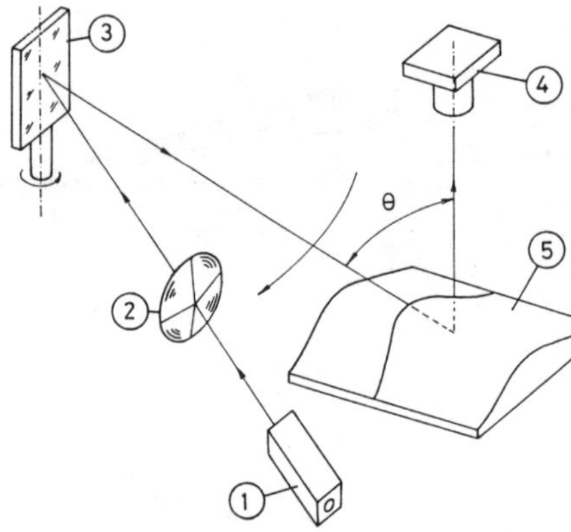

Fig. 2.39. Meso–optical light sectioning device with scanning mirror: 1 — laser, 2 — axicon, 3 — rotating scanning mirror, 4 — CCD camera, 5 — object's relief.

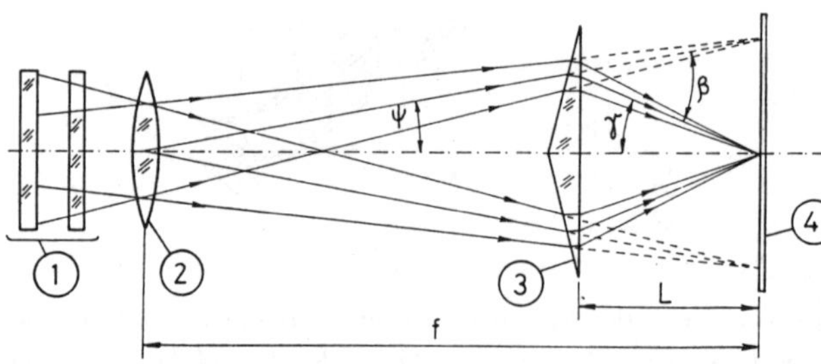

Fig. 2.40. Axicon-scanned Fabry–Perot spectrometer: 1 — Fabry–Perot etalon, 2 — imaging lens, 3 — convergent axicon, 4 — photodetector.

polarization effect, increases with wedge angle α. The small misalignment errors in this device can disturb severely the shape of the scanned spectral line.

2.9.6. Aberration free axicon-objective

A traditional kinoform-axicon [52] is made on a plane backing. Its working field of view is very small because of severe coma aberrations [38]. Therefore this kinoform-axicon cannot replace a traditional objective, having very small depth of focus. Meanwhile for observation of the object consisting of many point-like elements in $3D$ volume a principally new design is needed. This problem has arisen few years ago in high energy particle physics.

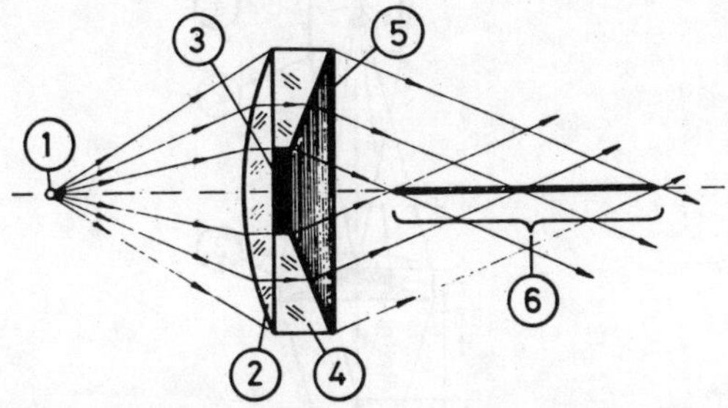

Fig. 2.41. Design of the aberration-free axicon-objective: 1 — point source, 2 — optical collimator, 3 — opaque disk and 4 — axicon objective on the conical backing 5, 6 — the focal region.

As will be explained in Chapter 3, the coma aberrations can be suppressed if we use a non-planar backing of the kinoform-axicon. Such an axicon-objective is shown in Fig. 2.41 [97, 98]. The spherical wave from a point source 1 traverses an optical collimator 2 and axicon-objective 4. The central part of the wave is absorbed by the opaque disk 3. The backing of this axicon-objective is a divergent cone. The secondary light rays are directed perpendicular to the generating line of the backing surface. These convergent light rays produce an extended meso-optical image of the point-like object. The light rays of zero diffraction order are directed away from the axis. The depth of focus of this axicon-objective does not depend on diffraction of light. However to see all working volume of the particle track chamber, the diameter of the axicon-objective must be chosen large (Fig. 2.42).

2.9.7. Axicons in the X-ray laser

For generation of coherent soft X-ray beam, a travelling wave excitation of the active medium is required. A time gradient must be about 10^{-9} s per 30 cm across the output aperture. For this aim a travelling axicon has been proposed [99]. A profile of this travelling wave axicon has been calculated in the Fresnel approximation for small numerical aperture. The radial derivation of the axicon profile $dT(r)/dr$ is defined by the equation

$$\frac{dT(r)}{dr} = -\frac{r}{f(n-1)}. \tag{2.87}$$

For uniform incident plane wave the radial profile $T(r)$ and the focal length of this axicon with uniform on-axis light intensity distribution are equal to

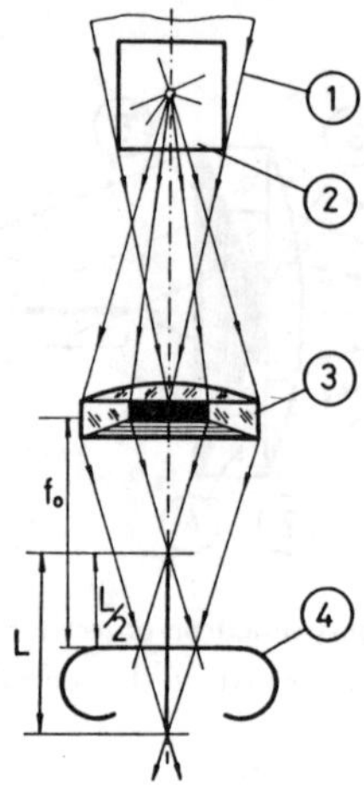

Fig. 2.42. Track chamber photoregistrator provided with aberration-free axicon-objective: 1 — convergent light beam, 2 — working volume of the track chamber, 3 — axicon-objective, 4 — photoregistrator, L — the length of the depth of view.

$$T(r) = T_\circ - \frac{1}{(n-1)} \frac{r^2}{2f_\circ} \left(1 - \frac{2\pi C r^2}{3f_\circ}\right) \tag{2.88}$$

and

$$f(r) = f_\circ \left(1 + \frac{4\pi C}{3f_\circ} r^2\right), \tag{2.89}$$

where C — constant, $f_\circ$ is the focal length of the equivalent optical lens.

The time delay as a function of r is defined by

$$t(r) = \frac{1}{C} \{n T(r) + [T_\circ - T(r)] + l(r)\} \tag{2.90}$$

where $l(r)$ is the length of the light ray from axicon to the z-axis. The travelling velocity of the focused light spot is equal to

$$v(r) = \frac{df(r)}{dt} = \frac{C}{df/dr} \cdot \frac{1}{(n-1)\frac{dT(r)}{dr} + \frac{dl(r)}{dr}}. \tag{2.91}$$

For

$$l \simeq f + \frac{1}{2}\frac{r^2}{f} \tag{2.92}$$

the travelling velocity

$$v = \frac{c}{1 - \frac{1}{2}(r/f)^2} \simeq c[1 + \frac{1}{2}(r/f)^2] > c \tag{2.93}$$

is approximately equal to the light velocity c.

To concentrate the pump energy over the small area and to produce a travelling wave pumping of the active medium, a pump pulse slanted to the direction of propagation was suggested [8]. The on-axis velocity of the pump pulse is equal to Lc/x (Fig. 2.43). Therefore X-ray photon, emitted at the near end of the medium in the direction towards the far end, will always find a medium with high excitation density. The X-ray laser beam is emitted from the far end of the medium.

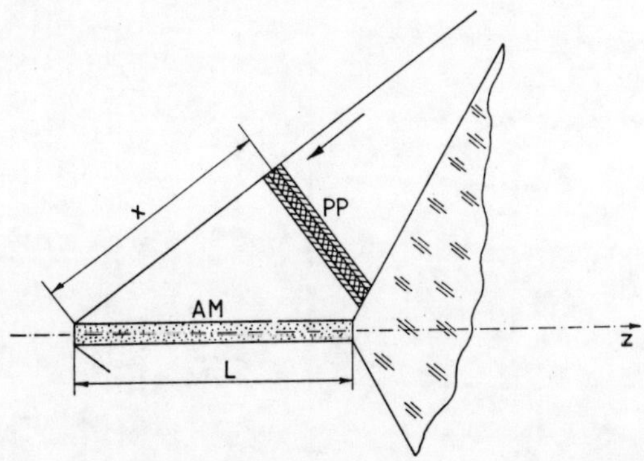

Fig. 2.43. Travelling wave pumping of the active medium AM with pulse conical wave PP.

Chapter 3

MESO-OPTICAL ELEMENTS WITH RING RESPONSE

3.1. Introduction

Meso–optical elements with ring response were described in Chapter 2 and some examples of its realization were given in Fig. 2.3 and Fig. 2.23. A point source on the optical axis is converted by this element into a narrow ring in the plane which is perpendicular to the optical axis. There is a principle difference between meso–optical element with line response and meso–optical element with ring response. The first one can be effectively used in some imaging or quasi-imaging systems. As it was demonstrated in Section 2.9.6, a kinoform-axicon on conical backing can be inserted as a working component into the objective for imaging of restricted class of objects. This possibility stems from the fact that radial dimension of the central spot produced by an axicon is equal to or even smaller than the radial dimension of the central spot produced by a lens with full aperture. A quite different situation occurs for meso–optical element with ring response. This element does not focus the light in the vicinity of the optical axis, and the dimension of the narrow ring is of the macroscopic scale, at least $10^4 - 10^5$ times of the radial dimension of the central spot of an axicon. Obviously, this feature of the meso–optical element with ring response does not satisfy the requirements of traditional imaging devices. It was widely believed that meso–optical element with ring response had no chances to be used in the imaging systems. Therefore sometimes this element has been considered as a toy among other meso–optical elements. Nobody had disproved this belief until a principally new meso–optical device has been suggested and constructed. This device, named "Meso–optical Fourier transform microscope", provides spatial resolution of the order of 2 μm with diameter of the meso–optical ring being equal to 160 mm.

In this Chapter we explain in detail the properties of the meso–optical elements with ring response and present various designs of the Meso–optical Fourier transform microscope (MFTM). This microscope was suggested to solve the depth-of-focus problem in light microscopy [2, 129, 130]. This device sees only straight line objects of a width comparable with the wavelength. Each such object is converted by the MFTM into **two** meso–optical images which are displayed on the ring of the annular shape [2]. The operation of depth scanning is entirely absent in the MFTM, although the information about z-coordinate of the microscopic line object is not lost. It should be pointed out that we cannot find any counterparts of this device in the traditional light microscopy. The MFTM offers the following principally new features. It performs a depth viewing without operation of depth scanning and presents a stereo-effect simultaneously for all orientations of the microscopic straight line objects. The MFTM demonstrates a meso–optical analogue of the classical moiré-effect and performs an instant angular selectivity without any adjustment or turning operations. Besides it put forward a new technique of small bending measurement which can be used for quasi-straight line microscopic objects. No parts of the MFTM are moving during these measurements.

As will be explained below the information about line object is displayed at the optical output of the MFTM in the extremely compressed form. The function of the photodetector system attached to the MFTM is to read out this information and to put it into a memory block. As was proved experimentally, the MFTM admits some new useful algorithms of the on-line and off-line processing of this compressed information. This feature of the MFTM opens new scopes of the noise suppression and new technique of the object intensity evaluation.

3.2. Meso–optical Fourier Transform Microscope (MFTM)

A principal scheme of the MFTM is shown in Fig. 3.1 [2, 129, 130]. A convergent beam of light produced by the Fourier transform lens 1 illuminates a working volume 2 of depth h in which microscopic line objects with small vertical inclination angles are located. This light beam produces a crossover in the vicinity of the meso–optical mirror with ring response 3 [2, 116, 129]. Each line object generates a far-field diffraction pattern, having a form of a narrow and long bright strip and crossing the optical axis of the MFTM. The width of this strip is equal to the diameter of the convergent beam crossover ($\sim 50\,\mu$m). The non-diffracted light goes through the central hole in the meso–optical mirror 3. Each line object produces **two** meso–optical images on the opposite sides of the focal ring: the left one, L, and the right one, R. Thus a line object is mapped by the MFTM into itself and is twice multiplexed.

The generating line of the axial symmetrical meso–optical mirror with ring response 3, shown on the cover of this book, is an arc of the ellipse. One focus of this ellipse is located in the median plane of the working volume 2 and the second focus is lying on the focal circle. The problem of the depth-of-focus is solved in the MFTM without loss of the radial resolution in the meridional cross-section of the MFTM. For this aim the profile of the meso–optical mirror 3 has been made of many, ~ 600, straight line segments with end points lying on the above described ellipse. Thus the real meso–optical mirror 3 performs a conversion of the point source on the optical axis into a shirt-shaped surface of very small height. This class of transformation can be written as $0D \rightarrow 2D$. If the length of each line segment of the generating line-polygon is chosen of the order of the depth of the working volume, $\sim 100\,\mu$m, the line objects lying at various depths of this volume should produce meso–optical images without substantial defocusing and without

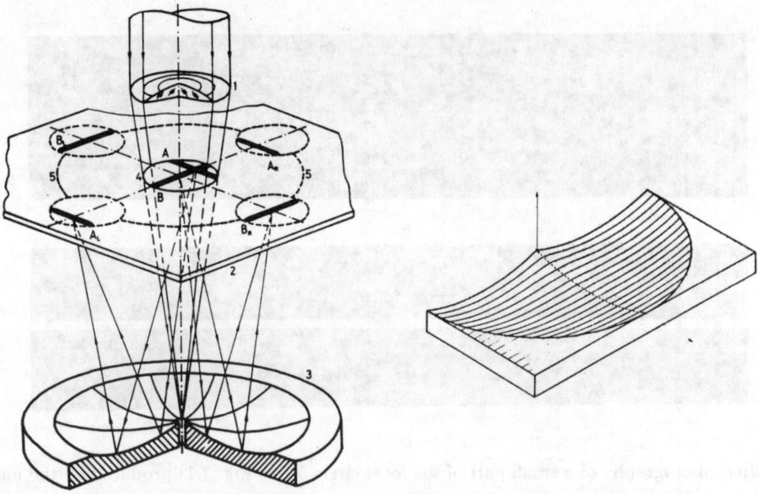

Fig. 3.1. Left: Principal scheme of the Meso–optical Fourier Transform Microscope (MFTM): 1 — Fourier transform lens of the illuminating system, 2 — working volume of the thickness h, 3 — meso–optical mirror with ring response, 4 — field of view, 5 — focal circle, A and B — two straight line objects, A_L and B_L — left meso–optical images of the objects A and B, A_R and B_R — right meso–optical images of the objects A and B.

Right: The view of the meso–optical mirror with ring response.

loss of spatial resolution. A small fragment of the focal circle produced by the meso–optical mirror with ring response 3 is shown in Fig. 3.2. The effective width of this focal circle is equal to $1.5\,\mu$m, whereas the diffraction limited width [2] is equal to $1.4\,\mu$m.

The diameter of the focal circle containing two meso–optical images of the line objects is changing with z-coordinate of the line object as is shown in Fig. 3.3. Due to this the information about z-coordinate of the line object can be estimated by comparing the positions of **two** meso–optical images without any operation of depth scanning. The only scanning operations, which are to be performed in the MFTM, are oriented perpendicular to or along to the expected orientations of the microscopic line objects to be searched for. The dip angle of the long line object can be measured in the MFTM as well. This feature of the MFTM is based on the fact that two end points of the inclined line object are lying on two different focal circles with unequal radii.

The MFTM has been designed for observation of straight line particle tracks in nuclear research emulsion [2]. Each particle track represents a line chain of small, $\sim 0.5\,\mu$m, silver grains. Due to the Babinet's principle [141], each meso–optical

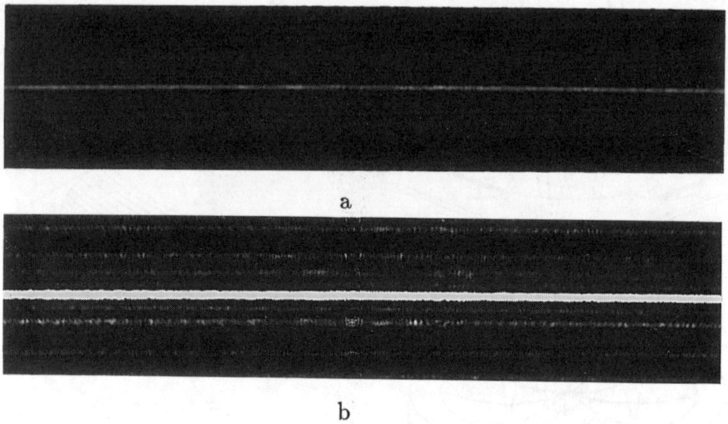

Fig. 3.2. Microphotography of a small part of the focal circle 5 (see Fig. 3.1) produced by the meso-optical mirror with ring response 3 (in Fig. 3.1): a) with underexposition, b) with overexposition.

image of this absorbent object represents a bright spot on semidark background. The line connecting two meso–optical images of the same line object, a left one L and a right one R, is perpendicular to orientation of the particle track. Two line objects, A and B, which cross each other within the illuminated region (Fig. 3.1), are converted in the MFTM into two pairs of meso–optical images, A_L and A_R for object A, and B_L and B_R for object B, on different parts of the focal circle. This automatic spreading of the meso–optical images in accordance with the orientation angle of these line objects represents the main feature of the MFTM. The locus of all meso–optical images has a form of an annular shaped ring of width D and of average radius $R_\circ$ (Fig. 3.4).

To explain the quantitative relations in the MFTM we consider first a line object which goes parallel to the median plane of the working volume. The position of this object is defined by the orientation angle θ_{xy}, by the distance $\rho_\circ$ to the optical axis and by $z_\circ$-coordinate with $z = 0$ for line object in the median plane of the working volume. From Fig. 3.5 we see that

$$\left.\begin{aligned} \rho_\circ &= \frac{\rho_1 - \rho_2}{2\cos\alpha_{1/2}}, \\ z_\circ &= \frac{\rho_1 + \rho_2}{2\sin\alpha_{1/2}}, \end{aligned}\right\} \tag{3.1}$$

where ρ_1 and ρ_2 are the "apparent" distances to the point 0 for the left and for the right observers, respectively, and $\alpha_{1/2}$ is the angle between the marginal optical axis and the main optical axis of the MFTM. If the line object has a small vertical inclination, the dip angle θ_z of the line object can be estimated if we measure z-coordinates of this line object in several adjacent fields of view (Fig. 3.6).

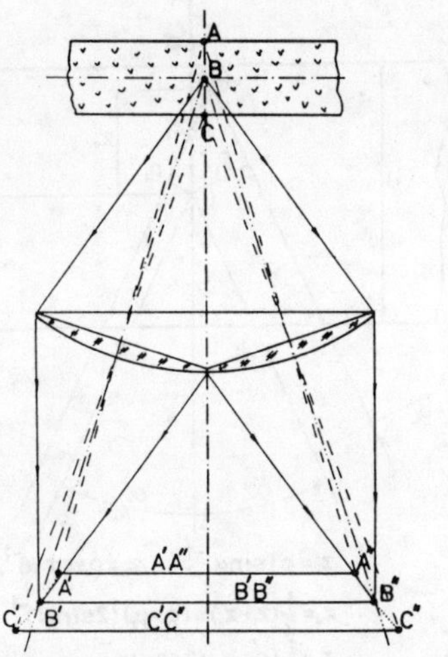

Fig. 3.3. The meridional cross-section of the refractive equivalent of the MFTM, shown in Fig. 3.1, which explains the main feature of the MFTM: the diameter of the focal circle, containing **two** meso–optical images of the straight line object, is changing with z-coordinate of this straight line object. These objects A, B and C are going perpendicular to the plane of the figure, $A'A''$, $B'B''$ and $C'C''$ are end points of the corresponding focal circles.

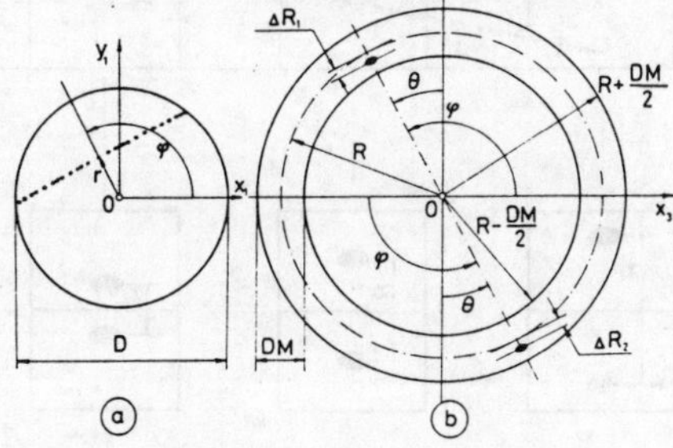

Fig. 3.4. Positions of two meso–optical images of the straight line object over the annular focal ring: a — position of the straight line object in the field of view of the MFTM, where r and φ are polar coordinates, b — annular focal ring with two meso–optical images, where φ and $\Delta R_1 = \rho_1$ are coordinates of the left meso–optical image, whereas $(\varphi + \pi)$ and $\Delta R_2 = \rho_2$ are coordinates of the right meso–optical image, ΔR_1 and ΔR_2 being taken with respect to the central focal circle of the radius R, D is the diameter of the field of view, and M is the linear magnification of the MFTM.

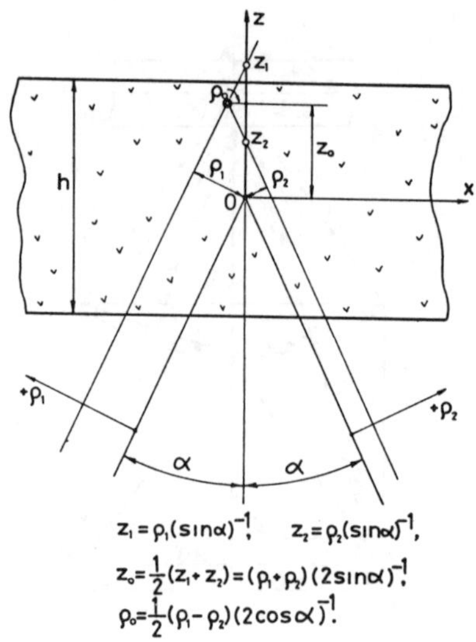

Fig. 3.5. Microscopic structure of the working volume with straight line object, oriented perpendicular to the figure: O — central point in the median plane of the working volume on z-axis, ρ_1 and ρ_2 are the apparent distances of the straight line object to the central point O, ρ_o is the distance from the optical axis to the straight line object, z_1 and z_2 are apparent z-coordinates of the straight line object, z_o is its true z-coordinate, h is the thickness of the working volume.

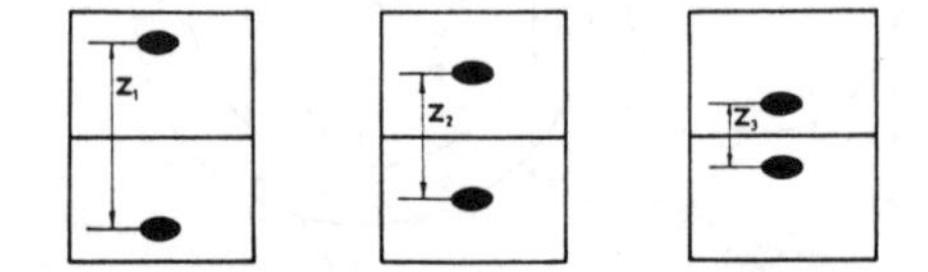

Fig. 3.6. Conventional method of estimation of the dip angle θ_z of the straight line object. Top — cross-section of the working volume with straight line object, bottom — positions of two meso–optical images on the screen of the CCD-photodetector.

The dip angle θ_z of the inclined line object can be measured by another technique without involving any movement of parts of the MFTM with respect to the working volume [129]. To understand the influence of the dip angle θ_z on the position of the meso-optical images of a given line object, let us consider first the end points

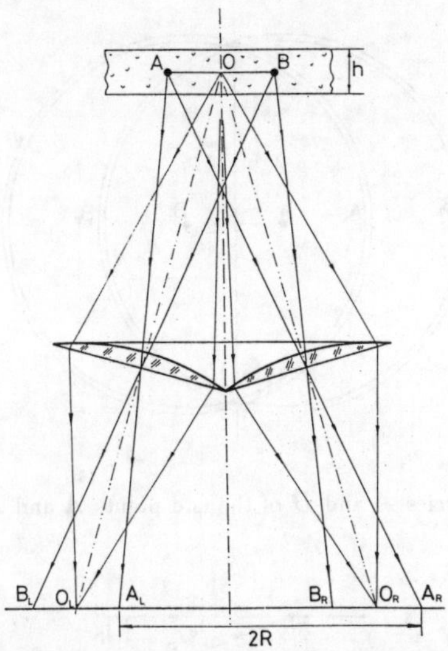

Fig. 3.7. Sagittal cross-section of the working volume with horizontal straight line object AB and the end points of the annular circles in the MFTM: A_L and A_R are the left and the right images of the end point A, B_L and B_R are the left and the right images of the end point B.

of horizontal line object (Fig. 3.7). Each end point goes into its own focal circle, and the cross points of two focal circles are on the line which goes through the optical axis of the MFTM (Fig. 3.8). If line object has a dip angle $\theta_z \neq 0$ in respect to the median plane of the working volume (Fig. 3.9), the radii of two focal circles of two end points of this inclined line object will be different (Fig. 3.10), and the line connecting two meso–optical images of the given line object does not cross the optical axis of the MFTM (Fig. 3.11), though it goes parallel to the line which connects meso–optical images of another line object with the same orientation angle θ_{xy} but with dip angle $\theta_z = 0$. The dip angle θ_z is defined by the divergent angle Φ [126].

If $h/H \ll 1$, where H is the distance between object and meso–optical mirror with ring response, we have [126]

$$\sin\Phi = \tan\theta_z \cdot \tan\alpha_{1/2}\,. \tag{3.2}$$

It should be noted that line object with very high dip angle can never produce meso–optical images. There is a critical dip angle θ_z^{crit} for which the focal circles have only one common cross point [126]. The critical dip angle is given by the equation

$$\theta_z^{crit} = \tan^{-1}[\cot\alpha_{1/2}] = \frac{\pi}{2} - \alpha_{1/2}\,. \tag{3.3}$$

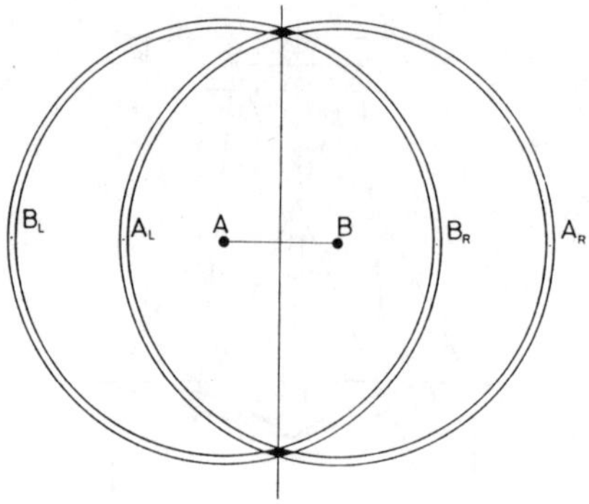

Fig. 3.8. View of two focal circles A and B of the end points A and B of the horizontal straight line object $A\,B$.

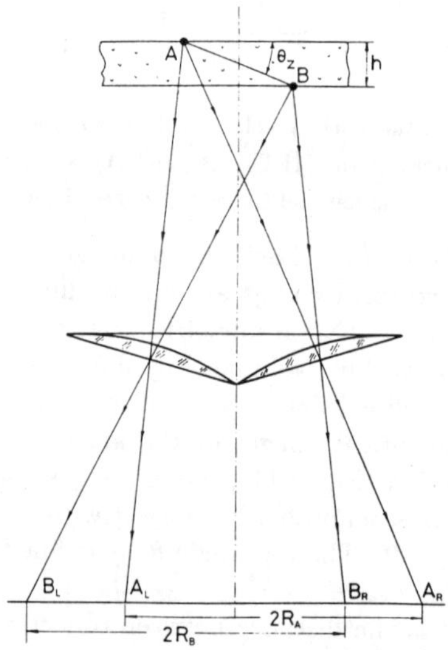

Fig. 3.9. Sagittal cross-section of the working volume with inclined straight object $A\,B$ and the end points of the annular circles in the MFTM: A_L and A_R are the left and the right images of the end point A, B_L and B_R are the left and the right images of the end point B, the diameters of the focal circles are different: $2\,R_A \neq 2\,R_B$.

To illustrate this feature of the MFTM two experiments have been performed [126]. In the first one the Fourier transform of the test line object in the form of a test-cross has observed at several dip angles: $\theta_z = 0°$, $45°$, and $80°$ (Fig. 3.12). The diffraction picture includes both Fraunhofer and Fresnel parts (Fig. 3.13). The information about the dip angle θ_z and its sign is defined by the curvature radius

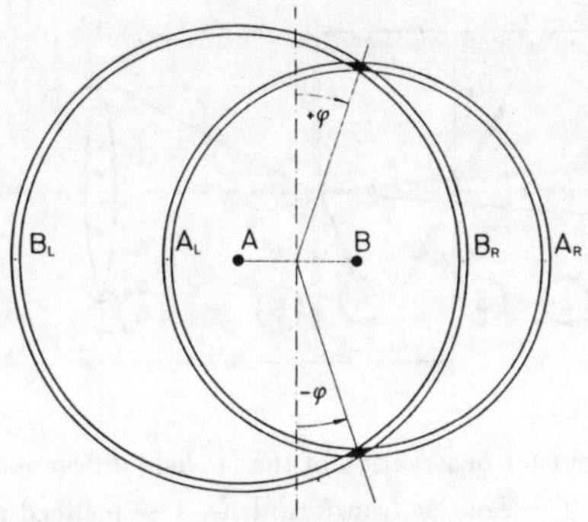

Fig. 3.10. View of two focal circles A and B of the end points A and B of the inclined line object AB, φ is the divergent angle.

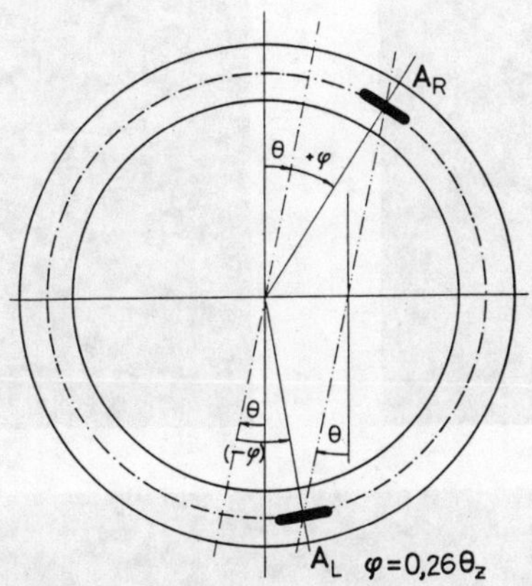

Fig. 3.11. Relation between the orientation angle θ, divergent angle φ and the dip angle θ_z of the straight line object observed in our design of the MFTM.

and by the curvature sign of this diffraction picture. In the second experiment [126] two meso–optical images of the test object have been observed. For this aim a kinoform with ring response [119] was used. In Fig. 3.14 we see two meso–optical images for different dip angles θ_z, and also the blurred diffraction picture. The later was seen because the kinoform with ring response had low diffraction efficiency ($\sim 60\%$). The results of the comparison of the theory with experiment are presented in Fig. 3.15 for test object on the glass backing. The calculated curve is given by the equation:

$$\sin\Phi \;=\; \tan\alpha_{1/2}\cdot\tan\left(\theta_z\,\frac{n-1}{n}\right),\;\; \theta_z < 10^\circ\,. \tag{3.4}$$

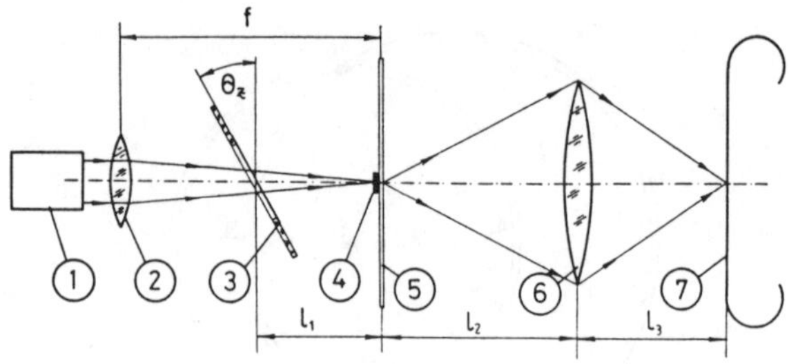

Fig. 3.12. The experimental layout for observation of the far field diffraction picture of the inclined straight line objects: 1 — laser, 2 — Fourier transform lens, 3 — inclined test object, 4 — central stop, 5 — screen, 6 — imaging objective, 7 — film.

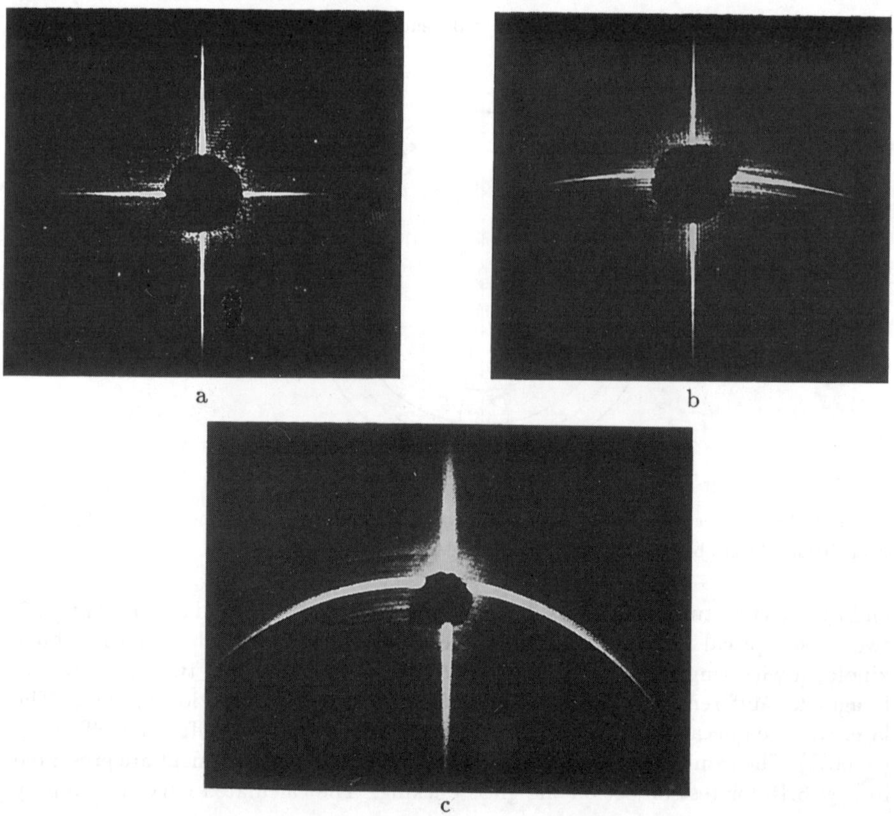

Fig. 3.13. Far field diffraction picture of the inclined test object, observed for different dip angles θ_z: a — 0°; b — 45° and c — 80°.

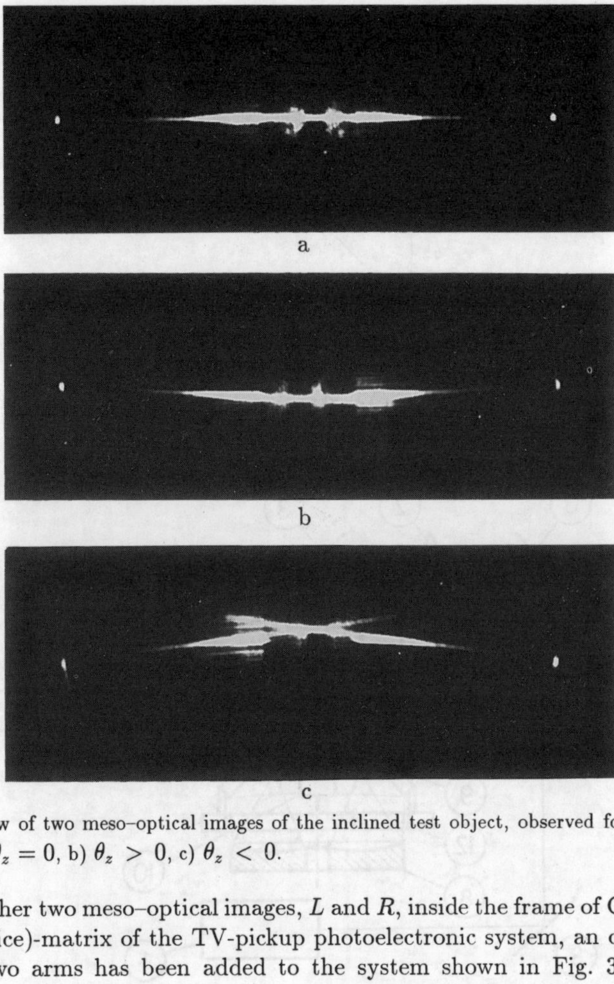

Fig. 3.14. View of two meso–optical images of the inclined test object, observed for different dip angles θ_z: a) $\theta_z = 0$, b) $\theta_z > 0$, c) $\theta_z < 0$.

To put together two meso–optical images, L and R, inside the frame of CCD(charge-coupled device)-matrix of the TV-pickup photoelectronic system, an optical interface with two arms has been added to the system shown in Fig. 3.1 [2]. The functional scheme of the computer controlled MFTM with one CCD-matrix [103] is given in Fig. 3.16 [2].

The light source 1 is a 5 mW He–He laser. The lens 2 with focal length $f = 400$ mm accomplishes $2D$ spatial Fourier transformation of the line objects in the working volume. The positive cylindrical lens 4 decreases the width of the illuminated region and thus diminishes the noise from randomly distributed silver grains. The dimensions of the rectangular field of view were $700\,\mu$m $\times$ $50\,\mu$m. The negative cylindrical lens 10 increases the width of the meso–optical images from $2\,\mu$m to $50 - 200\,\mu$m and thus matches the high spatial resolution of the MFTM,

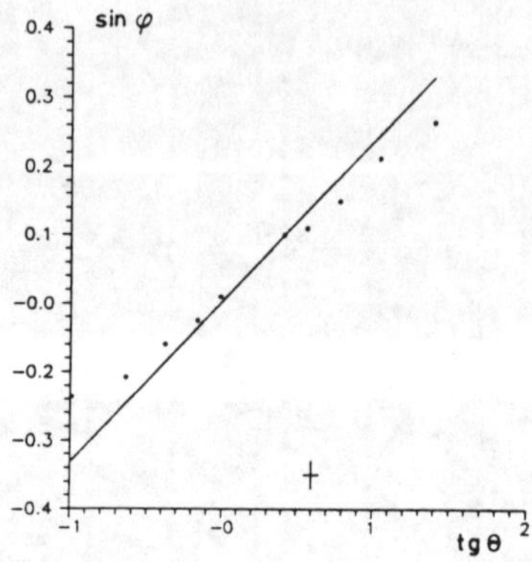

Fig. 3.15. Relation between $\sin\varphi$ and $\tan\theta_z$ for straight line test object in air: points — the results of the experiment, line — theory.

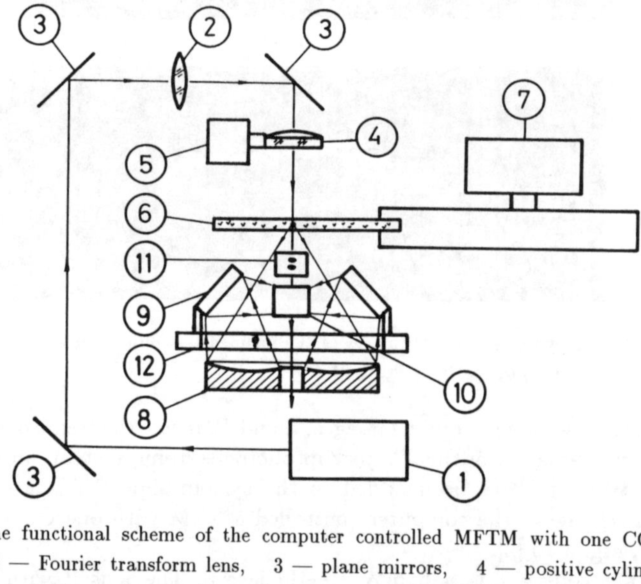

Fig. 3.16. The functional scheme of the computer controlled MFTM with one CCD matrix: 1 — laser, 2 — Fourier transform lens, 3 — plane mirrors, 4 — positive cylindrical lens, 5 — rotating lens holder, 6 — working volume, 7 — $(X-Y)$-stage equipped with position indicators, 8 — meso–optical mirror with ring response, 9 — plane mirrors of the optical interface, 10 — negative cylindrical lens, 11 — CCD-TV-camera, 12 — rotating platform of the optical interface.

$\sim 2\,\mu$m, with dimensions of the picture element (pixel) of the photosensitive detector [123, 133], $\sim 30\,\mu$m. Astigmatically deformed two meso–optical images, L

and R, are transferred by plane mirrors 9 and negative cylindrical lens 10 onto the common CCD-matrix of overall dimensions [2] 3×5 mm with 232×144 pixels.

Four degrees of freedom which possess two meso–optical images of the line object within CCD-matrix are explained in Fig. 3.17. There is one-to-one correspondence between these four degrees of freedom and four coordinates which define the orientation and the spatial position of the microscope line object: orientation angle θ_{xy}, dip angle θ_z, transverse coordinate x and depth coordinate z. There are four combinations of the coherent moving modes of two meso–optical images in the frame of the CCD-matrix. Each such moving mode corresponds to one of these four coordinates.

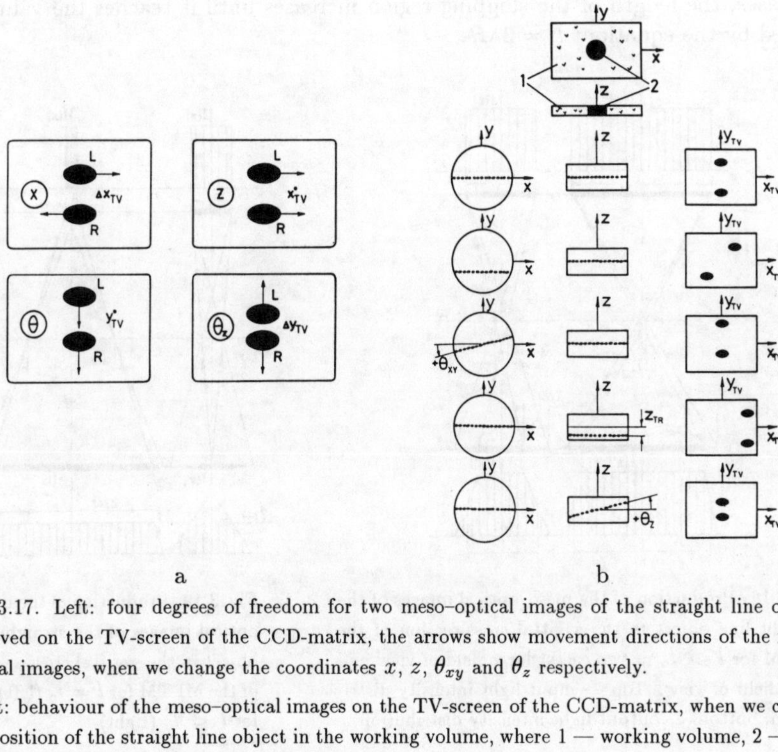

a b

Fig. 3.17. Left: four degrees of freedom for two meso–optical images of the straight line object, observed on the TV-screen of the CCD-matrix, the arrows show movement directions of the meso–optical images, when we change the coordinates x, z, θ_{xy} and θ_z respectively.
Right: behaviour of the meso–optical images on the TV-screen of the CCD-matrix, when we change the position of the straight line object in the working volume, where 1 — working volume, 2 — field of view, x, y, z are coordinates of the points in the working volume, x_{TV}, y_{TV} are coordinates of the points on the TV-screen, the position of the straight line object in the working volume is defined by four coordinates: x, z, θ_{xy}, θ_z.

3.3. Sagittal Section of the MFTM

The MFTM in a sagittal section can be considered as one-dimensional $1D$ "pinhole" camera with a virtual transmitting slit oriented perpendicular to the given line object [120]. The diffracted light rays in the sagittal section of the MFTM are shown in Fig. 3.18 for line object of the length l. The diameter of the crossover of the convergent light beam is equal to $\Delta\omega = \lambda H/l$. The light intensity distribution $I(x')$ along meso–optical image in the sagittal section can be approximated for $l > \Delta\omega$ by a trapezium with small slopping regions on both sides. As the length l decreases, the length of the slopping region increases until it reaches the value $l_\circ$, defined by the equation: $l_\circ^2 = 2\lambda H$.

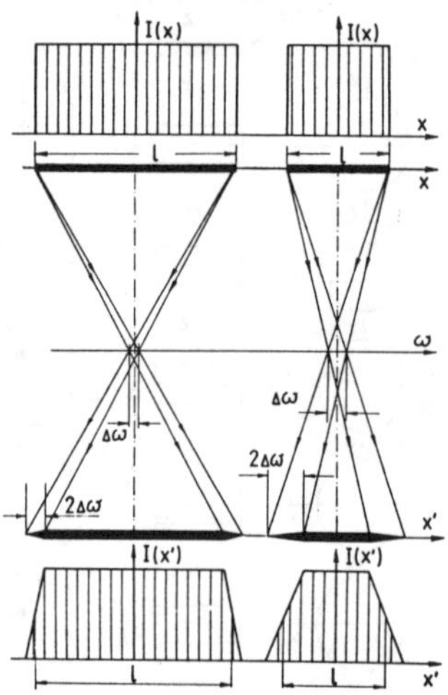

Fig. 3.18. Production of the meso–optical images of the straight line object in the sagittal cross-section of the MFTM for $l > l_\circ$ in two cases: large field of view and small field of view. Top — input light intensity distribution, bottom — output light intensity distribution.

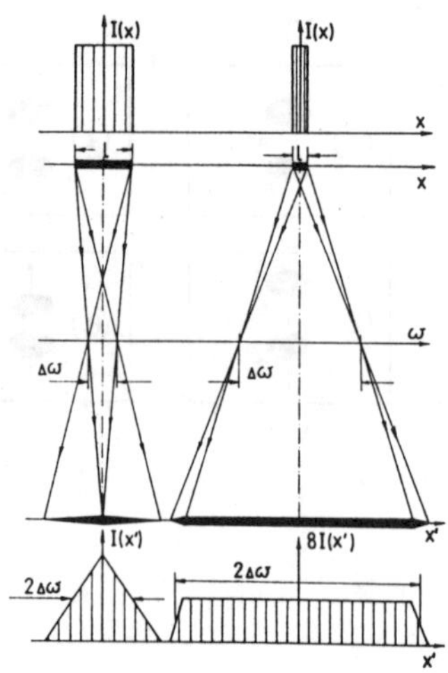

Fig. 3.19. Production of the meso–optical images of the straight line object in the sagittal cross-section of the MFTM for $l = l_\circ$ (left) and for $l < l_\circ$ (right).

For $l < l_\circ$ the diameter of the crossover exceeds the length of the line object. The function $I(x')$ is spread out over the length $2\Delta\omega$, and the intensity of the light drops to the value $l/2\Delta\omega$ (Fig. 3.19). Two experiments have been performed [120]. In the first one we have chosen $l > l_\circ$ (Fig. 3.20), and in the second one we had $l < l_\circ$ (Fig. 3.21). In the case of $l < l_\circ$, the meso–optical image of the line object assumes the form of an arc with angular spread $\Delta\theta \approx \delta/l$, where δ is the width of the line object. These experiments support the suggestion that MFTM in the sagittal section should be considered as a one-dimensional "pinhole" camera with

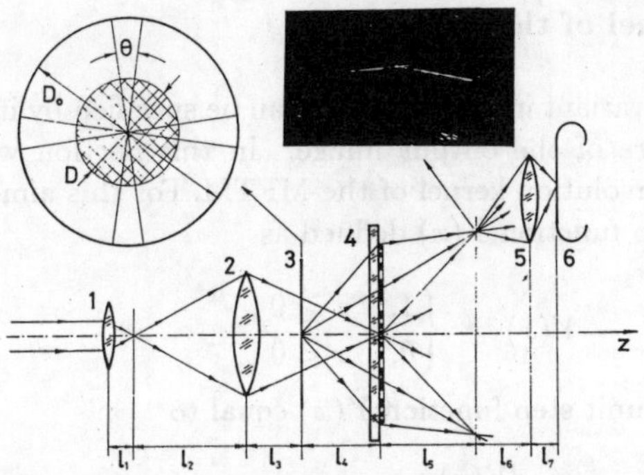

Fig. 3.20. The object in the form of a double cross, the experimental layout for observation of the left meso–optical images of two straight lines for $l > l_o$ and view of these two (left) meso–optical images.

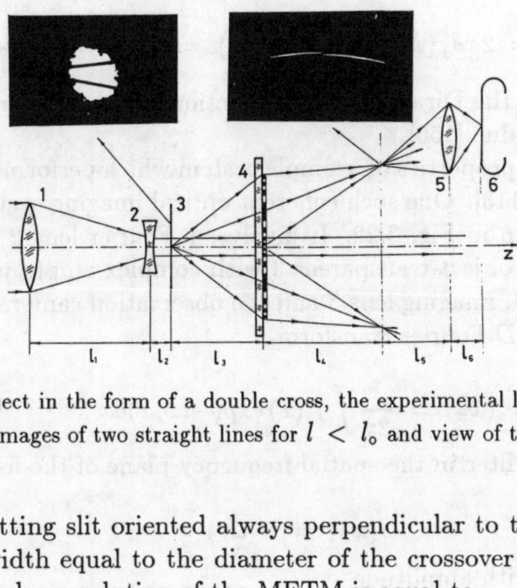

Fig. 3.21. The object in the form of a double cross, the experimental layout for observation of the left meso–optical images of two straight lines for $l < l_o$ and view of these two (left) meso–optical images.

virtual transmitting slit oriented always perpendicular to the observed line object and is of the width equal to the diameter of the crossover of the convergent light beam. The angular resolution of the MFTM is equal to

$$\Delta\theta_{xy}^{eff} = \frac{\lambda H}{\sqrt{2}\,DR}, \tag{3.5}$$

or $\Delta\theta_{xy}^{eff} \approx 1'$ for $\lambda = 0.63\,\mu$m, $D = 3$ mm, $H = 150$ mm, and $R = 80$ mm.

3.4. Convolution Kernel of the MFTM

Any coherent space-invariant imaging system can be specified by its kernel $h(x)$, which define the structure of the output image. In this Section we deduce the analytical form of the convolution kernel of the MFTM. For this aim we recall the properties of the unit step function $Y(x)$ defined as

$$Y(x) = \begin{cases} 1, & x \geq 0\,, \\ 0, & x < 0\,, \end{cases} \tag{3.6}$$

and the derivative of the unit step function $Y(x)$ equal to

$$\frac{dY(x)}{dx} = \delta(x)\,, \tag{3.7}$$

where $\delta(x)$ is Dirac delta function. The Fourier transform of the unit step function $Y(x)$ is given by

$$\hat{\mathcal{F}}[Y(x)] = 2\pi\delta_+(x), \quad \text{with} \quad \delta_+(x) = \frac{1}{2}\,\delta(\omega) + \frac{1}{2\pi i\omega}\,. \tag{3.8}$$

The second term of the Dirac delta-plus function $\delta_+(\omega)$ is taken in the sense of the Cauchy principal value [156].

First we deduce the properties of a simple system which performs a spatial filtering of the input scene [113]. One such coherent optical imaging system with spatial frequency filter is shown in Fig. 3.22. It involves a Fourier lens 2 for producing a convergent light beam, object-transparent 3 with complex amplitude transparency $f_\circ(x)$, knife-edge filter 4, imaging lens 5 and the observation camera 6. We consider $1D$ spatial filter with $1D$ Fourier transform

$$F_\circ(\omega_x) = \frac{1}{2\pi}\int f_\circ(x)\exp(-i\omega_x x)dx\,. \tag{3.9}$$

The $1D$ knife-edge filter in the spatial frequency plane of the form

$$G(\omega_x) = Y(\omega_x) \tag{3.10}$$

generates a wavefield with amplitude

$$F(\omega_x) = F_\circ(\omega_x)\cdot Y(\omega_x)\,. \tag{3.11}$$

After inverse Fourier transformation we get the light field with amplitude

$$f_\Phi(x) = \frac{1}{2}\,[f_\circ(x) + i\,\chi(x)]\,, \tag{3.12}$$

where $\chi(x)$ is the Hilbert transform [157] of the function $f_\circ(x)$.

As an another example we consider a phase-edge filter with amplitude transparency [158]

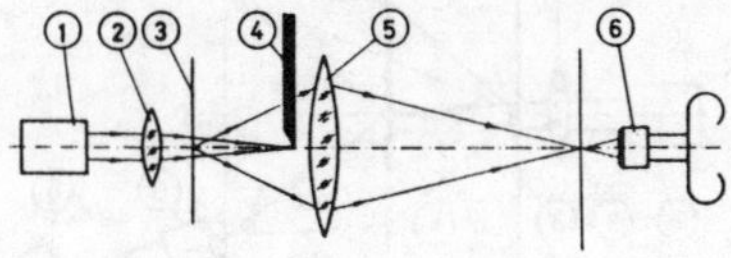

Fig. 3.22. Coherent optical imaging system which has a spatial frequency filter $G(\omega)$: 1 — laser, 2 — Fourier transform lens, 3 — object, 4 — spatial frequency filter, 5 — imaging lens, 6 — camera.

$$G(\omega) \;=\; i\,\mathrm{sgn}\,\omega \;=\; i\begin{cases} 1, & \omega \geq 0\,, \\ -1, & \omega < 0\,, \end{cases} \tag{3.13}$$

where the sign function $\mathrm{sgn}\,\omega$ is related to the unit step function $Y(x)$ by the equation

$$\mathrm{sgn}(\omega) \;=\; 2\,Y(\omega) \;-\; 1(\omega)\,. \tag{3.14}$$

The Fourier transform of the *sgn* function is equal to

$$\hat{\mathcal{F}}[\mathrm{sgn}(\omega)] \;=\; \hat{\mathcal{F}}[2\,Y(\omega) \;-\; 1(\omega)] \;=\; 2\,\delta_+(x) \;-\; \delta(x) \;=\; \frac{1}{\pi\,i\,x}\,. \tag{3.15}$$

Of a particular interest for us is the spatial filter in the form of the diffraction grating with phase jump [159]. The distance between two adjacent central grooves of this grating is equal to $a/2$, where a is the grating period. The line of this phase jump is located on the line $\omega_x = 0$. The amplitude of the filtered field observed in the first diffraction order is equal to the Hilbert transform $\chi(x)$ of the input field with amplitude $f(x)$ [157]. In the general case the distance between two adjacent central grooves can be made not equal $a/2$, and the intensity of the signal in the first diffraction order will be given by the equation

$$I(x) = |\cos\alpha \,\cdot\, \chi(x) + \sin\alpha \,\cdot\, f(x)|^2\,, \tag{3.16}$$

where α is the difference angle which is equal to zero for perfect diffraction grating with phase jump and is equal to $\pi/2$ for ordinary diffraction grating without any phase jump. In the general case the convolution kernel of this "wrong" filter is equal to

$$\delta_n(x,\alpha) \;=\; \sin\alpha \,\cdot\, \delta(x) \;+\; \cos\alpha \,\cdot\, \frac{1}{\pi\,x}\,. \tag{3.17}$$

The inherent property of the convolution kernel, defined by Eq.(3.17), is that two components are in phase, whereas for Dirac delta-plus function these two components are in quadrature.

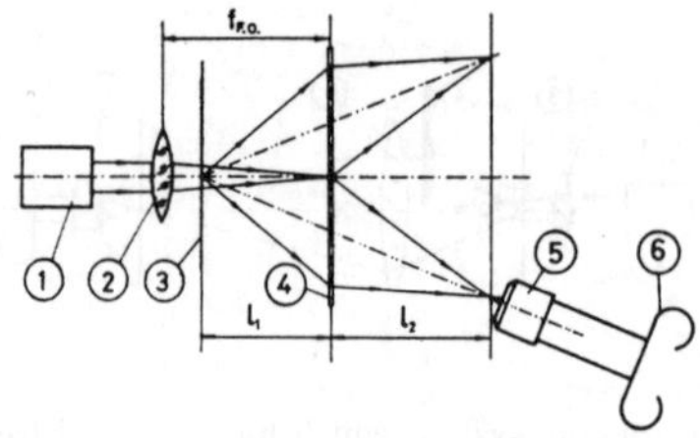

Fig. 3.23. Coherent optical imaging system which contains the meso–optical element with ring response: 1 — laser, 2 — Fourier transform lens, 3 — object, 4 — meso–optical element with ring response, 5 — observation camera, 6 — screen.

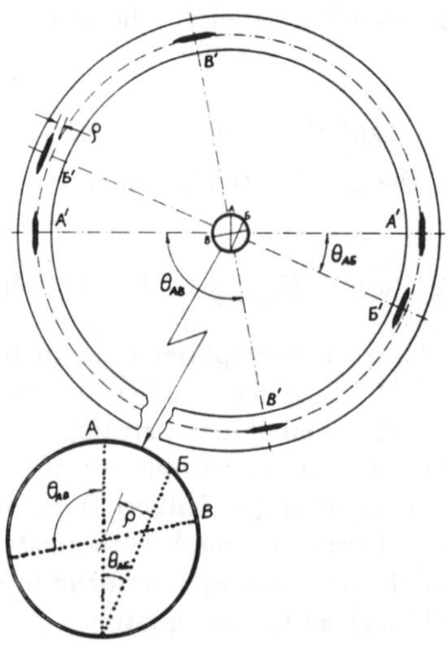

Fig. 3.24. Structure of the transformed object (three straight line segments) observed in the meso–optical imaging system, shown in Fig. 3.23.

Now let us consider the meso–optical counter-part of the imaging system shown in Fig. 3.22, with meso–optical filter with ring response (Fig. 3.23). We know that position of the meso–optical images of the line object on the focal ring is defined by the orientation angle θ_{xy}, by the distance from the center of the field of view ρ_o and by its z-coordinate [3]. An example of three line object A, B, C is shown in Fig. 3.24. Three pairs of meso–optical images are located within the focal ring.

Now we prove that the line object is subjected in the meridional section of the MFTM by the spatial transformation which is described by the Dirac delta-plus function. To prove this theorem we note that in the MFTM a knife-edge filter with amplitude transparency defined by Eq.(3.10) is present virtually [113]. This follows from the fact that each meso–optical image of the line object is produced only by the positive (or negative) spatial frequencies and that the edge of this virtual knife-edge filter is always oriented parallel to the orientation of the line object. Unlike the traditional Hilbert filter in optics, each line object has its own virtual knife-edge filter no matter how many line objects are present in the field of view of the MFTM. Furthermore, the light which produces the meso–optical image, is gathering over a narrow sector of the meso–optical element with ring response. From this fact it follows that Fourier transform of the line object is subjected to multiplication by the function $i\omega$, or spatial differentiation of the filtered spatial function. Hence we may conclude that the initial optical field $f_\circ(x)$ is subjected in the MFTM to the following sequence of transformations [113]:

$$f_\circ(x) \overset{\hat{\mathcal{F}}_x}{\rightarrow} F_\circ(\omega_x) \overset{Y(\omega_x)}{\rightarrow} F_\circ(\omega_x) \cdot Y(\omega_x) \overset{i\,\omega_x}{\rightarrow}$$

$$\rightarrow i\,\omega_x\, F(\omega_x)\, Y(\omega_x) \overset{\hat{\mathcal{F}}_x^{-1}}{\rightarrow} \frac{d}{dx}\left[\delta_t(x)\, A\, f_\circ(x)\right] =$$

$$= \frac{d}{dx}\left[f_\circ(x) + i\chi(x)\right] = \frac{d}{dx}\, f_\circ(x) + i\,\frac{d}{dx}\,\chi(x)\,, \tag{3.18}$$

for the left meso–optical image, and equals to $= \frac{d}{dx}\, f_\circ(x) - i\,\frac{d}{dx}\,\chi(x)$ for the right meso–optical image. Thus, the convolution kernel of the MFTM in the meridional section is equal to the first spatial derivative of the Dirac delta-plus and delta-minus functions.

To verify the prediction of this theory an experiment has been performed [107]. We used a kinoform with ring response of the focal length $f = 202\,\text{mm}$ and with focal ring of the width $10\,\mu\text{m}$. A fragment of the meso–optical image of the object in the form of the spectral slit of width $40\,\mu\text{m}$ is shown in Fig. 3.25. The effect of spatial differentiation of the object is clearly visible. In the same fashion the meso–optical images of the open circles, rectangles, and a semiplane shutter were observed. An object in the form of an open circle is transformed by the MFTM into two narrow concentric circles.

3.5. Meso–optical Moiré-effect

If we bring into a close contact two regular diffraction gratings, a classical moiré-effect is observed. It involves difference and sum fringes [142], which are oriented, respectively, perpendicular and parallel to the bisector of the angle α between grooves of two crossed diffraction gratings. In the case when the angle α is very small, the period of the sum fringes is equal to the period of the diffraction grating. The observer does not notice the sum moiré-effect on the background of the well visible difference moiré-effect [142]. If one of the diffraction grating is moving with respect to another one in a direction perpendicular to the orientation of the grooves of the second diffraction grating, the fringes of the difference moiré-effect are moving in the direction perpendicular to the relative velocity vector. The fringe period of the difference moiré-effect is $1/\alpha$ times the period of the diffraction grating, and the velocity of the linear movement of these fringes is $1/\alpha$ times the relative velocity of these diffraction gratings.

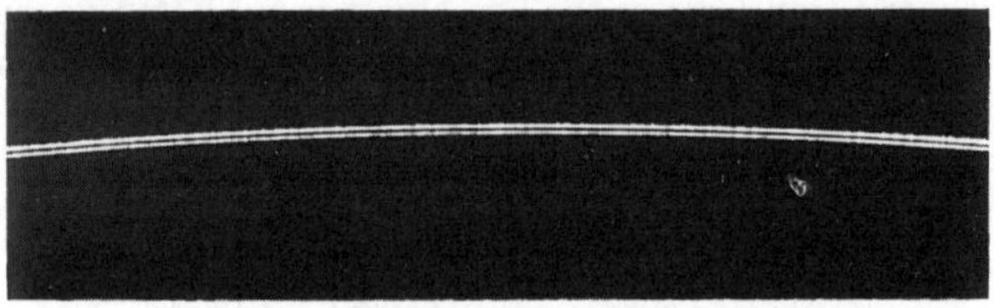

Fig. 3.25. Meso–optical image of the open spectral slit of the width 30 μm. The transverse spatial resolution is equal to 10 μm. The effect of the spatial differentiation is clearly visible.

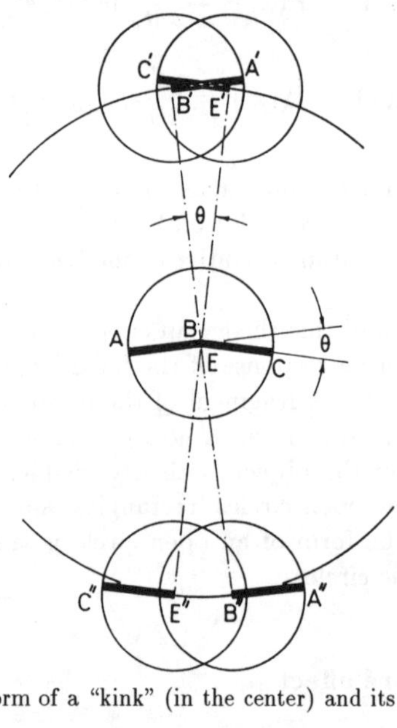

Fig. 3.26. The event in the form of a "kink" (in the center) and its meso–optical images (top and bottom).

A meso–optical analogue of the classical moiré-effect observed in the MFTM involves the following phenomena [111]. Let the external tip of the right half of the line object be subjected to rotation around the center of the field of view in the direction perpendicular to the orientation of the line object. As a consequence both meso–optical images of this line half-object will be moved along initial meso–optical images, that is perpendicular to the velocity vector of the external tip of the right half-object. This is the first characteristic of the meso–optical moiré-effect which is in close analogy with corresponding characteristic of the classical moiré-effect. The linear velocity of this movement is R/D times of the linear velocity of the external tip of the right line half-object. The same qualitative relation holds for absolute displacement.

To explain clearly the meso–optical analogue of the moiré-effect observed in the MFTM let us choose an event in the form of the "kink" shown in Fig. 3.26. A particle track AB produced by some primary particle undergoes a scattering at small angle θ and produces a secondary particle track EC with $B = E$. We see that the left and the right meso–optical images of this "kink" event are different in shape. The left observer sees the event ABC as a convex obtuse angle, whereas the right observer sees the same event CBA as a concave obtuse angle. So the shapes of the left and of the right meso–optical images of this event will be different.

For the left configuration two meso–optical images of the line objects $C'E'$ and $B'A'$ do overlap each other. As the angle θ increases the point B' moves along the meso–optical image $C'E'$ and there occurs an overlapping of two meso–optical images on the segment $B'E'$. This follows from the first characteristic of the meso–optical moiré-effect. In the right configuration we observe a gap between two meso–optical images $C'E'$ and $B'A'$. The length of the gap is equal to the length of the overlapping region.

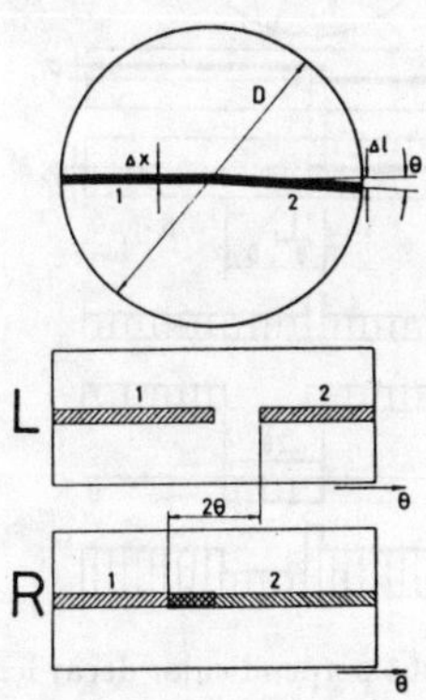

Fig. 3.27. The "kink-event" with small scattering angle θ: top — the object, bottom — the left (L) and the right (R) meso–optical images.

This meso–optical moiré-effect can be used in the MFTM for estimation of very small scattering angle θ. The left and the right meso–optical configurations are shown in Fig. 3.27. The light intensity distributions $I_L(\theta)$ and $I_R(\theta)$ versus angular coordinate θ are presented in Fig. 3.28 for both meso-optical configurations together with difference signal $\Delta(\theta) = I_L(\theta) - I_R(\theta)$ and sum signal $\sum(\theta) = I_L(\theta) + I_R(\theta)$. To estimate small angle θ it is sufficient to measure the length of the rectangular pulse in the signal $\Delta(\theta)$ or $\sum(\theta)$.

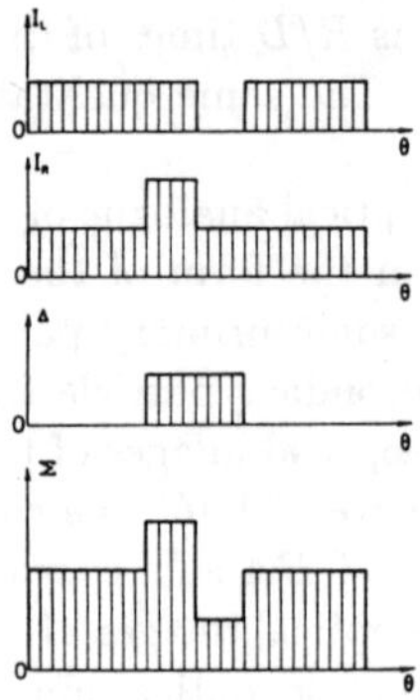

Fig. 3.28. Light intensity distributions I_L and I_R versus angle θ, and functions $\Delta(\theta)$ and $\sum(\theta)$.

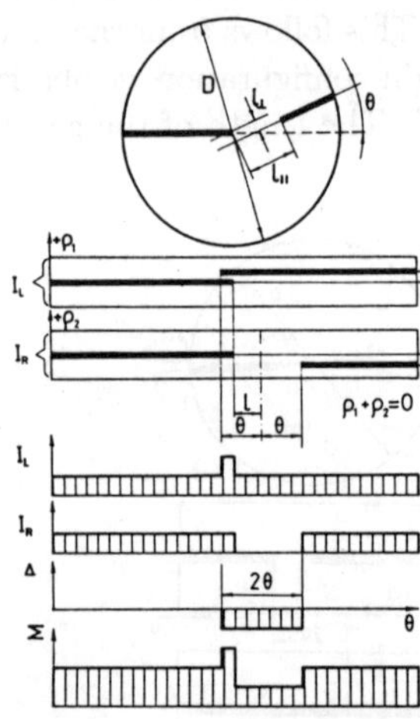

Fig. 3.29. Technique used for estimation of the perpendicular decay length $l_\perp$ and of the longitudinal decay length $l_{||}$.

The same approach can be used for more complex object, for example, when one of particle track is located eccentrically so, that it does not cross the end point of the primary particle track. In this case a gap between extreme points of these two particle tracks occurs (Fig. 3.29). Our aim is to estimate the lengths of the perpendicular $l_\perp$ and of the longitudinal $l_{||}$ components of the decay length l of the eccentrically going particle track. The meso–optical configurations for such a complex event and the light intensity distributions I_L, I_R, Δ and $\sum$ are presented in Fig. 3.29. The visible distances ρ_1 and ρ_2 for two meso–optical configurations define the perpendicular component of the decay length

$$l_{\perp} = \rho_{\circ} = \frac{|\rho_2 - \rho_1|}{2cos\alpha_{1/2}} . \tag{3.19}$$

The longitudinal component $l_{\|}$ of the decay length is defined by the length of the gap G in the signal $I_R(\theta)$, where

$$G = \theta \cdot R + l_{\|} . \tag{3.20}$$

The angle θ is determined by the length of the rectangular pulse in the difference signal $\Delta(\theta)$.

In our design of the MFTM [2, 103, 109] the radius of the focal circle $R = 80$ mm, the diameter of the field of view $D = 1$ mm, and the distance between the line object and the meso–optical mirror with ring response $H = 145$ mm. The linear magnification factor M of the meso–optical moiré-effect in this design of the MFTM is equal to $M = R/D = 80 : 1$. So if we shift the first tip of the second particle track C in Fig. 3.26 over the distance 1 μm, the end points of the meso–optical images are rotated giving rise to the linear shift of 80 μm. The length of the rectangular pulse of the difference signal $\Delta(\theta)$ will be equal to 160 μm. The measurement error is of the order of $1'$.

The most interesting conclusions we may draw by considering the object of three line segments in the form of a polygon (Fig. 3.30). It may be thought of as three tangents to the common circle arc. It is seen that the main difference between the left and the right meso–optical configurations consists in that the effective length of the left meso–optical image is larger than the effective length of the right meso–optical image. From this observation we deduce a new technique of the particle track curvature measurement which can be realized by means of the MFTM [114]. To estimate the curvature radius of the curved particle track it is sufficient to measure **two** linear quantities, the effective lengths of the left and of the right meso–optical images of this curved particle track. The principle of this technique is explained in Fig. 3.31. The effective lengths of two meso–optical images, l_L and l_R, of the curved particle track are coupled by the equation

$$\frac{l_L}{\rho + R} = \frac{l_R}{\rho - R} , \tag{3.21}$$

where ρ is the curvature radius and $2R$ is the diameter of the focal ring in the MFTM. The curvature radius is equal to

$$\rho = R\,\frac{l_L + l_R}{l_L - l_R} \approx \frac{2RL}{\xi} , \tag{3.22}$$

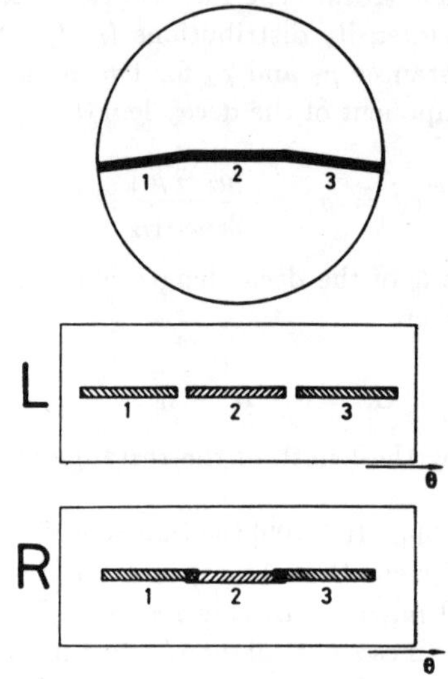

Fig. 3.30. The object of three line segments "1", "2" and "3" of the polygon; the left (L) and the right (R) meso–optical images of this object; the functions I_L, I_R, $\Delta(\theta)$ and $\sum(\theta)$ for this object.

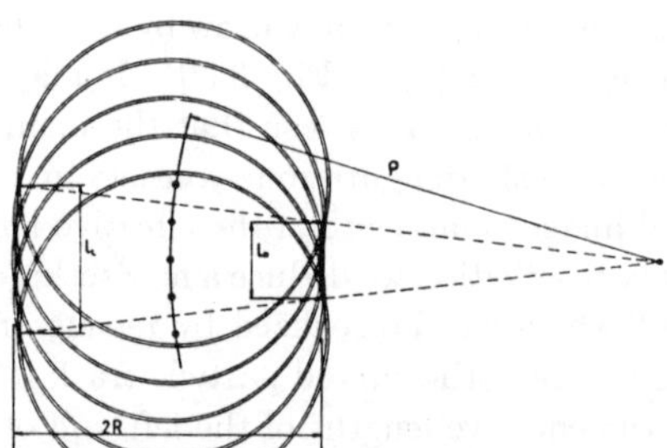

Fig. 3.31. Technique of the line curvature estimation in the case when curvature radius $\rho > R$, where R is the radius of the focal circle in this MFTM.

where $\xi = l_L - l_R$ is the difference length and L is the length of the particle track. The ratio of the difference length ξ to sagitta S of the curved particle track of the length L, where $S = L^2/8\rho$, is equal to

$$\frac{\xi}{S} = \frac{16R}{L}. \tag{3.23}$$

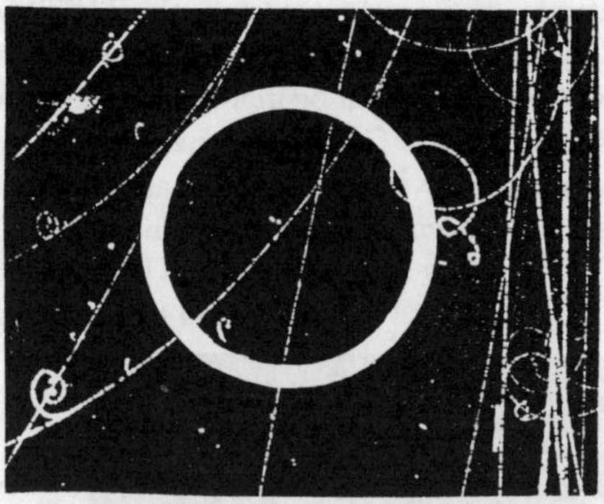

Fig. 3.32. The object of two lines with various curvature (inside the circle).

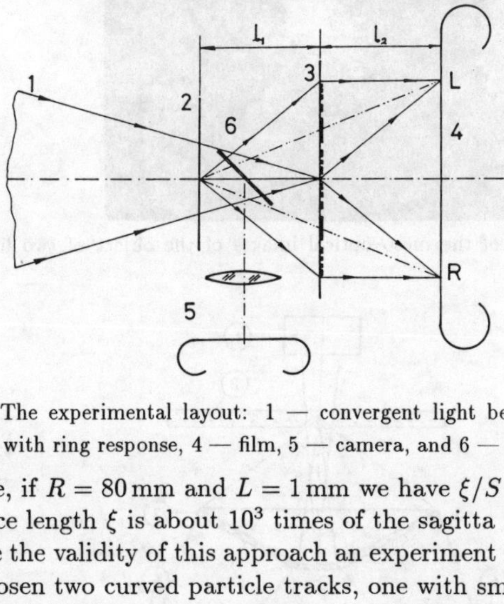

Fig. 3.33. The experimental layout: 1 — convergent light beam, 2 — object–transparency, 3 — kinoform with ring response, 4 — film, 5 — camera, and 6 — plane mirror.

For example, if $R = 80\,\text{mm}$ and $L = 1\,\text{mm}$ we have $\xi/S = 1.28 \times 10^3$. We see that the difference length ξ is about 10^3 times of the sagitta S(!).

To prove the validity of this approach an experiment have been performed [114]. We have chosen two curved particle tracks, one with small and another with large curvature radii. This object is presented in Fig. 3.32. The experimental setup is shown in Fig. 3.33. Two pairs of the meso–optical images of these two curved particle tracks are given in Fig. 3.34. This new technique offers high linear magnification of the sagitta value of the curved particle track and does not involve any mechanical displacements of the microscope parts or object itself.

3.6. MFTM with One-channel Photodetectors

The searching for particle tracks in nuclear emulsion by means of the MFTM is performed without depth scanning, and the angular selectivity is achieved in time instantly. Due to these features, the productivity of the MFTM is at least two orders of magnitude higher than in the systems with traditional optical microscope. Meanwhile the productivity of the MFTM can be increased additionally, by reducing the scanning time. For this aim the $2D$ CCD-matrix should be replaced by only two one-channel photodetectors. This innovation is based on the fact that the information about line object is concentrated in two small spots. The scanning time of each picture in CCD-matrix is about 20 ms. The scanning time of one-channel photodetector with time constant 1.0 ms is 100 times smaller.

Fig. 3.34. Photo of two pairs of the meso–optical images of the object of two lines, shown in Figure 3.32.

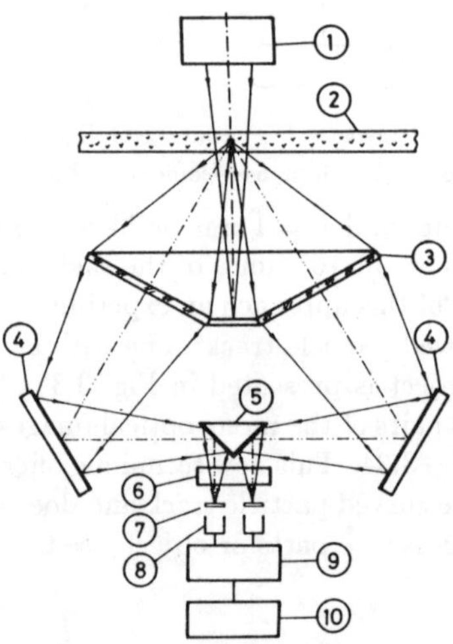

Fig. 3.35. Design of the MFTM with one-channel photodetectors: 1 — laser, 2 — working volume, 3 — meso–optical element with ring response, 4 — plane mirrors, 5 — mirror prism, 6 — negative cylindrical lens, 7 — narrow slit, 8 — photoresistor, 9 — electronics and 10 — computer.

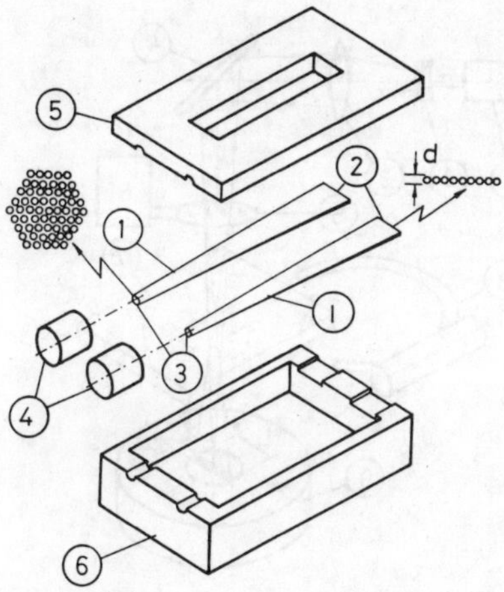

Fig. 3.36. Light guide comb, which serves as a narrow slit: 1 — light guide comb, 2 — input faces, 3 — output faces, 4 — photoresistors, 5 — top deck, 6 — case.

The principle of the MFTM with one-channel photodetector [127, 128] in each arm of the MFTM is explained in Fig. 3.35. Before entering into input face of the light-guide comb 7, the diffracted light rays are reflecting from plane mirror 4 and double-mirror 5, and pass through the negative cylindrical lens 6. The output faces of the light-guide comb (Fig. 3.36) are in optical contact with photoresistors 4. The diameter of each light-guide is equal to 15 μm.

The essential parts of this version of the MFTM are the system of the illumination and the rotating fork (Fig. 3.37). The convergent light beam has two $1D$ crossovers, one in the plane of the nuclear emulsion 2 and another in the vicinity of the meso–optical mirror with ring response 3. With this design we can produce various illumination configurations [127], four of them are shown in Fig. 3.38. In Fig. 3.38 (a) (b) each one-channel photodetector sees the whole illuminated volume of various width, $\sim 100\,\mu$m in (a) and $\sim 20\,\mu$m in (b). The slit 7 is absent. No information about z_o-coordinate of the particle track is available. In the configuration, presented in Fig. 3.38 (c), a narrow slit 7 composed of many light guides is placed in front of each photodetector (see Fig. 3.36). Only two narrow crossing regions of the illuminated volume are seen in this system. The width of the illuminated region is chosen so that $\Delta x_o \geq h \cdot \sin \alpha_{1/2}$ ($\approx 80\,\mu$m). The meso–optical signals from one particle track are detected in **two** arms of the MFTM at different moments of time at different positions of $(X - Y)$ stage. The difference between these moments of time is the measure of z_o-coordinate of the particle track. The time diagram is shown in Fig. 3.39. The spatial resolution is about 5 μm. The illumination configuration with very narrow illumination region, $\sim 2\,\mu$m, given in Fig. 3.38 (d), can be realized

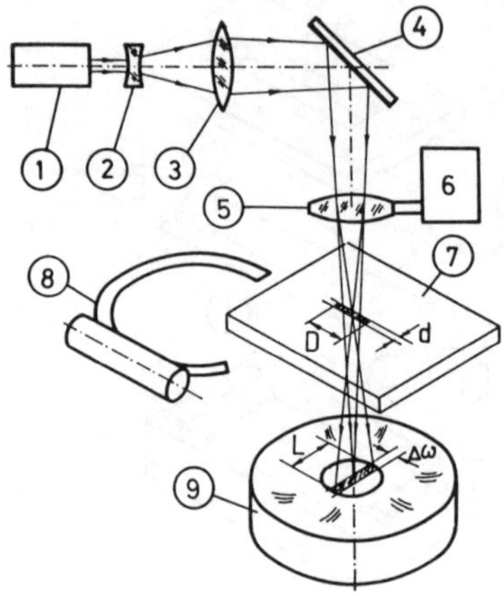

Fig. 3.37. The illuminating system of the MFTM with one-channel photodetectors: 1 — laser, 2 — negative lens, 3 — Fourier transform lens, 4 — plane mirror, 5 — positive cylindrical lens, 6 — micrometer screw, 7 — working volume, 8 — rotating fork, 9 — meso-optical mirror with ring response.

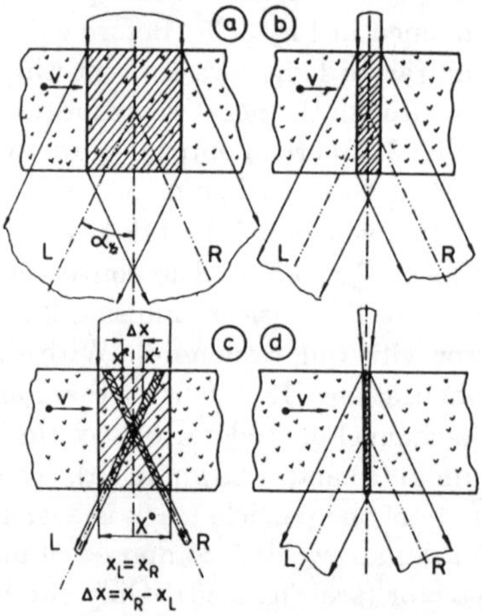

Fig. 3.38. Four basic illumination configurations: a) broad illuminated region $\sim 100\,\mu$m, b) narrow illuminated region, $\sim 20\,\mu$m, c) width $\gtrsim 80\,\mu$m, with the light guide combs in two arms of the MFTM, d) very narrow illuminated region, $\sim 2\,\mu$m.

only by means of the meso–optical illumination system. The later will be explained in Chapter 8. No information about z_o-coordinate of the particle track is available, also in the illumination configurations (a) and (b). The signal-to-noise ratio is very

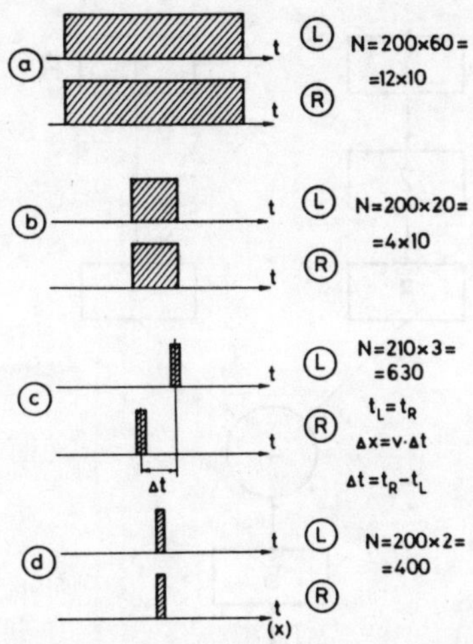

Fig. 3.39. The time diagram of video signals versus x (or t) coordinate for illumination configurations presented in Fig. 3.38.

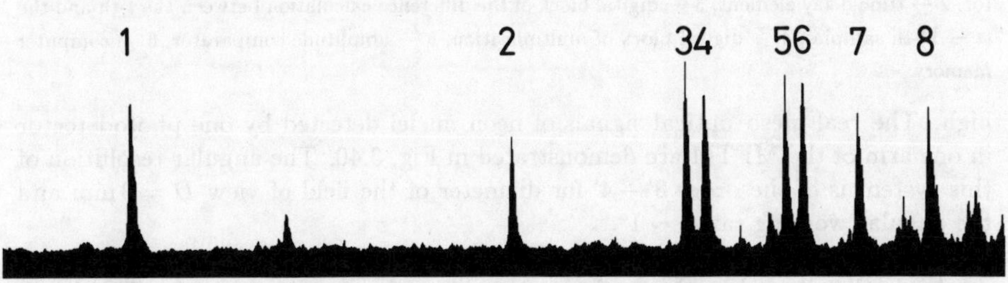

Fig. 3.40. Top: typical video signals, detected in the illumination configuration (b) (see Fig. 3.38) from accelerated relativistic neon nuclei and from some "gray" particle tracks in the nuclear emulsion. Bottom: real position of the particle tracks which induced the video signals shown top, with respect to the marked grid of period 1 mm.

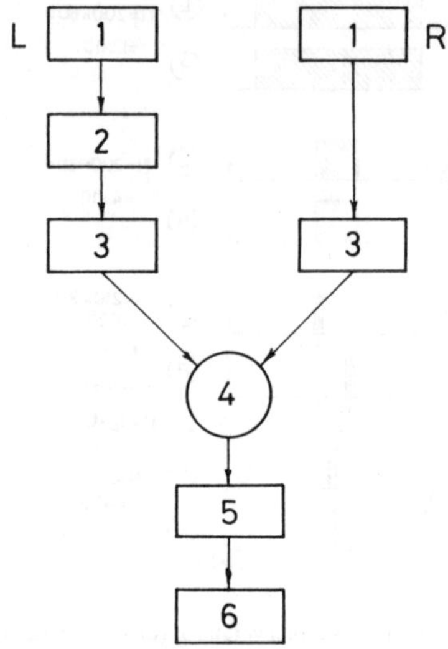

Fig. 3.41. General algorithm of the signal-to-noise ratio enhancement: 1 — one-channel photodetector, 2 — time delay element, 3 — digital block of the difference calculation between the i-th and the $(i-1)$-th samples, 4 — digital block of multiplication, 5 — amplitude comparator, 6 — computer memory.

high. The real meso–optical signals of neon nuclei detected by one photodetector in one arm of the MFTM are demonstrated in Fig. 3.40. The angular resolution of this system is of the order $3' \div 4'$ for diameter of the field of view $D = 3\,\text{mm}$ and the angular working range $\sim 1°$.

The noise, or background, is produced in the nuclear emulsion by silver grains randomly distributed in the working volume and by non-line objects. This background makes difficult the observation of the particle tracks of very low linear density of silver grains, ~ 25 per $100\,\mu\text{m}$. The dramatic compression of information which offers the MFTM with one-channel photodetectors opens new scopes of the noise suppression. New algorithms of the noise reduction have been suggested [124]. One of this algorithm involves the differentiation of the $1D$ meso–optical signals.

One such algorithm, which can be accomplished in any illumination configuration given in Fig. 3.38, is presented in Fig. 3.41. The block scheme of this algorithm is shown in Fig. 3.42 in the form of the electronic-computer system. The results of the computer simulation of this noise suppression are presented in Fig. 3.43. For illumination systems shown in Fig. 3.38 (a), (b), and (d), a second stage of the algorithm can be added. At this stage a product of two derivatives in both arms of the MFTM is calculated for each position of the $(X-Y)$ stage. It has been proved that with this processing algorithm of the $1D$ meso–optical signals we can clearly detect the particle track which have an initial signal-to-noise ratio $S/N = 1 : 7.5$.

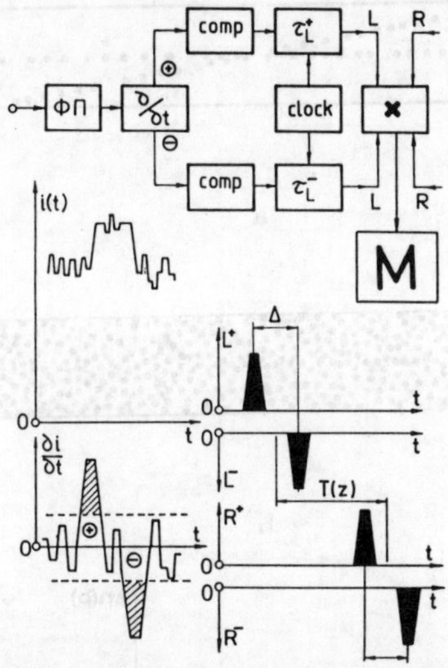

Fig. 3.42. Block scheme of the processing algorithm of the MFTM with one-channel photodetectors in the illumination configuration (c) (see Fig. 3.38): PD — photodetector, "comp" — comparator block, "clock" — generator of the clock signals for sampling, × — digital multiplication, M — memory, $i\,(t)$ — input signal, $\frac{\partial i}{\partial t}$ — digital differentiation, Δ — time delay between rising and falling parts of the electronic pulse, $T\,(z)$ — time delay between left and right arms of the scheme, which defines the z-coordinate of the particle track.

The ability of the MFTM to compress picture information about line objects in $3D$ space opens new scopes to perform some traditional measurements in another, new, fashion. As an example we consider the problem of the measurement of the linear density of silver grains in the given particle track. As will be explained below this measurement can be performed automatically in the MFTM according to non-traditional scheme [129].

Each $1D$ signal at the output of the MFTM with one-channel photodetectors represents a $1D$ function $f(x)$ of the spatial coordinate (x). As we use one-mode laser radiation, the function $f(x)$ can be approximated by the Gaussian

$$f(x) = A\,\exp(-kx^2)\,, \quad \text{with} \quad k = 1/\sigma^2\,, \tag{3.24}$$

where A is a constant defined by the linear density of silver grains belonging to the particle track, and σ is the effective width of the illuminated region. Two photoelectric signals, $f_L(x)$ and $f_R(x)$, are detected in the MFTM on a quasi-constant noise background (see Fig. 3.40). The noise components in both arms of the MFTM with one-channel photodetectors can be considered as statistically independent.

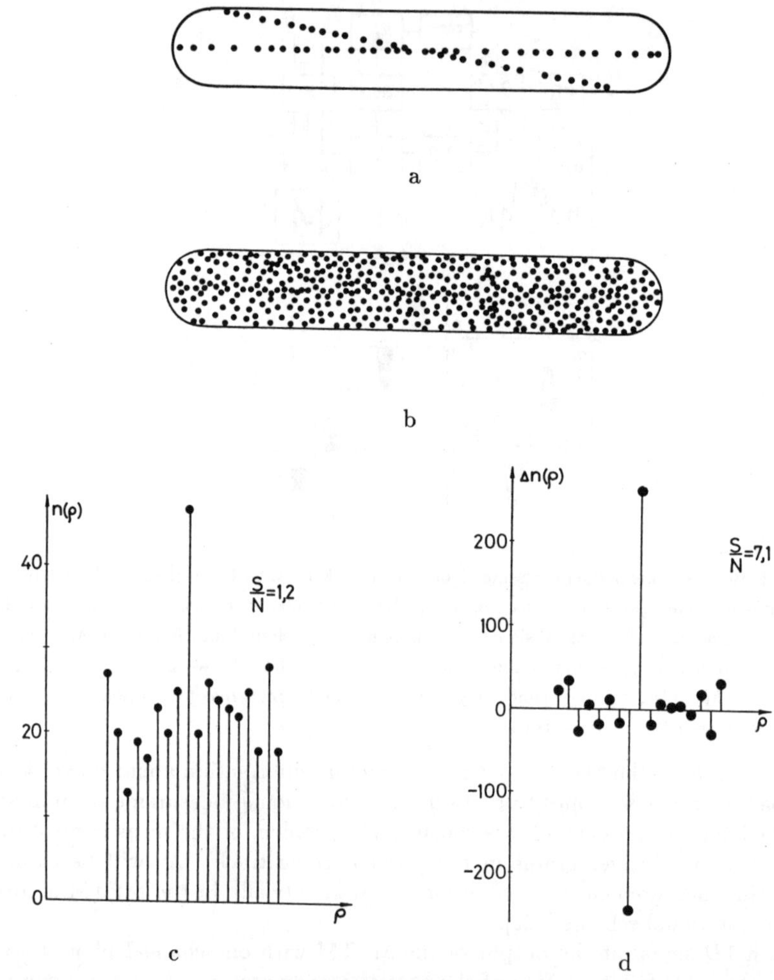

Fig. 3.43. Computer simulation of the signal differentiation: a) signals without noise, b) signals plus noise, c) photoelectronic signals in the MFTM versus transversal coordinate ρ for signals plus noise $(S/N = 1.2)$, d) the results of the differentiation $(S/N = 7.1)$.

On the first stage we detect two functions of the spatial coordinate x, $f_L(x)$ and $f_R(x)$. On the second stage we estimate the surfaces under these functions by means of the smoothing algorithm, S_L and S_R, where

$$S_L = \int_{-\infty}^{\infty} f_L(x)dx = A\int_{-\infty}^{\infty} exp(-ikx^2)dx = A\sqrt{\frac{\pi}{k}} \tag{3.25}$$

with $A_L = A_R = A$ for zero noise, and calculate the product

$$S_L \times S_R = \frac{\pi A^2}{k} = \pi A^2 \sigma^2 \,. \tag{3.26}$$

To eliminate the effective width of the illuminated region σ, we involve first derivations of the functions $f_L(x)$ and $f_R(x)$ and fix their extreme values. We have

$$\frac{df(x)}{dx} = -2kx \cdot \exp(-kx^2) \tag{3.27}$$

and the extreme values are equal to

$$\frac{df(x)}{dx}|_{EXTR} = \pm A\sqrt{\frac{2k}{e}} \tag{3.28}$$

at points $x_\pm = \pm 1/\sqrt{2k}$. Now we calculate the second product

$$\left.\frac{df_L(x)}{dx}\right|_{EXTR_{AV}} \times \left.\frac{df_R(x)}{dx}\right|_{EXTR_{AV}} = A^2\frac{2k}{e} \,, \tag{3.29}$$

where "AV" denotes the averaging of the absolute values of $|df(x)/dx|$ at points $x_\pm = \pm 1/\sqrt{2k}$. To estimate the geometrical parameter $k = 1/\sigma^2$ we must calculate the overall product

$$G = S_L \times S_R \times \left.\frac{df_L(x)}{dx}\right|_{EXTR_{AV}} \times \left.\frac{df_R(x)}{dx}\right|_{EXTR_{AV}} = \frac{A^4}{e}\,2\pi \tag{3.30}$$

from which it follows the estimation of the constant A:

$$A = \sqrt[4]{\frac{Ge}{2\pi}} \,. \tag{3.31}$$

3.7. Some Other Designs of the MFTM

3.7.1. FT-microscope of the direct observation

The information about the charged particle tracks in nuclear emulsion is of three-dimension character. During a long time the direct application of the picture processing algorithms developed for plane objects [160] to $3D$ objects was considered as questionable. Only in 1981 a new technique of the nuclear emulsion viewing and processing was proposed [143] with the aim to increase the productivity of the searching for rare particle tracks of known orientation in space. For this purpose a nuclear emulsion layer is illuminated by the convergent light beam with its focus in the plane where FT-pattern of the $3D$ object is produced [100, 101, 102]. This far field diffraction pattern can be observed directly or viewed by means of suitable imaging system. With this new technique the scanning operation in depth proved to be completely unnecessary. Besides, due to inherent properties of the FT-pattern, the $(x-y)$ scanning can be performed with nuclear emulsion in motion. New scopes of the automation of the searching for rare particle tracks were opened by these features of the FT-microscope of the direct observation [131]. The term "direct observation" means that no imaging elements are present in this nontraditional microscope and no images are viewed by the operator or by the camera. The only source of the information about the particle track in the nuclear emulsion is $2D$ FT-pattern of the straight line particle track.

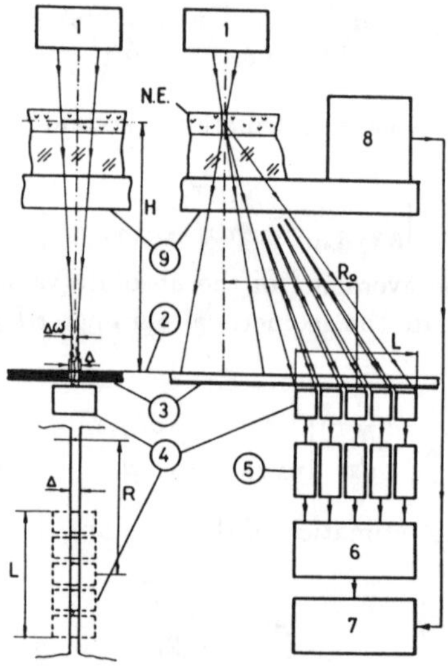

Fig. 3.44. Fourier-transform microscope of the direct observation: 1 — source of the convergent light beam, 2 — Fourier-transform plane, 3 — transmitting slit, 4 — array of five photodetectors, 5 — array of five signal amplifiers, 6 — interface, 7 — computer, 8 — $(X-Y)$-stage with position indicator, 9 — rotating fork.

This device is shown in Fig. 3.44. The signals from array of m photodetectors 4 disposed just behind the narrow transmitting slit 3 undergo two processing algorithms: a) multiplication of all m photosignals, which gives $Pr(x)$; b) estimation of the m-th root of $Pr(x)$, which gives $\sqrt[m]{Pr(x)}$; c) averaging of photosignals, which gives $M(x)$. These operations we accomplish for each x with moving step equal to $10\,\mu$m. The product $Pr(x)$ is compared with the given comparison level P_o. If the product $P_r(x)$ is smaller than the comparison level P_o, the output signal will be equal to P_o. If the product $Pr(x)$ exceeds the comparison level P_o, the real value $Pr(x)$ is stored in the memory block. Between the adjacent scanning operations the operator changes manually the angular position of the rotating fork 9 over given angular increment $\Delta\theta_{xy}$. The output signals $I(x)$, $Pr(x)$, and $M(x)$ form an array of $1D$ scans into the frame $(x,\ \theta_{xy})$. The typical number of X-scans is about 200 and the number of discrete θ_{xy}-orientations is about 20.

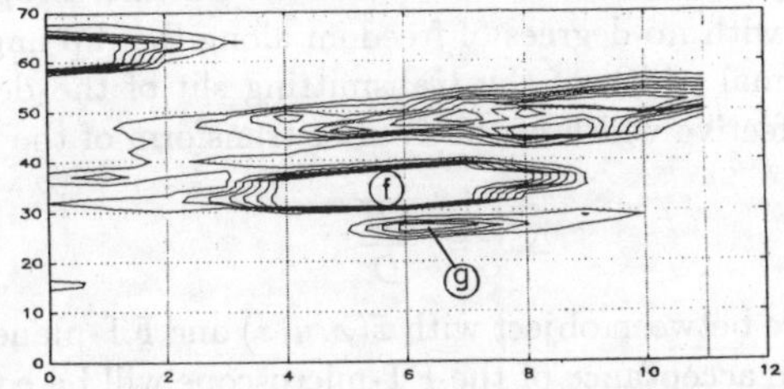

Fig. 3.45. $2D$ plot of the signals from proton track "f" with linear density of silver grains 120 per 100 μm and from proton track "g" with linear density of silver grains 38 per 100 μm, presented in the frame $(x,\ \theta_{xy})$.

Under a real condition the linear density of the silver grains is equal to $n = 30$ per 100 μm, and the noise N from the whole illuminated region is equivalent to 10^5 silver grains. For the length of the illuminated region $D = 3\,\text{mm}$ we have $S = 900 \pm 30$, and $N = 10^5 \pm 300$. The experiments with FT-microscope of the direct observation have shown that a proton track with linear density of 40 silver grains per 100 μm can be detected on the $2D$ plot in the $(x,\ \theta_{xy})$ frame (Fig. 3.45) with initial S/N ratio equal to $S/N = 1/3$. The same experiment was performed for 500 proton tracks with energy of 250 GeV with illuminated region of the width 18 μm. The average linear density of silver grains in these particle tracks was equal to 37.5 ± 2.5 per 100 μm. The spatial resolution of the system is equal to 20 μm and the angular resolution $\sim 10'$ [131].

3.7.2. *Acceptance of the FT-microscopes*

To compare the performance of different designs of the MFTM a new term: "acceptance of the Fourier transform microscope" for nuclear emulsion has been introduced [132]. As will be explained below this term is of the universal feature and can be applied to various constructions of these microscopes. The acceptance of any Fourier transform microscope $d\Omega$ is defined as a product of the spatial $d\Omega_{\vec{x}}$ and of the angular $d\Omega_{\vec{\theta}}$ parts:

$$d\Omega = d\Omega_{\vec{x}} \cdot d\Omega_{\vec{\theta}}\,, \tag{3.32}$$

and is referred only to the line objects. The acceptance $d\Omega$ represents a volume element in the 5-dimensional space which can be factorized into 3-dimensional spatial space $\vec{x}(x,y,z)$ and 2-dimensional angular space $\vec{\theta}(\theta_{xy},\theta_z)$. The spatial part of the acceptance represents a volume of the illuminated region of the working volume of the width a, of the depth h and of the length D or the corresponding volume of the observation zone with transversal dimension Δx, where Δx is the spatial resolution of the FT-microscope.

To explain the properties of the acceptance of the FT-microscope, we consider a horizontal line object with no degrees of freedom along the dip angle θ_z, for which $d\omega_{\vec{\theta}} = d\theta_{xy}$. For optimal width of the transmitting slit of the detector [132], Δ, chosen equal to the effective width of the Fourier transform of the line object, $\Delta\omega$, with

$$\Delta\omega = \frac{\lambda H}{D}, \tag{3.33}$$

where H is the distance between object with $\vec{x}(x,y,z)$ and FT-plane with $\vec{\omega}(\omega_x,\omega_y)$, the angular part of the acceptance of the FT-microscope will be equal to

$$d\omega_{\vec{\theta}} = d\theta_{xy} = \frac{\Delta\omega}{R} = \frac{\lambda H}{DR_\circ}, \tag{3.34}$$

where $R_\circ$ is the average distance from the transmitting slit to the optical axis. The spatial part of the acceptance is equal to

$$d\Omega_{\vec{x}} = a \cdot h \cdot D, \tag{3.35}$$

and the total acceptance of the detected system is given by

$$d\Omega = d\Omega_{\vec{x}} \cdot d\omega_{\vec{\theta}} = a \cdot h \cdot D \cdot \frac{\lambda H}{DR_\circ} = \frac{a\lambda hH}{R_\circ}. \tag{3.36}$$

We see that the acceptance of the FT-microscope does not depend on the length of the illuminated region D. For $h = 0.2\,\mathrm{mm}$, $\lambda = 0.6\,\mu\mathrm{m}$, $H = 150\,\mathrm{mm}$, $R_\circ = 80\,\mathrm{mm}$, and $a = 20\,\mu\mathrm{m}$ we have $d\Omega = 4.5 \cdot 10^{-6}\,\mathrm{mm{\cdot}rad}$.

The spatial part of the acceptance of the MFTM with one-channel photodetectors and with transmitting slit, set in front of the meso–optical mirror with ring response [132], is given by

$$d\Omega_x^\circ = \frac{\lambda}{\sin\alpha_{1/2}} \cdot \frac{a}{\tan\alpha_{1/2}} \cdot D. \tag{3.37}$$

For $\alpha_{1/2} = 4/15$ rad, $d\Omega_x^\circ = 8.3 \cdot 10^{-3} \cdot aD$. The total acceptance of this MFTM, with one-channel photodetectors and with transmitting slit, is equal to $d\Omega = 1.86 \cdot 10^{-7}\,\mathrm{mm{\cdot}rad}$ or ~ 23 times smaller than for the FT-microscope of the direct observation. To explain this difference let us consider a cross section of the illuminating system of the FT-microscope of the direct observation [132] and a cross-section of the observation zone of the MFTM with one-channel photodetectors

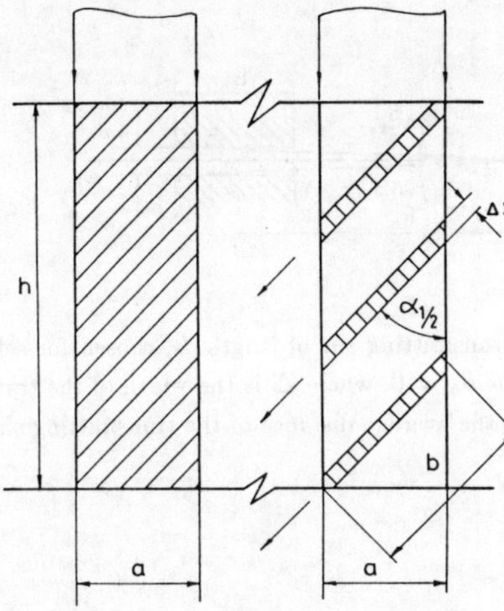

Fig. 3.46. The structure of the illuminated region of the width "a" (left) and of the observation zones (right) of width Δx inside the transversal cross-section of the nuclear emulsion layer of the thickness h in the FT-microscope of direct observation; $\alpha_{1/2}$ is the angle between the marginal axis and the main optical axis of the FT-microscope of the direct observation.

(Fig. 3.46)(right). We see that the spatial selectivity of the first system is k times higher than that of the second system, where

$$k = \frac{a \cdot h}{b\Delta x} = \frac{h}{\lambda} \sin\alpha_{1/2} \tan\alpha_{1/2} \,. \tag{3.38}$$

At the extreme condition with $\alpha_{1/2} = 1$ rad, $k \approx 670 : 1$.

To give insight into the properties of the acceptance of the FT-microscope, let us consider a line object with given dip angle θ_z. What will be the information properties of the transmitting slit in these conditions? The feature of the transmitting slit shown in Fig. 3.47, is defined by its width Δ, by the length of the working part L and by the average distance $R_\circ$ from the center of the working part to the optical axis of the system. The position of the transmitting slit in the $\vec{\omega}$-plane is changed by means of the rotating fork [131] (see Fig. 3.37). In the optimal condition the width of this transmitting slit Δ should be chosen equal to

$$\Delta_{opt} = \frac{\Delta\omega}{2} = \frac{\lambda H}{2D} \,. \tag{3.39}$$

The interval $\Delta\theta_{xy}$ of the orientation angles θ_{xy} which are picked up by this transmitting slit is equal to

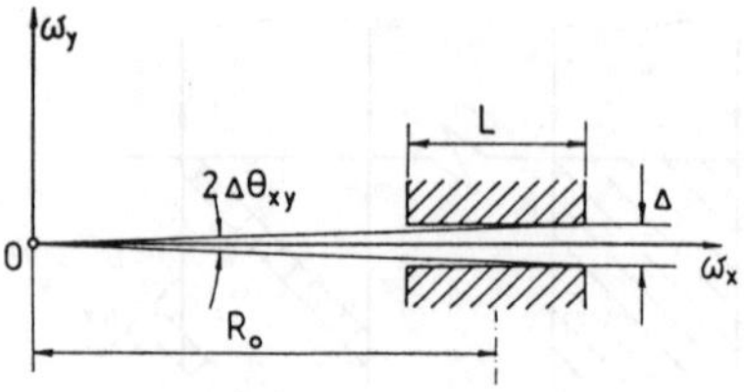

Fig. 3.47. The position of the transmitting slit of length L, chosen for selective detection of the straight line object with dip angle $\theta_z = 0$, where Δ is the width of the transmitting slit, $\Delta\,\theta_{xy}$ is its angular selectivity, and $R_\circ$ is the average distance of the transmitting slit from optical axis.

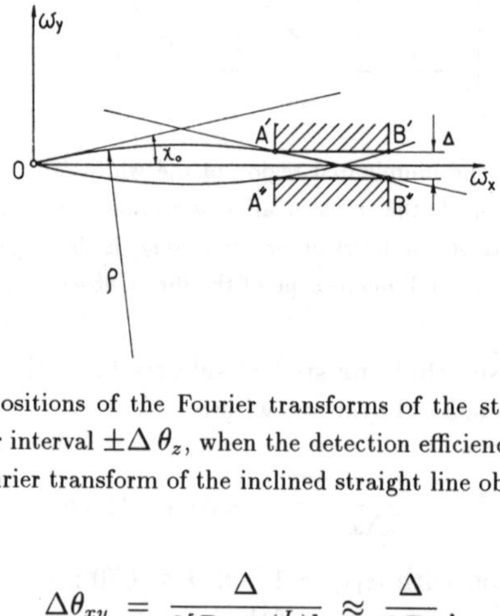

Fig. 3.48. Two extreme positions of the Fourier transforms of the straight line object with dip angle θ_z within the angular interval $\pm\Delta\,\theta_z$, when the detection efficiency is equal to 1, ρ being the curvature radius of the Fourier transform of the inclined straight line object.

$$\Delta\theta_{xy} = \frac{\Delta}{2[R_\circ + (\frac{L}{2})]} \approx \frac{\Delta}{2R_\circ}\,, \tag{3.40}$$

(Fig. 3.48). This transmitting slit sees also line objects with dip angles θ_z. To obtain high efficiency of the detection the transmitting slit should be shifted up as shown in Fig. 3.49 in the position with an eccentricity ε. The angle $\chi_\circ$ shown in Fig. 3.48, defines the angular interval $\Delta\theta_z$ of the dip angles around $\theta_z^\circ = 0$ with eccentricity $\varepsilon = 0$. The extreme circle which goes through the origin 0 and the points A' and B'' of the transmitting slit (Fig. 3.50), is described by the equation

$$\begin{gathered}(\omega_x - a)^2 + (\omega_y - b)^2 = \rho^2\,, \\ \text{with} \quad a \cong \frac{R_\circ}{2}\,, \quad b \cong -\frac{LR_\circ}{2\Delta}\,, \quad \rho \cong \frac{LR_\circ}{2\Delta}\,.\end{gathered} \tag{3.41}$$

The detecting efficiency of the transmitting slit is equal to zero with $\Delta\theta_z = \Delta\theta_{xy} = 0$ for $\rho = \rho_{min}$, when the circle arc crosses the points A, D and B in

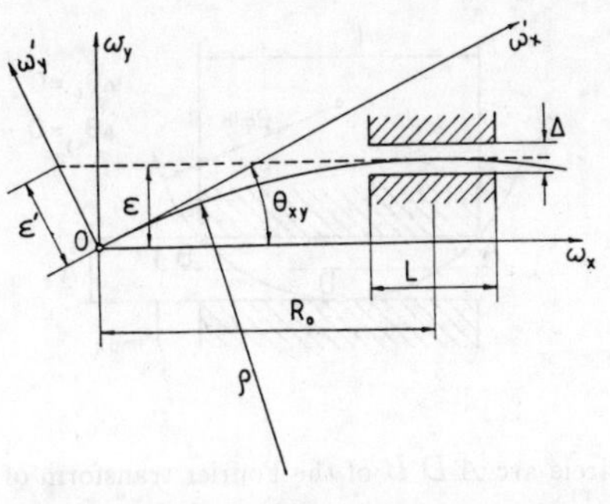

Fig. 3.49. The position of the transmitting slit with lateral eccentricity ε for selective detection of the straight line object with given dip angle θ_z and with orientation angle θ_{xy}.

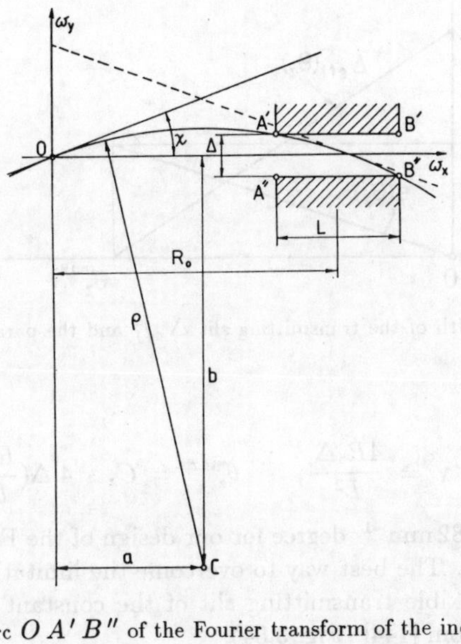

Fig. 3.50. The circle arc $O\,A'\,B''$ of the Fourier transform of the inclined straight line object at the extreme position without lateral eccentricity of the transmitting slit.

Fig. 3.51. The effective width of the transmitting slit $\Delta_{eff}(\theta_z)$ and the required eccentricity $\varepsilon(\theta_z)$ versus θ_z are presented in Fig. 3.52.

Fig. 3.53 shows the angular acceptance of the FT-microscope of the direct observation in the form of rectangles versus angles θ_{xy} and θ_z. The external points of this diagram correspond to the design of the FT-microscope with translationally moving transmitting slit. The moving of the transmitting slit along ω_x-axis induces a change of the tuned optimal orientation angle θ_{xy}. Due to this the center of the rectangular acceptance will be on the line $C'C''$, and

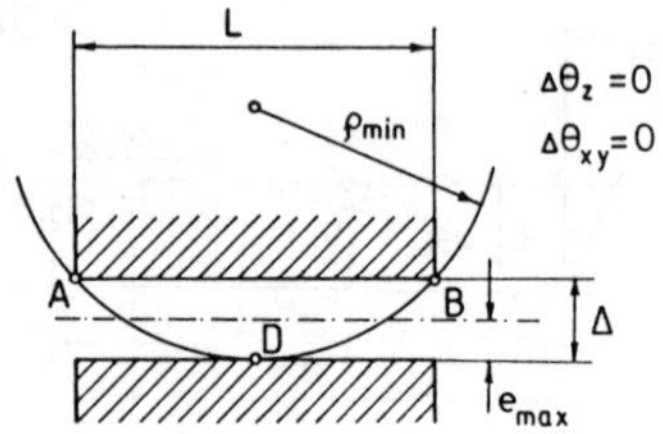

Fig. 3.51. The position of the circle arc $A\,D\,B$ of the Fourier transform of the straight line object at the extreme position with zero angular intervals $\Delta\,\theta_z = \Delta\,\theta_{xy} = 0$.

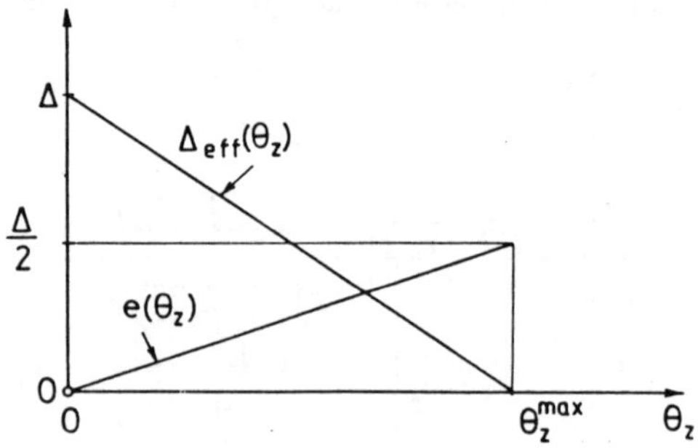

Fig. 3.52. The effective width of the transmitting slit Δ_{eff} and the parameter e (see Fig 3.51), versus dip angle θ_z.

$$\theta_{xy}^{max} = \chi \cong \frac{4R_\circ\Delta}{L^2}\,, \qquad \theta_z^{max} = C_z \cdot 4\,\Delta(\frac{R_\circ}{L})^2\,, \tag{3.42}$$

where constant $C_z = 2.82\,\text{mm}^{-1}\cdot$ degree for our design of the FT-microscope of the direct observation [131]. The best way to overcome the limitation on the dip angle θ_z is to introduce a flexible transmitting slit of the constant clearance and with bending degree of freedom [144] (Fig. 3.54).

The structure of the signals at the output of the FT-microscope of the direct observation has different form for different line objects. For line objects of very high linear density of silver grains we have $S \gg N$, or $N \approx 0$. For line object with very low linear density of silver grains we have $N \gg S$. In the first case the product signal $Pr = S^m$ is maximal if the line object goes parallel to the symmetry axis of the transmitting slit (Fig. 3.55). For dip angle $\theta_z \neq 0$ when FT-pattern does not cross at least one photodetector, the product signal $Pr = 0$. In the second case the normalized signal in each photodetector will be equal to $1+\alpha$, with $\alpha = S/N \ll 1$. After multiplication we have $Pr = (1+\alpha)^k \approx 1 + k\alpha$, where k is the number of photodetectors which are illuminated by the FT pattern, $k \leq m$. The intensity of

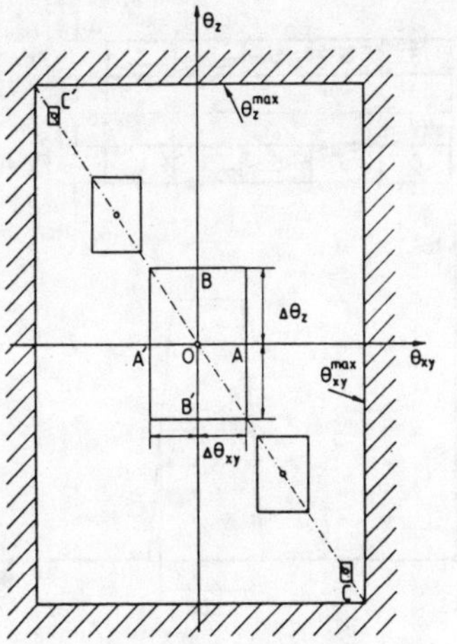

Fig. 3.53. The overall dependence of the angular acceptance $4\,\Delta\,\theta_{xy}\cdot\Delta\,\theta_z$ in the Fourier transform microscope of the direct observation versus angles θ_{xy} and θ_z.

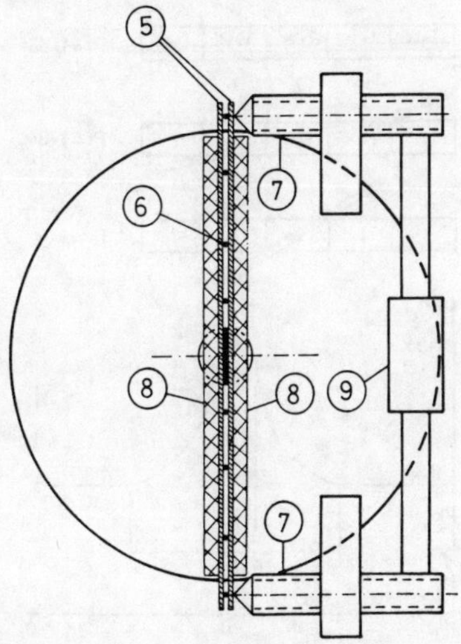

Fig. 3.54. General design of the system, containing the flexible transmitting slit of constant clearance and with bending degree of freedom: 5 — flexible plate, 6 — clearance rods, 7 — fixing screw, 8 — rubber support of the flexible plate, and 9 — screw driver. The orientation angle θ_{xy} is defined by means of the rotating fork (see Fig. 3.37).

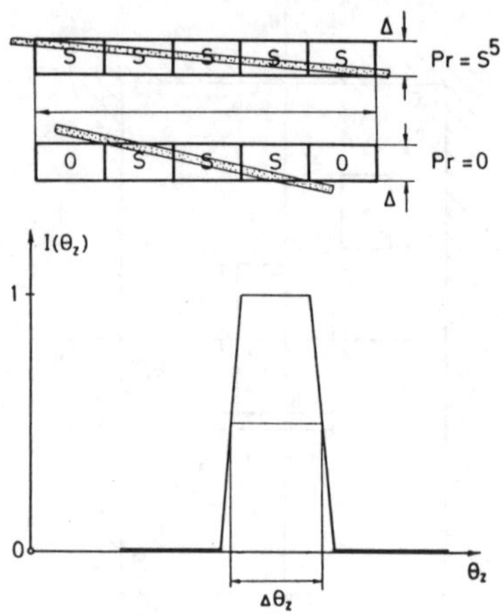

Fig. 3.55. The structure of the output product signal in the Fourier transform microscope of the direct observation for straight line objects of high signal-to-noise ratio.

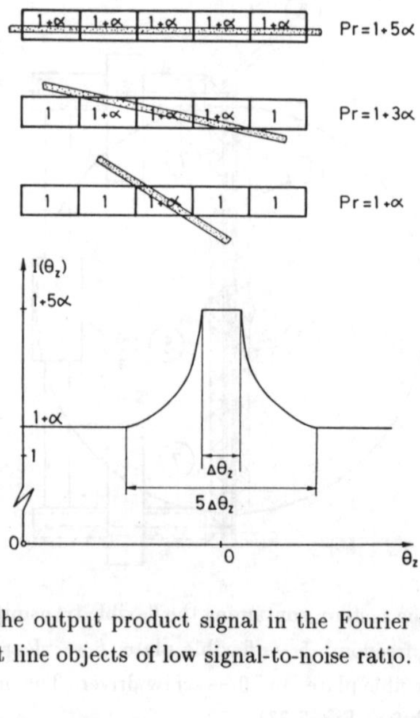

Fig. 3.56. The structure of the output product signal in the Fourier transform microscope of the direct observation for straight line objects of low signal-to-noise ratio.

the product signal Pr is ranged from $Pr^{min} = 1 + \alpha$ to $Pr^{max} = 1 + m\alpha$ (Fig. 3.56). In our design of the FT-microscope of the direct observation we have

$$\Delta\theta_z = 16.9' ; \qquad \Delta\theta_{xy} = 4.4' \tag{3.43}$$

with $\theta_z^{max} = 5.5°$ and $\theta_{xy}^{max} = 1.43°$.

3.7.3. MFTM with double focusing

In the MFTM described in Section 3.2, the meso–optical mirror with ring response produces two meso–optical images of the line object. These signals are on the opposite sides of the focal ring. The focusing of the diffracted light is taking place only in the meridional section along the radial coordinate. In the sagittal section of the MFTM there is no focusing at all and as was explained before [134] (see Section 3.3), the MFTM in the sagittal section can be considered as one-dimensional pinhole camera. Therefore the length of the meso–optical images in the sagittal section [118] is equal to the length (diameter) of the field of view of the MFTM. This property of the MFTM with single focusing enables one to perform some rough metrical measurements such as to localize spatially the end or the beginning of the line segment inside the field of view of the MFTM. The error of these measurements in the sagittal section is defined by the length of the field of view. To prove this two experiments have been performed [118]. In the first one the opaque screen shut off two strokes of the cross (Fig. 3.57). From Fig. 3.58 we see that the opaque screen does act upon each meso–optical image independently. The blurring of the extreme points is increasing as the opaque screen is shutting off the strokes. In the second experiment an object in the form of three optical grids has been imaged by means of the MFTM. The distance between two grids of width 20 μm was (80 ± 5) μm (Fig. 3.59). One of the meso–optical images of this object is shown in Fig. 3.60. We see that the length of the blurred parts of the image is of the order of 30 μm in accordance with estimations made earlier [118].

The function of the meso–optical mirror with ring response can perform an meso–optical assembly which involves three meso–optical lenses with each lens having virtual ring response. To suppress the spherical aberrations of this meso–optical assembly we must fulfill three necessary conditions: a) the internal surface of each meso–optical lens should be conical one with generating line perpendicular to the marginal axis of the system, b) the external surface of each meso–optical lens should be spherical one with center in the median plane of the working volume on the optical axis of the system, c) the number of these meso–optical lenses with virtual ring response should be equal or more than three. The spherical aberration of this meso–optical assembly is 10 times smaller than in the system with one optical component of very high index of refraction n and of small radius of the spherical surface.

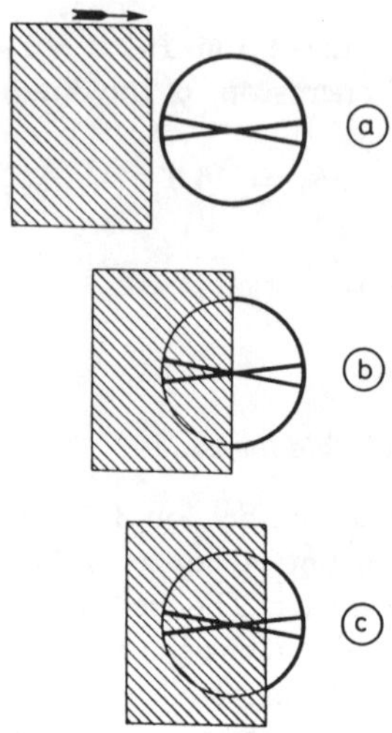

Fig. 3.57. Positions of the opaque screen relative to the object in the form of a cross: a) opaque screen is out of the field of view, b) opaque screen shuts off a half of the field of view, c) opaque screen shuts off 3/4 part of the field of view.

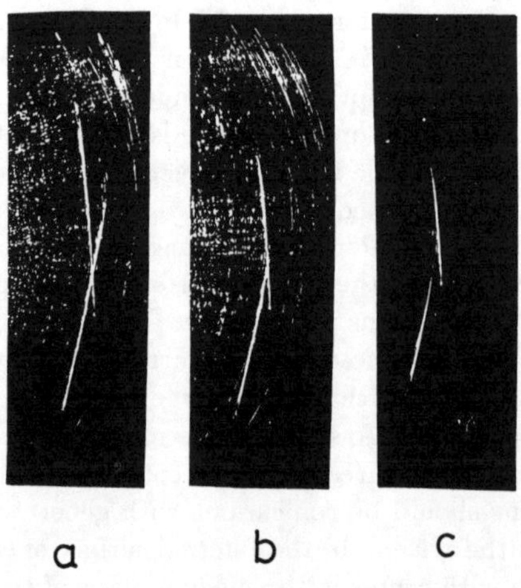

Fig. 3.58. Meso–optical images of two straight lines of a cross, shown in Fig. 3.57, in three positions of the opaque screen a), b), c), respectively.

Fig. 3.59. View of the three optical grids with gap 80 μm and of the width 20 μm.

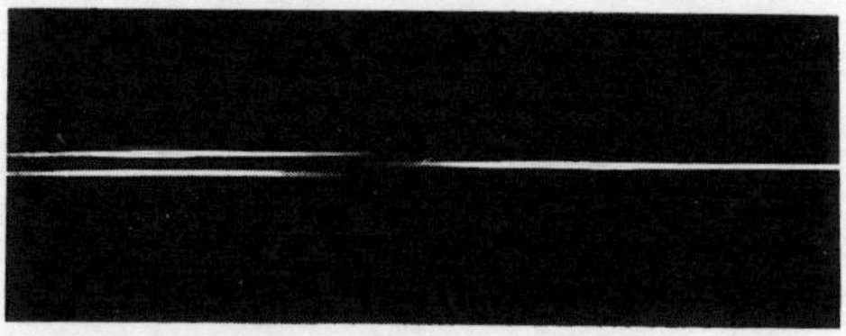

Fig. 3.60. Meso–optical image of the object shown in Fig. 3.59.

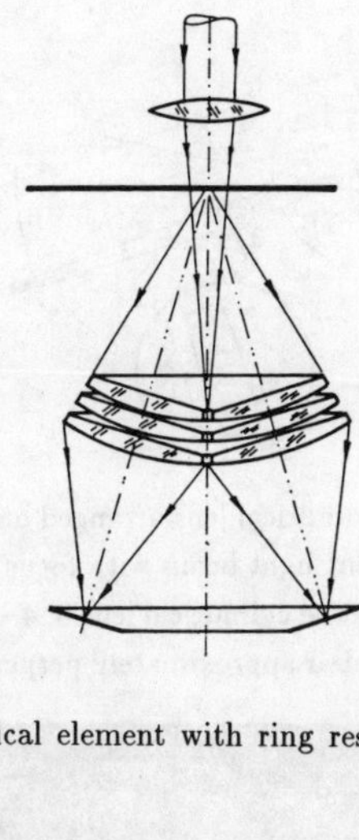

Fig. 3.61. MFTM with meso–optical element with ring response made of three coaxial axiconic lenses.

The diameter of the MFTM can be made be 10 times smaller than that in the system with large spherical aberrations. The MFTM with three optimal meso–optical lenses is shown in Fig. 3.61.

In the case when the required numerical aperture of the system is small, of the order of $\alpha \approx 10^{-2}$ rad, the design of the MFTM can be simplified additionally (Fig. 3.62). The angle $\alpha_{1/2}$ between marginal and central optical axes is kept constant and each meso–optical lens is converted into traditional cylindrical lens.

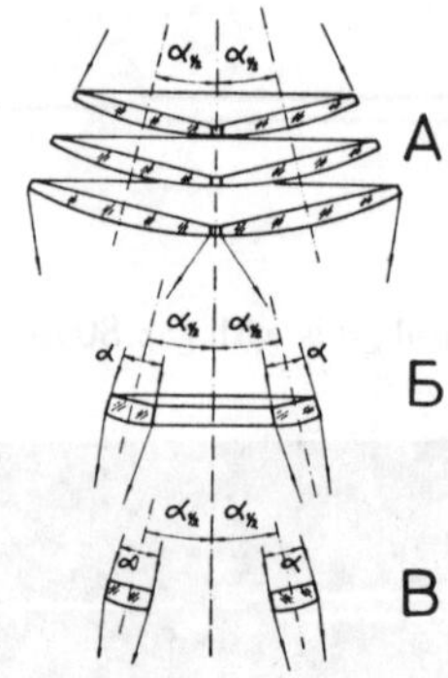

Fig. 3.62. Various designs of the meso-optical element with ring response: a) with three axiconic lenses, b) with one narrow ring-shaped axiconic lens, c) with two narrow cylindrical lenses, put along the marginal axes of the MFTM.

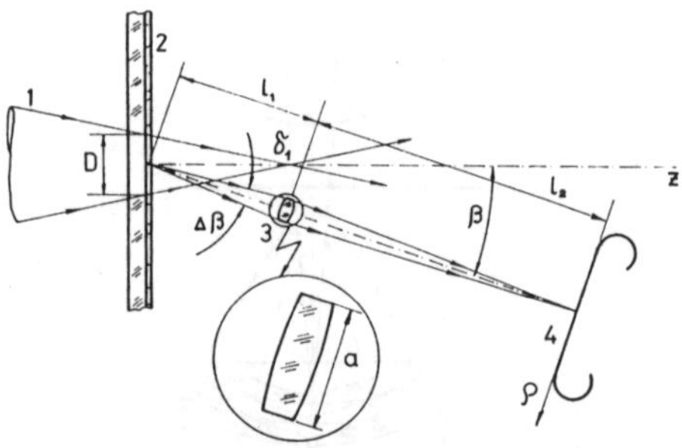

Fig. 3.63. The first experiment with cylindrical lens arranged on the marginal axis of the MFTM, as a meso-optical element: 1 — convergent light beam with focus at the point δ_1, which is located at the same distance from the object 2, as the cylindrical lens 3, 4 — detection plane with meso-optical images of the straight line object oriented approximately perpendicular to the figure plane.

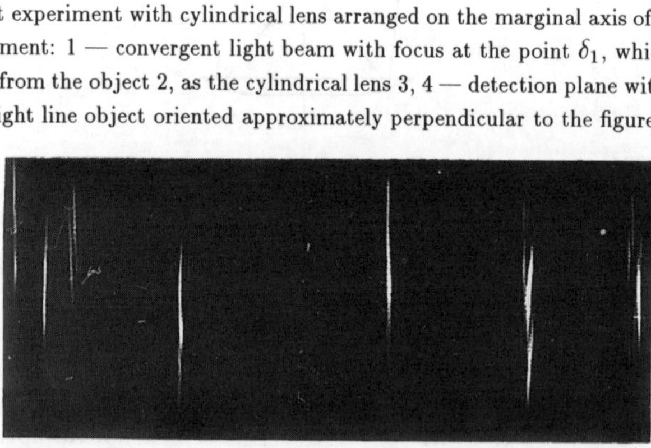

Fig. 3.64. Photo of the meso-optical images of the straight line objects observed in the first experiment (see Fig. 3.63).

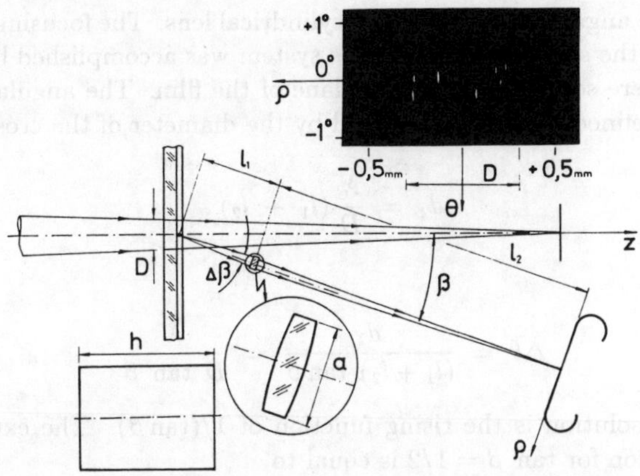

Fig. 3.65. The second experiment with cylindrical lens arranged on the marginal axis of the MFTM as a meso–optical element, and the meso–optical images of the straight line objects going approximately perpendicular to the figure plane; the focus of the convergent light beam is located at the distance $l_1 + l_2$ from the working volume, where l_1 is the distance between the working volume and the cylindrical lens, and l_2 is the distance from the cylindrical lens to the detection plane.

Such a design is dedicated to observe line objects with small spread of their orientation angles, as for these objects we need only small part of the whole focal ring and small arc of the focal circle can be replaced by the tangent. Here we use a rotating fork [131] to orient the searching for line object with any known orientation angle along symmetry axis of the illumination region. The tolerated uncertainty in the orientation angle is of the order of $\pm 1°$.

Two experiments were performed to test this idea[118]. In the first one (Fig.3.63) the meso–optical element with ring response in the form of two simple cylindrical lenses were placed near the crossover of the convergent light beam on the corresponding marginal axes of the system. The view of the meso–optical image of the line object in one arm of the system is shown in Fig. 3.64. The focusing along radial coordinate performs a cylindrical lens, but the sagittal focusing is absent. The length of the meso–optical image is equal to $D \cdot M$, where D is the diameter the length of the MFTM field of view and $M = l_2/l_1$ is linear magnification of the system, where l_1 and l_2 are distances between the object and the cylindrical lens and between the cylindrical lens and the meso–optical image, respectively.

In the second experiment the crossover of the convergent light beam was in the plane of the film (Fig. 3.65). The radial resolution of such a configuration was equal to

$$\Delta\rho = \frac{\lambda}{\Delta\beta}, \tag{3.44}$$

where $\Delta\beta$ is the angular aperture of the cylindrical lens. The focusing along angular coordinate θ in the sagittal section of the system was accomplished by the FT lens. Both focuses were set in the common plane of the film. The angular resolution of the system is defined by the angle β and by the diameter of the crossover

$$d_2 = \frac{\lambda}{D}(l_1 + l_2)\,, \tag{3.45}$$

namely,

$$\Delta\theta = \frac{d_2}{(l_1 + l_2)\,\tan\beta} = \frac{\lambda}{D\,\tan\,\beta}\,. \tag{3.46}$$

The angular resolution is the rising function of $1/(\tan\beta)$. The extreme value of angular resolution for tan $\beta = 1/2$ is equal to

$$\Delta\theta^{EXTR} \approx \frac{2\lambda}{D}\,. \tag{3.47}$$

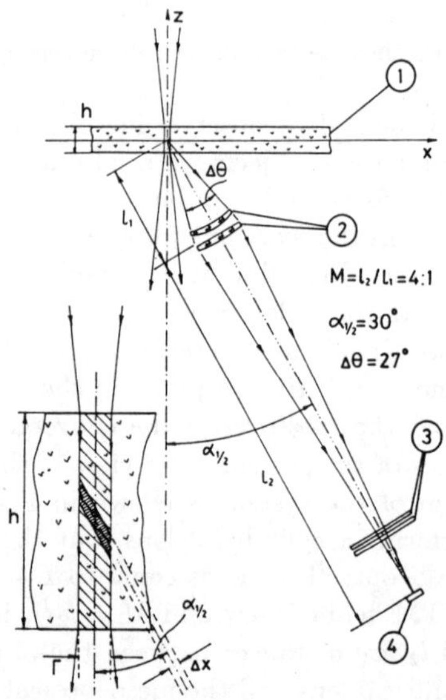

Fig. 3.66. Schematic diagram of the one-arm MFTM with double focusing: 1 — working volume, 2 — cylindrical objective, 3 — neutral light absorber, 4 — CCD-matrix of the TV-camera. Below to the left — meridional cross-section of the working volume with single hatched illuminated region and with double hatched region as a part of the illuminated region, which is seen by one pixel of the CCD-matrix.

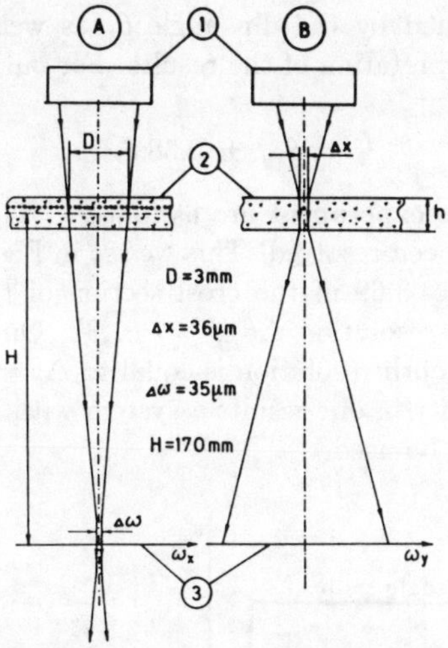

Fig. 3.67. Light rays of the illumination system of the MFTM with double focusing: 1 — source of the astigmatic convergent light beam, 2 — working volume, 3 — spatial frequency plane, "A" — in the sagittal cross-section, "B" — in the meridional cross-section of the MFTM with double focusing.

To get extremely high concentration of information about line objects, a MFTM with double focusing has been build [134]. Its schematic diagram is given in Fig. 3.66. The convergent illuminated light beam has an astigmatic structure. The focus in the meridional section is in the medium plane of the nuclear emulsion layer 1, and the focus in the sagittal section is in the plane with pickup matrix of the CCD-TV-camera. The distance between the center of the CCD matrix and the optical axis of the system is R. The distance between the nuclear emulsion layer and Fourier transform plane is H. The width of the illuminated region was $d = 36\,\mu$m and its length $D = 2.7$ mm. The diffracted light is converted by the cylindrical objective 2 into meso–optical image which is picked up by CCD-matrix of the TV-camera 4. The concentration of the diffracted light into meso–optical spots was so high that a neutral absorber filter 3 was inserted before CCD-TV-camera.

The cylindrical objective of this device was constructed as a projected system of two components with fixed distances $l_1 = 32$ mm and $l_2 = 128$ mm. The linear magnification of this system was $M = 4:1$. The expected spatial resolution in the object space was $2\,\mu$m.

The rays of the illuminated light in the MFTM with double focusing are shown in Fig. 3.67 in the meridional section. In the accomplished experiments only one cylindrical objective has been used. An effective orientation angle θ_{xy}^{eff} detected by one-arm system is defined by the dip angle θ_z as well, so we have two-fold ambiguity [134] in the interpretation of the results. For our design we have

$$\theta_{xy}^{eff} = \theta_{xy} \pm 0.58\,\theta_z\,. \tag{3.48}$$

The main results of this experiment are as follows. The $C'C''$-doublet of two parallel particle tracks has been resolved. This we see in Fig. 3.68 in the coordinate system $(z,\ \theta_{xy}^{eff})$ and in Fig. 3.69 in the cross-section of the $C'C''$-doublet along z-coordinate. The angular resolution $\Delta\theta_{xy}^{eff} = \pm 1.3'$. One pixel of CCD-matrix corresponds to $0.54'$. The depth resolution is equal to $\Delta z = 10\,\mu$m. To remove the mentioned two-fold ambiguity of our results a system with two arms and with one common CCD-matrix must be used.

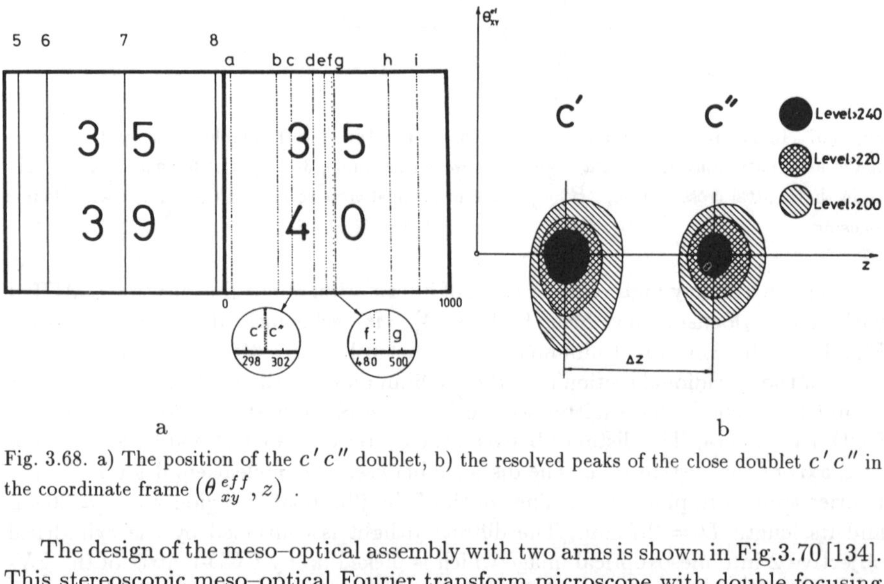

Fig. 3.68. a) The position of the $c'c''$ doublet, b) the resolved peaks of the close doublet $c'c''$ in the coordinate frame (θ_{xy}^{eff}, z) .

The design of the meso–optical assembly with two arms is shown in Fig.3.70 [134]. This stereoscopic meso–optical Fourier transform microscope with double focusing involves the frontal spherical immersion lens 1, two cylindrical objectives 2, plane mirrors 3, double mirror 4, plane mirrors 5, and photosensitive matrix 6 of the CCD-TV-camera.

Two cylindrical objectives 2 operate as two independent $1D$ projection systems. The diffracted light rays gathered by the cylindrical objective 2 are focused in the meridional cross-section into the bright spot on the photosensitive matrix 6 of the CCD-TV-camera. To bring two meso–optical spots onto one CCD-matrix, an optical interface with plane mirrors 3 and 5, and with double mirror 4 was used. The geometrical dimensions of two meso–optical images, $\Delta\rho$ and $\Delta\omega$, define the resolution of the device along transversal x-coordinate (1 pixel $= 0.9\,\mu$m) and the

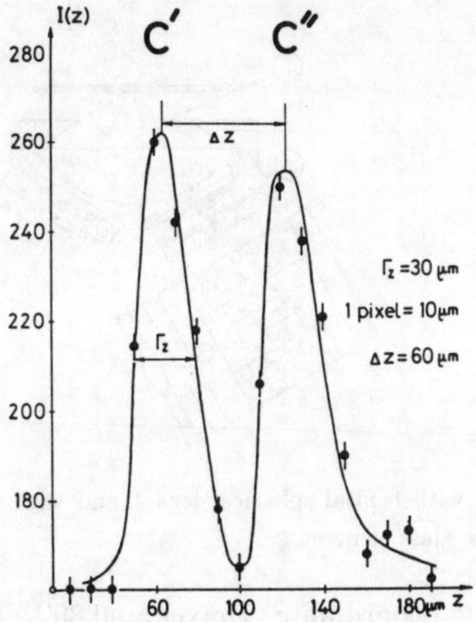

Fig. 3.69. The cut of the signals from $c'\,c''$ doublet, shown in Fig. 3.68(b), along z-coordinate.

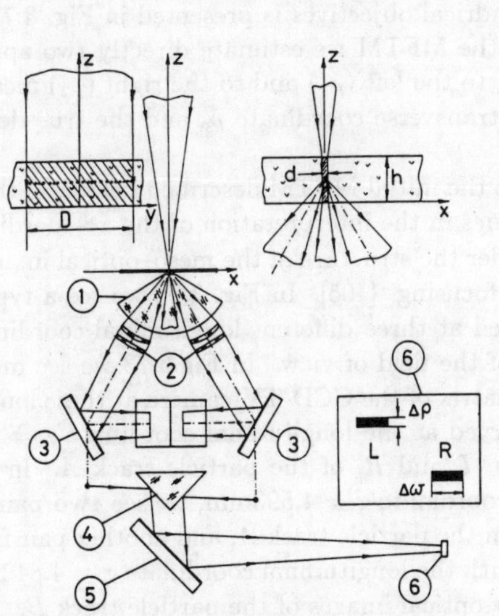

Fig. 3.70. Stereoscopic meso-optical Fourier transform microscope with double focusing: 1 — frontal spherical immersion lens, 2 — cylindrical objectives, 3 — plain mirrors, 4 — double mirror, 5 — plain mirror, 6 — photosensitive matrix of the CCD-TV-camera, L and R — the left and the right meso-optical images (spots), d — the width of the illuminated region, D — the length of the illuminated region, and h — the depth of the working volume.

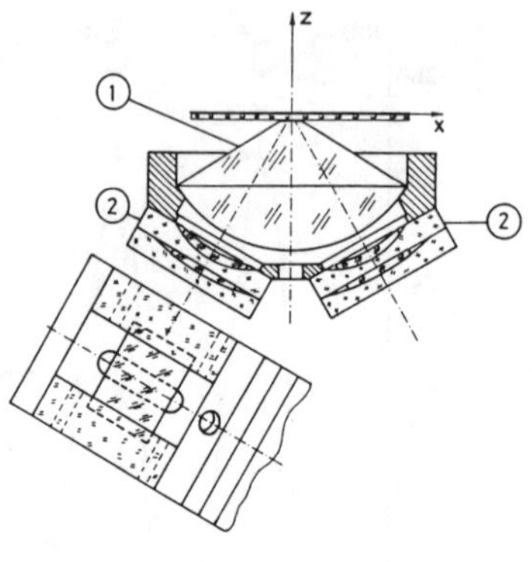

Fig. 3.71. The optical block with frontal spherical lens 1 and with two cylindrical objectives 2, mounted rigidly to the organic glass element.

resolution along angular θ_{xy}-coordinate (1 pixel = 0.89′). The linear magnification of the whole system of the light transportation (optical interface) was equal to 4 : 1.

The real construction of the optical block with frontal spherical immersion lens 1 and with two cylindrical objectives is presented in Fig. 3.71 [405]. In the stereoscopic prototype of the MFTM we estimate directly two apparent distances from the particle track O, to the left (ρ_1) and to the right (ρ_2) meso–optical images (see Fig. 3.6). The true transverse coordinate $\rho_\circ$ and the true depth coordinate $z_\circ$ are defined by Eq. 3.1.

Thus, opposed to the initial MFTM described before, in this stereoscopic device we have no ambiguities in the interpretation of the xz-coordinates.

Now let us consider the structure of the meso–optical images in the stereoscopic system with double focusing [405]. In Fig. 3.72 we see a typical small angle scattering event, observed at three different longitudinal coordinates y of the MFTM stage in the frame of the field of view. In Fig.3.73 we see meso–optical images on the photosensitive matrix of the CCD-TV-camera at these longitudinal coordinates. In the picture, observed at the longitudinal coordinate $x = 4.413$mm, we see two meso–optical images, L and R, of the particle track A. In the picture, emerged at the longitudinal coordinate $x = 4.593$mm, we see **two** pairs of the meso–optical images, one pair from the particle track A, and another pair from the particle track B. In the position with the longitudinal coordinate $x = 4.842$mm we see again only one pair of the meso–optical images of the particle track B.

The results of the measurements of this small angle scattering event, performed by means of the stereoscopic MFTM, are shown in Fig. 3.74, as orientation angle θ_{xy} for different longitudinal coordinates y over the length $\approx$ 3mm. We see that the small angle scattering event manifests itself in our MFTM as a local jump in this frame.

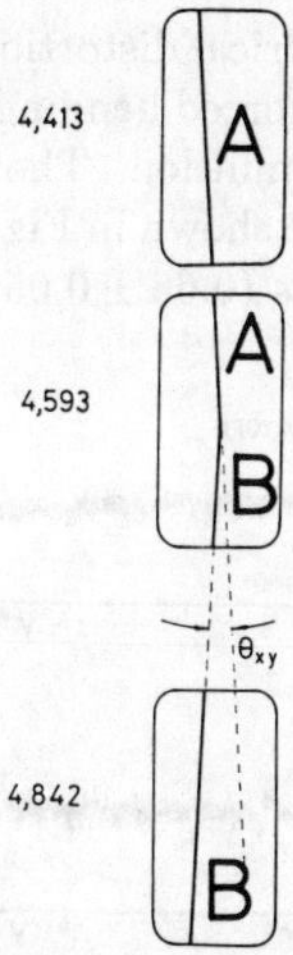

Fig. 3.72. Small angle scattering event $A \to B$, observed in the nuclear emulsion at three different longitudinal coordinates of the MFTM stage: $y = 4.313$; 4.593; 4.842 mm.

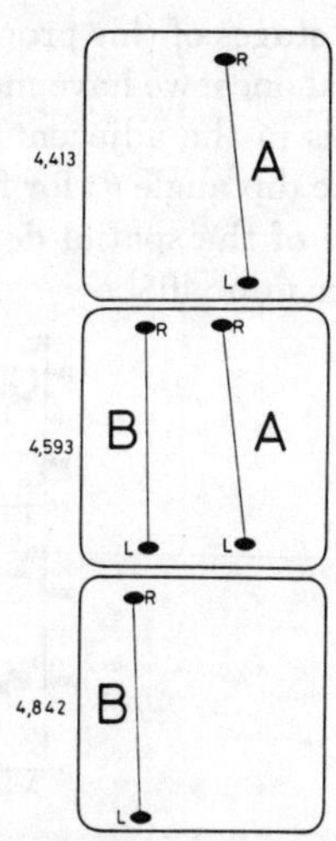

Fig. 3.73. Three pictures observed on the screen of the MFTM at the same longitudinal coordinates as in Fig. 3.71.

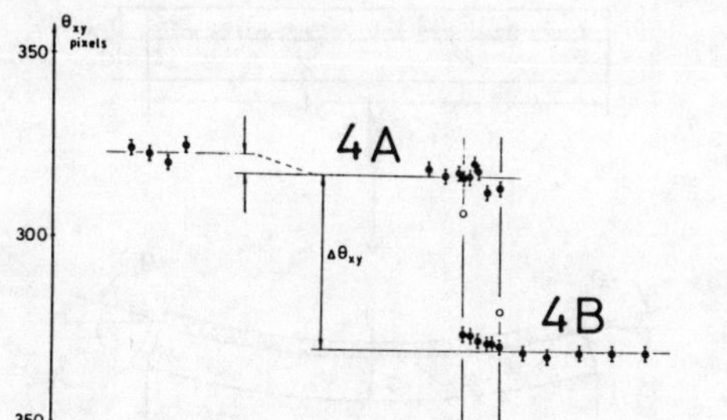

Fig. 3.74. Local jump of the orientation angle $\Delta\theta_{xy}$, observed at $y = 4.593$ mm, when the vertex of the scattering event is located in the center of the field of view.

The observed values of the dip angle θ_z can be approximated by the linear variation versus x over the distance 10mm. These variations of the dip angle θ_z can be explained by the forced bending of the glass support of the nuclear emulsion layer in the course of the drying stage of the whole complex photochemical treatment of the nuclear photoemulsion. In Fig. 3.76 we show schematically the initial and the final stages of this process. To prove that this geometrical distortions are indeed the global ones, we have measured the magnitude of the forced bending of other particle tracks in the adjacent regions of the nuclear photoemulsion. The linear variations of the dip angle θ_z for five particle tracks versus y are shown in Fig. 3.77. The mean value of the spatial derivative $d\theta_z/dy$ is estimated as (0.65 ± 0.05) pixels / mm or $2'$ per mm [408].

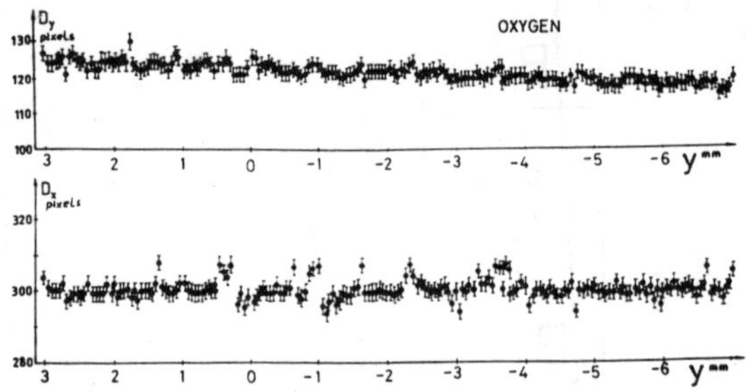

Fig. 3.75. The orientation angle θ_{xy} (expressed in pixels D_x) and the dip angle θ_z (expressed in pixels D_y) of the Oxygen nucleus, which were estimated for various longitudinal coordinate y over the length ~ 10 mm.

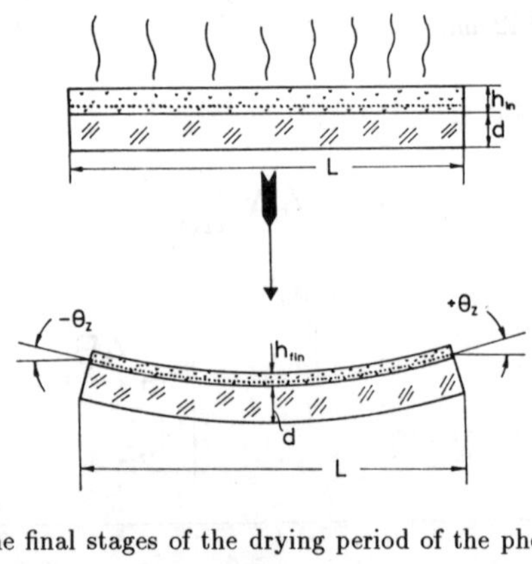

Fig. 3.76. The initial and the final stages of the drying period of the photochemical treatment of the nuclear photoemulsion layer on the glass backing.

3.8. Isoplanatic Systems in Meso–optics

It is well known that to suppress coma aberrations in the imaging system we must fulfill the Abbe sine condition [145–147] or the equivalent cosine condition [148, 149]. To suppress coma aberrations in holography we must choose non-plane backing of the holographic lens [147, 150, 151] or introduce refractive-diffraction elements [152], or optimize the parameters of the entrance or exit pupil of the system with holographic lens on the spherical backing [153]. To optimize the form of the hologram backing in general case the Welford theorem [151, 154] is to be used.

The problem of coma aberrations also arises in meso–optics. For example, the meridional coma aberrations restrict the usable diameter of the field of view of the MFTM [109]. Therefore, we consider here the problem of coma aberration in optics, in holography, and in meso–optics.

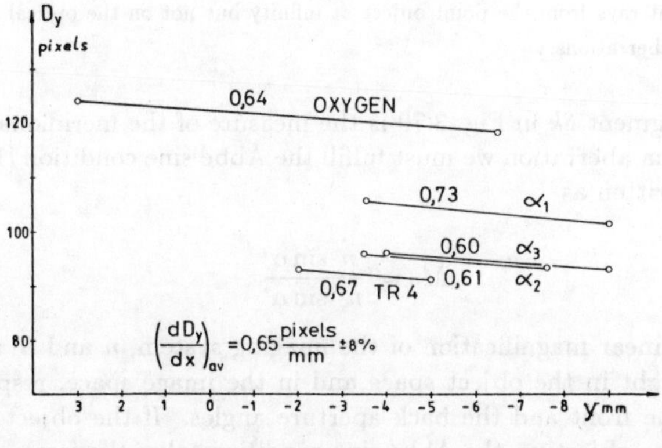

Fig. 3.77. The linear variations of the dip angle θ_z (expressed in pixels D_y) versus longitudinal coordinate y for five particle tracks in the same region of the nuclear emulsion.

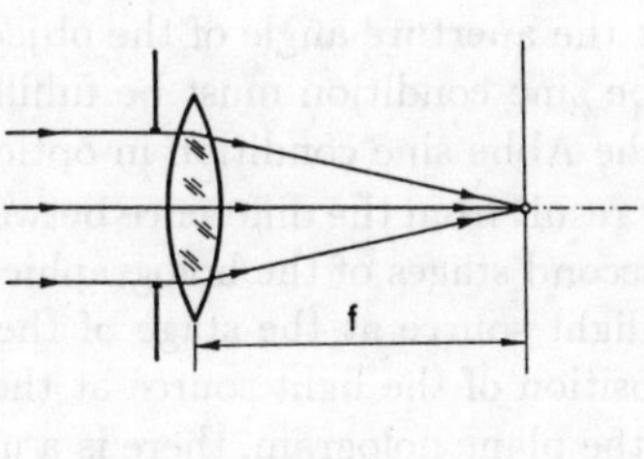

Fig. 3.78. The basic scheme of the imaging of the point object at infinity on the optical axis of a thin lens without spherical aberrations.

The light rays in the spherical aberration free lens are given in Fig. 3.78 for the point-like object on the optical axis of the lens. The main and the marginal light rays intersect in one point. If the point-like object is not on the optical axis of the lens, the point of intersection of two marginal light rays does not lie on the main light ray in the image space. As a result, the image of the point will be

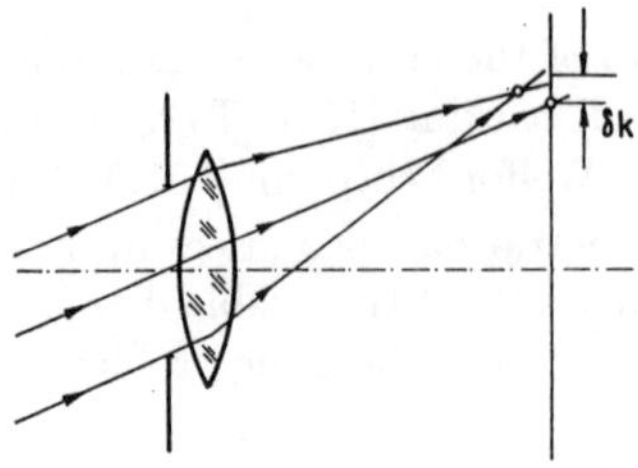

Fig. 3.79. The light rays from the point object at infinity but not on the optical axis of the lens without spherical aberrations.

blurred. The segment δk in Fig. 3.79 is the measure of the meridional coma [149]. To suppress coma aberration we must fulfill the Abbe sine condition [145, 147, 149], which can be written as

$$M = \frac{n \sin\alpha}{n' \sin\alpha'} . \tag{3.49}$$

Here M is the linear magnification of the imaging system, n and n' are refractive indices of the light in the object space and in the image space, respectively, and α and α' are the front and the back aperture angles. If the object is at infinite distance from the objective, the Abbe sine condition takes the form

$$h = f' \sin\alpha' , \tag{3.50}$$

where h is the collision parameter of the light ray and f' is the back focal distance of the objective. The larger is the aperture angle of the objective, the higher is the precision with which the Abbe sine condition must be fulfilled. There are several techniques for realization of the Abbe sine condition in optics [148, 149].

Aberrations in holography result from the difference between the wave lengths of light on the first and on the second stages of the holographic process and also from the fact that position of the light source at the stage of the hologram production does not coincide with the position of the light source at the reconstruction stage. As was shown for the case of the plane hologram, there is a unique relative position of reference and object light waves, for which the coma aberration is minimized. In this configuration the point-like object and the convergent focus of the reference wave must be mirror images of each other in the plane hologram as a mirror. For arbitrary configuration of the reference and the object light waves the hologram backing is to be chosen as a sphere with curvature radius R defined by the equation

$$\frac{1}{R} = \frac{1}{l_1} + \frac{1}{l_2} , \tag{3.51}$$

(Welford theorem) [151, 154], where l_1 and l_2 are the distances from the reference point-source to the hologram and from the hologram to the object-source, respectively, the signs of l_1 and l_2 being chosen in accordance with rules accepted in optics (Fig. 3.80). For $l_2 \to \infty$, $R \to l_1$ the curvature radius of the hologram backing is equal to the distance between hologram and the reference light source. For $l_1 = -l_2$, $R \to \infty$, and hence the hologram backing must be plane. Finally, if $l_2 \gg l_1$, then $R \approx l_1$.

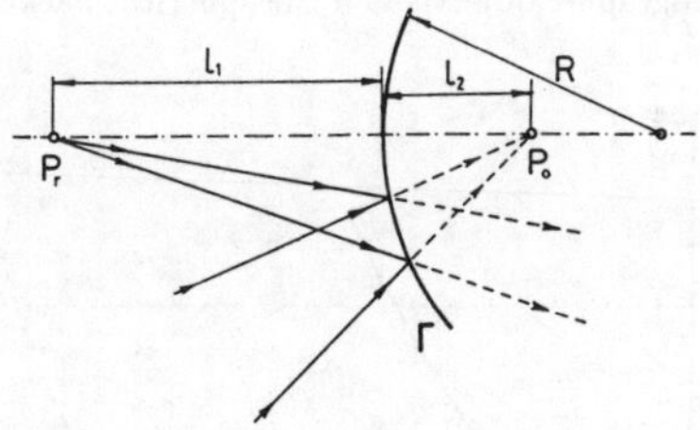

Fig. 3.80. Aplanatic spherical backing of the hologram for arbitrary position of the object and reference beams of light with respect to the hologram.

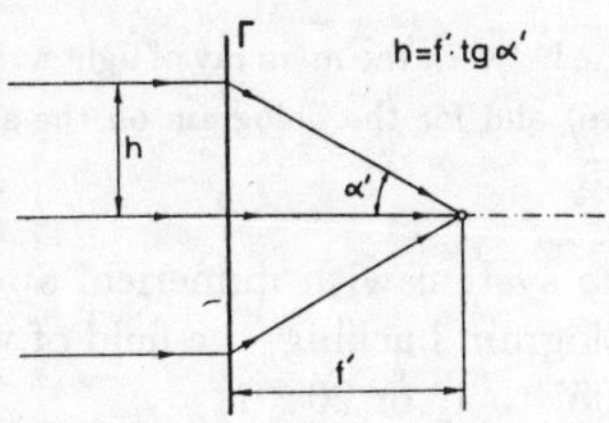

Fig. 3.81. The light rays in the holographic objective on the plane backing at the reconstruction stage.

The design of the aberration-free holographic lens with high numerical aperture has been suggested [155]. The reconstruction stage with plane holographic lens is shown in Fig. 3.81. Here

$$h = f' \tan \alpha' , \tag{3.52}$$

and the Abbe sine condition, Eq. (3.50), is violated. For hologram on the spherical backing, shown in Fig. 3.82, we have

$$h = f' \sin \alpha' , \tag{3.53}$$

that is, the Abbe sine condition is fulfilled. To illustrate the effect of the coma aberration suppression we present the dependence of the coma aberration versus

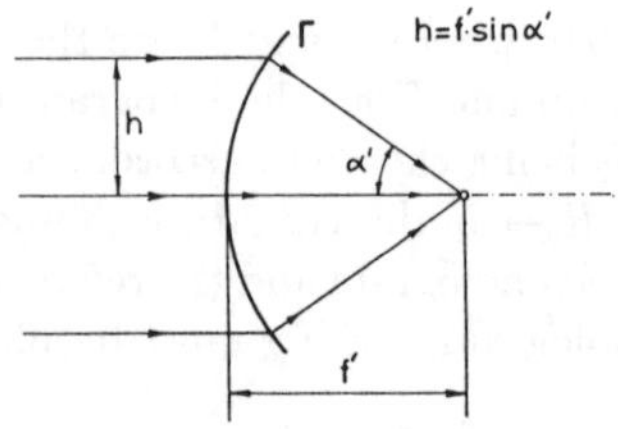

Fig. 3.82. Aberration-free holographic objective on the spherical backing.

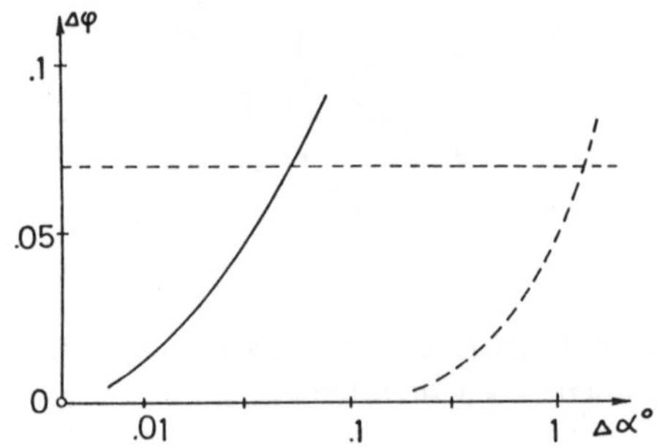

Fig. 3.83. Aberrations versus angle between the main ray of light and the optical axis of the hologram on the plane backing (solid curve) and for the hologram on the spherical backing (dashed curve) with numerical aperture $A = 0.5$.

the observation angle in the system with numerical aperture $A = 0.5$ (Fig. 3.83). By shaping properly the hologram backing, the field of view of the holographic lens has been increased from 0.05° to 1° or 20 : 1.

The problem of coma aberration arises in meso–optics as well [109]. The light rays in the meridional section of the MFTM are presented in Fig. 3.84. The convergent light beam 2 illuminates the line object 3 in the object space 1. The diffracted light rays 6, produced by the line object 3 oriented perpendicular to the figure plane, illuminate the meso–optical mirror with ring response 5 over the narrow strip with width of the order of 60 μm. One of two meso–optical images of the line object 3 is in the point 7 inside the image space 8.

The effect of coma aberration in the MFTM is illustrated graphically in Fig. 3.85 via diffracted light rays for three positions of the line object, shown in Fig. 3.85-A. For the line object "1" these light rays are presented in Fig. 3.85-B, for line object "2" in Fig. 3.85-C, and for line object "3" in Fig. 3.85-D [109]. We see that for positions of the line object "2" and "3" there is caustic, distinction for meridional coma. The magnitude of the meridional coma is no more than 1 μm for field of view of dimensions 0.4×0.4 mm.

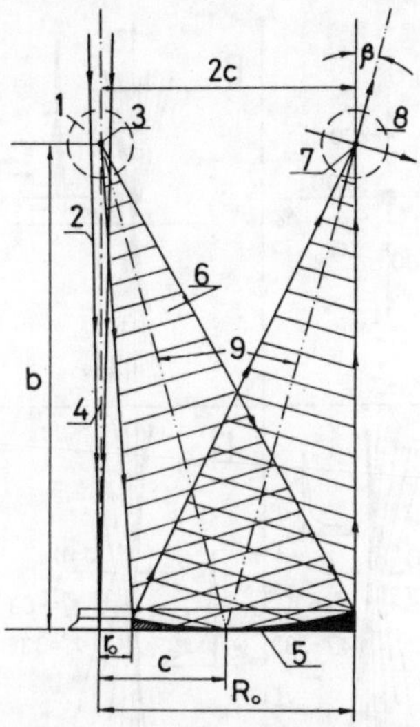

Fig. 3.84. The light rays in the meridional cross section of the MFTM: 1 — object space, 2 — convergent light beam, 3 — straight line particle track, 4 — optical axis of the MFTM, 5 — meso–optical mirror with ring response, 6 — diffracted light rays, 7 — meso–optical image of the straight line particle track, 8 — image space, 9 — marginal optical axis of the MFTM, $\alpha_{1/2}$ — angle between the main optical axis and the marginal optical axis of the MFTM.

To suppress coma aberrations in the MFTM and to decrease the overall dimensions of the meso–optical element with ring response, we use the Welford theorem, concerning the curvature radius of the hologram backing, and make meso–optical element in the form of a hologram on the spherical backing with curvature radius chosen in accordance with Eq. 3.51. The center of this circle 0 is located on the marginal optical axis of the MFTM as shown in Fig. 3.86. The lengths l_1 and l_2 have the same meaning as the lengths l_1 and l_2 in Fig. 3.81.

In the design with high geometrical magnification, when $L_2 \gg L_1$ and $R \approx L_1$ the center of the circle arc as generating line of the meso–optical hologram backing is located approximately in the position of the line object in the median plane of the working volume (Fig. 3.87). The overall dimensions of the meso–optical element with ring response can be reduced by this method from 160 mm to 40 mm.

To fabricate such meso–optical element-hologram on the spherical backing we can use a method suggested in [38] for manufacturing of the circular diffraction grating. A linear diffraction grating is placed in mechanical contact with opaque screen and with wedge-shaped aperture of very small central angle. The illuminated sector of the grating is rotating around the center of the wedge sector by means of

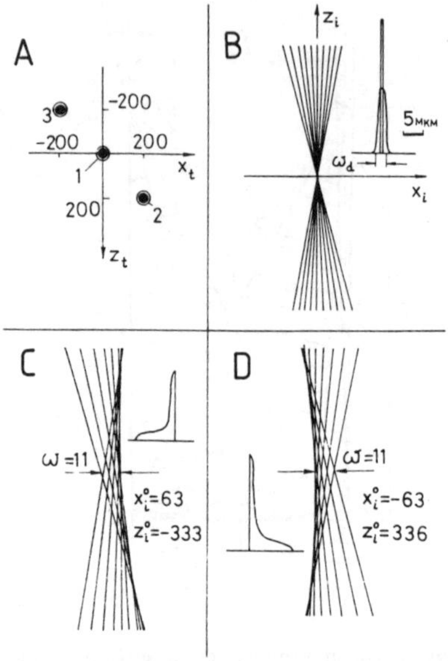

Fig. 3.85. The light rays which produce one of two meso–optical images of the straight line particle track in the MFTM for three positions of the straight line particle track and the light intensities near the crossover.

the tool spindle. For manufacturing of the plane circular hologram another system should be used (Fig. 3.88). A convergent object light beam is produced by the point-like source 1, optical collimator 2, plane mirror 4 and the lens 6. A divergent reference light beam is produced by the point-like source 1, optical collimator 2, beam splitter mirror 3, plane mirror 4, micro-objective 5 and by a small plane mirror 7. The opaque screen with sector aperture 8 is immobile. The photo plate 9 attached to the motor 10 is rotating during exposure. The center of the interference rings is on off-axis point at the distance $D_1/2$ from the optical axis of the circular hologram. The mutual arrangement of the sector wedge 8, of the interference fringes on the photo plate 9, and rotation axis of the motor 10 are shown in Fig. 3.89. The analogous rotating device for manufacturing of the meso–optical hologram with ring response on the spherical backing is given in Fig. 3.90: 1 — the source of the convergent light beam, 2 — the point-source of the divergent light beam, 3 — the spherical opaque screen with sector aperture, 4 — the photosensitive layer on the spherical backing 5, which accomplishes simultaneously the function of the immersion element in the MFTM.

Now we consider the design of the MFTM provided with *isoplanatic meso-optical mirror*. The generating line of this axially symmetrical mirror has the multielement structure, each element of which being a short arc of the *tautochronic ellipses* [407].

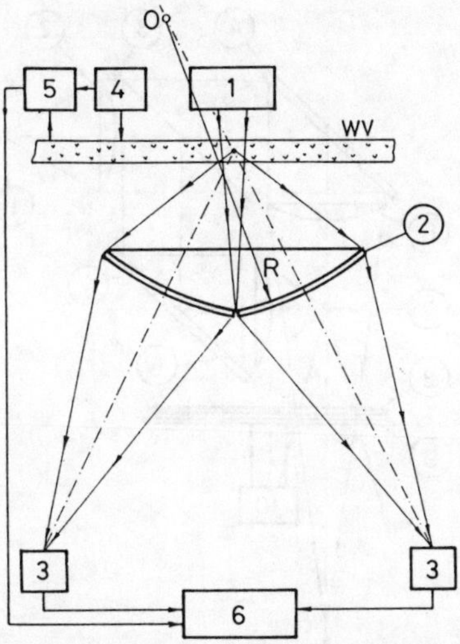

Fig. 3.86. Aplanatic meso–optical hologram with ring response, used in the MFTM: 1 — source of the convergent light beam, 2 — aplanatic meso–optical hologram on the nonplane backing, 3 — observation system, R — curvature radius of the hologram backing, where $R^{-1} = l_1^{-1} + l_2^{-1}$ and l_1 and l_2 are taken with allowance for the signs, O — point on the marginal optical axis of the MFTM in meridional cross section, on which the center of the circle of radius R lies.

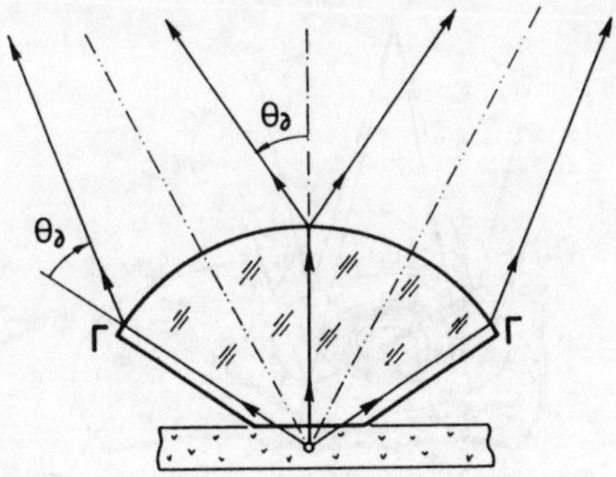

Fig. 3.87. Basic aplanatic meso–optical hologram Γ with ring response, made on the spherical backing of the frontal spherical immersion lens.

By definition, a family of the tautochronic ellipses is specified by two focuses. Two adjacent tautochronic ellipses are separated by the mutual distance defined by the wave length of light λ. To evaluate this distance, we consider the reflection of the light ray, going from the first focus F_1, at the points A and B, lying on two

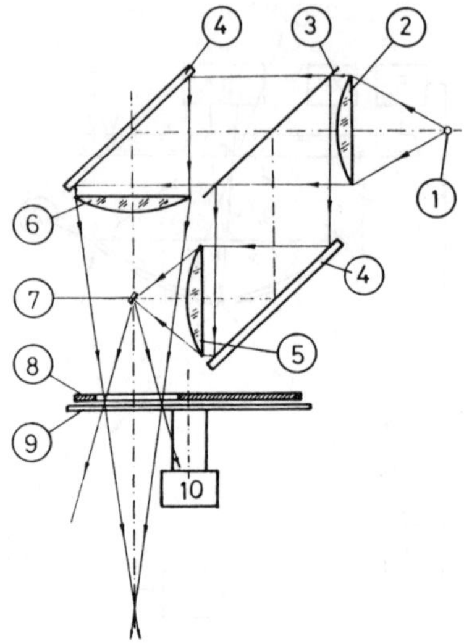

Fig. 3.88. Device for production of the meso-optical hologram with ring response on the plane backing: 1 — point light source, 2 — positive lens, 3 — beam splitter, 4 — plane mirror, 5 — positive lens, 6 — positive lens, 7 — microscopic plane mirror, 8 — shutter provided with narrow sector aperture, 9 — plane photosensitive layer, 10 — rotating unit.

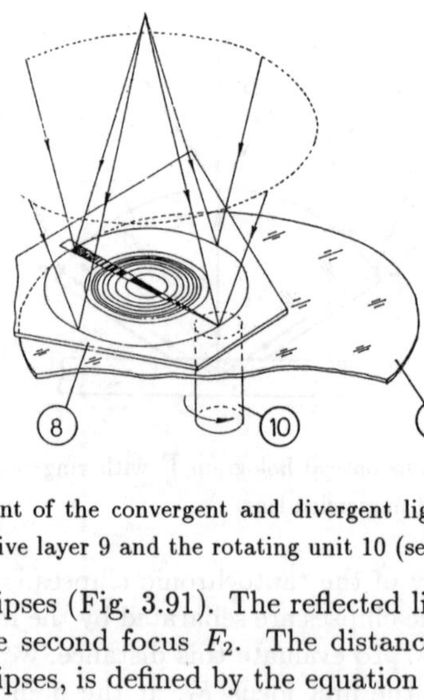

Fig. 3.89. Mutual arrangement of the convergent and divergent light beams, plane shutter with sector aperture 8, photosensitive layer 9 and the rotating unit 10 (see Fig. 3.88).

adjacent tautochronic ellipses (Fig. 3.91). The reflected light rays AC and BD are to added in phase at the second focus F_2. The distance d, which separates two adjacent tautochronic ellipses, is defined by the equation

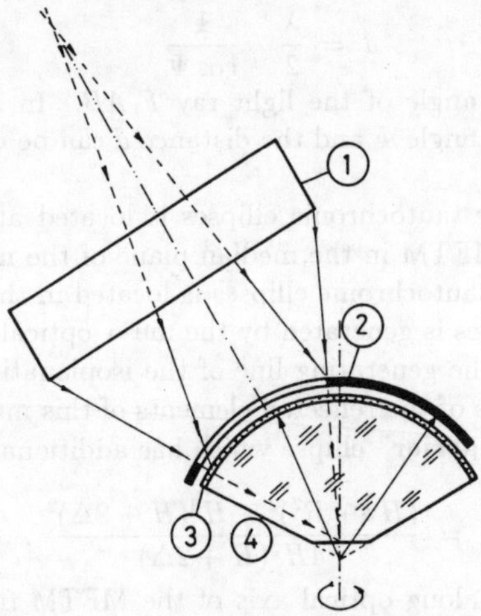

Fig. 3.90. Device for production of the aplanatic meso-optical hologram on the spherical backing: 1 — source of the convergent and divergent light beams, 2 — spherical shutter with open aperture in the form of a narrow sector, 3 — photosensitive layer mounted on the spherical surface of the frontal immersion lens.

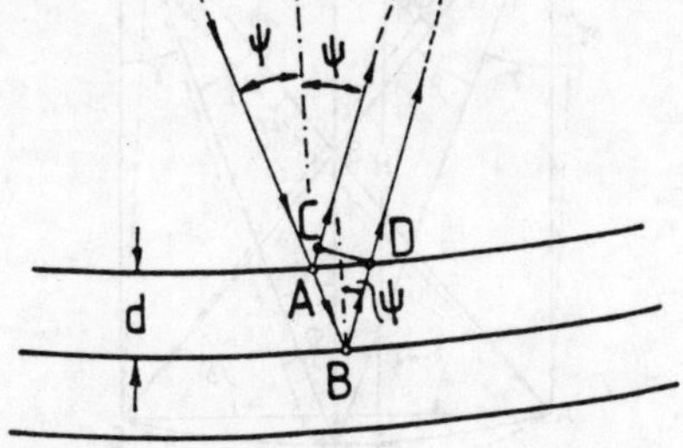

Fig. 3.91. Mutual arrangement of two adjacent tautochronic ellipses: F_1 and F_2 are two focuses of the given family of the tautochronic ellipses, ψ-incident angle of the light ray F_1AB, d - distance, which separates two adjacent tautochronic ellipses.

$$d = \frac{\lambda}{2} \cdot \frac{1}{\cos\Psi}\,, \tag{3.54}$$

where Ψ is the incident angle of the light ray F_1AB. In a small region of two tautochronic ellipses, the angle Φ and the distance d can be considered as constant factors.

The first focus of our tautochronic ellipses is located at the point F_1 on the optical axis F_1A of the MFTM in the median plane of the nuclear emulsion layer. The second focus of our tautochronic ellipses is located in the point F_2, where one of two meso–optical images is generated by the meso–optical mirror.

The main feature of the generating line of the isoplanatic meso–optical mirror is that the gravity centers of the reflected elements of this multielement generating line are located on the "**carrier**" ellipse which has additional sagitta (Fig. 3.92)

$$V = \frac{(H^2+R^2)^2 - H^2(H+2\Delta)^2}{4H^2(H+2\Delta)}\,, \tag{3.55}$$

where H is the distance along optical axis of the MFTM from the nearest point of the meso–optical mirror to the first focus of the tautochronic ellipses, R is the distance between two focuses of the tautochronic ellipses, and Δ is the sagitta of the traditional bulk meso–optical mirror, defined by the equation [407]

$$\Delta = \sqrt{\frac{H^2 + H\sqrt{H^2+R^2}}{2}} - H \tag{3.56}$$

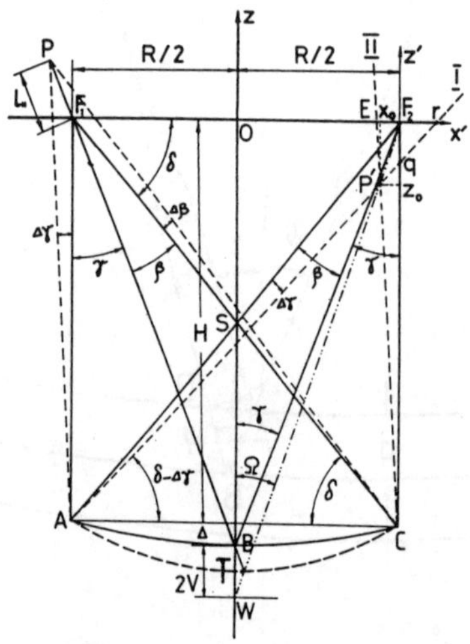

Fig. 3.92. Meridional cross section of the traditional meso–optical mirror (ABC) and of the multielement isoplanatic meso–optical mirror (ATC).

To prove Eq.(3.53), let us consider Fig. 3.92. The point P is the position of the particle track which is out of the center of the field of view of the MFTM and which is lying on the line segment PF_1BT. The point B is on the traditional continuos meso–optical mirror with ring response at the distance $R/2$ from the optical axis of the MFTM, γ is the angle AF_1B, β is the angle BF_1C, δ is the angle ACF_1, $\Delta\beta$ is the angle PCF_1, $\Delta\phi$ is the angle BF_2W, Ω is the angle OWF_2. The segment $\Delta = DB$ is the sagitta of the traditional meso–optical mirror with ring response, $2V = BW$.

The point P' is the image of the point P, produced by the isoplanatic meso–optical mirror with "carrier ellipse" ATC. The dashed line I is the line $AP'qr$. The dashed line II is the line $CP'E$.

The sine theorem for the triangle PF_1A gives:

$$\frac{\sin\Delta\gamma}{l_{\|}} = \frac{\sin(\gamma - \Delta\gamma)}{H}, \tag{3.57}$$

where $l_{\|} = pF_1$, $(\gamma - \Delta\gamma)$ is the angle PAP'. From Eq.(3.57) we get

$$\frac{\Delta\gamma}{l_{\|}} = \frac{(R/2)}{H\sqrt{(H+\Delta)^2 + (R/2)^2}} \tag{3.58}$$

The sine theorem for the triangle PF_1C gives:

$$\frac{\sin\Delta\beta}{l_{\|}} = \frac{\sin(\beta - \Delta\beta)}{\sqrt{H^2 + R^2}}, \tag{3.59}$$

where $(\beta - \Delta\beta)$ is the angle F_1CP.

If now we use the sine theorem for the triangle F_1BS

$$\frac{\sin\gamma}{\sqrt{(\frac{H}{2})^2 + (\frac{R}{2})^2}} = \frac{\sin\beta}{(\frac{H}{2} + \Delta)}, \tag{3.60}$$

then we have

$$\frac{\Delta\beta}{l_{\|}} = \frac{(R/2)(H + 2\Delta)}{(H^2 + R^2)\sqrt{(H+\Delta)^2 + (R/2)^2}} \tag{3.61}$$

Now let us calculate the coordinates $x_\circ$, $z_\circ$ of the cross point of two dashed lines I and II, defined by the equations:

$$\left.\begin{aligned} z &= x\,\tan(\delta - \Delta\gamma) - F_2q\,, \\ x &= -\Delta\beta \cdot H\,(1 + \tfrac{z}{H}) \end{aligned}\right\} \tag{3.62}$$

Also we have

$$\left.\begin{aligned} F_2q &= \Delta\gamma \cdot \tfrac{H^2+R^2}{R+H\Delta\gamma}\,, \\ F_2r &= F_2q \cdot \tfrac{1}{\tan(\delta-\Delta\gamma)} \end{aligned}\right\} \tag{3.63}$$

where

$$\tan(\delta - \Delta\gamma) = \frac{(\frac{H}{R}) - \Delta\gamma}{1 + (\frac{H}{R})} = \frac{HR - \Delta\gamma(H^2 + R^2)}{R^2} \tag{3.64}$$

From these equations we find:

$$\left.\begin{aligned} x_\circ &= -H \cdot \Delta\beta\,, \\ z_\circ &= -\Delta\beta \cdot H \cdot \tan(\delta - \Delta\gamma) - \Delta\gamma \tfrac{H^2+R^2}{R+H\Delta\gamma} \end{aligned}\right\} \tag{3.65}$$

Let us introduce the distance

$$\mathcal{L} = 2V + \Delta + H = \frac{R}{2}\cot\Omega_\circ\,, \tag{3.66}$$

where

$$\cot\Omega_\circ = \frac{z_\circ}{x_\circ} = -\frac{[(H^2 + R^2)^2 + H^4 + 2H^3\Delta]}{RH^2(H + 2\Delta)}\,. \tag{3.67}$$

We have then

$$\mathcal{L} = \frac{R}{2} \cdot \frac{[(H^2 + R^2)^2 + H^4 + 2H^3\Delta x]}{RH^2(H + 2\Delta)} = \frac{(H^2 + R^2)^2 + H^3(H + 2\Delta)}{2H^2(H + 2\Delta)}\,. \tag{3.68}$$

Finally we calculate the additional sagitta V :

$$2V = \mathcal{L} - (H + \Delta) = \frac{(H^2 + R^2)^2 + H^3(H + 2\Delta) - 2(H^4 + 3H^3\Delta + 2H^2\Delta^2)}{2H^2(H + 2\Delta)}\,. \tag{3.69}$$

or Eqs. (3.55) and (3.56).

The schematic diagram of the new design of the MFTM provided with isoplanatic meso–optical mirror is given in Fig.3.93. [407]. The convergent light beam from the light source 1 illuminates the nuclear emulsion layer and produces, in the vicinity of the meso–optical mirror with ring response 2, the Fourier transform of the straight line particle track. This meso–optical mirror 2 produces two meso–optical images of the particle track. One of them is picked up by the photosensitive matrix detector 4. The signals from the detector are fed to computer memory block 5. The unit 3 transports the nuclear emulsion layer and generates the position signals.

The multielement generating line of the isoplanatic meso–optical mirror is shown in detail inside the circle. Two adjacent elements of the whole generating line are connected by the auxiliary segment. Such jumps give rise to the diffraction orders in the place of the meso–optical images. To separate the diffraction order from the zero diffraction meso–optical image, we must choose the length $\Delta_\circ$ of each arc of the tautochronic ellipse in accordance with inequality

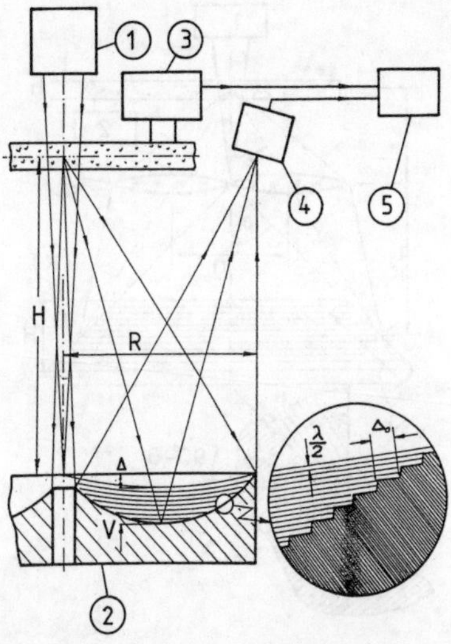

Fig. 3.93. Schematic diagram of the MFTM with multielement isoplanatic meso–optical mirror with ring response: 1 — source of the convergent light beam, 2 — isoplanatic meso–optical mirror with ring response, 3 — driver and position indicator, 4 — photosensitive matrix detector, 5 — computer memory block.

$$\frac{\lambda}{\Delta_\circ} \cdot l_2 > D\,, \tag{3.70}$$

where l_2 is the distance along light ray from the meso–optical mirror to the second focus F_2 of the tautochronic ellipses.

Now we present an example of realization of the MFTM, which has isoplanatic meso–optical mirror with ring response. For $R = 40\,\text{mm}$, $H = 0.3\,\text{mm}$, $\lambda = 0.6\,\mu\text{m}$, we have: $\Delta = 2.45\,\text{mm}$, $V = 9.05\,\mu\text{m}$, $l_2 \approx 77\,\text{mm}$, $\Delta_\circ = 100\,\mu\text{m} < \frac{\lambda l_2}{D} = 154\,\mu\text{m}$. The number of elements in this multielement meso–optical mirror with ring response $N = 400$. The jump between two adjacent elements $\Delta h = 22.6\,\mu\text{m}$.

3.9. Problem of Depth-of-focus in Meso–optics

The meso–optical element with shirt-shaped response described in Section 3.2 is used to solve the problem of depth-of-focus. Meanwhile it has a disadvantage: it produces intense side lobes in the meso–optical image. This restricts the dynamic range of the MFTM and provides only moderate values of the S/N ratio. To overcome this drawback a new design of the MFTM has been suggested [162]. The

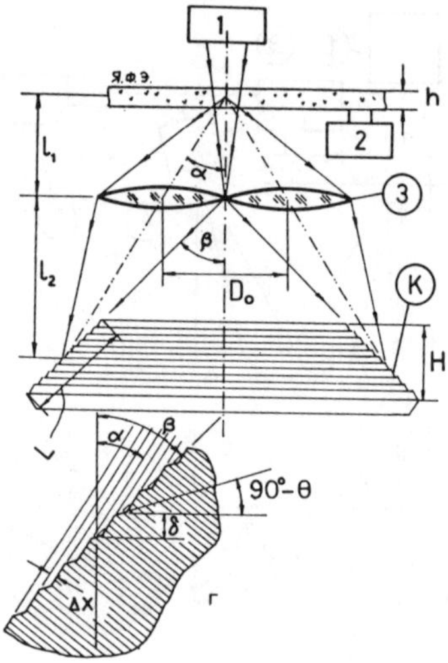

Fig. 3.94. General explanation of the MFTM with large depth-of-focus: 1 — source of the convergent light beam, 2 — $(X-Y)$ stage with position indicator, 3 — meso–optical element with ring response, K — multielement conical surface with generating line, consisting of N_z short microscopic blazed grooves, α — orientation angle of the auxiliary part of the blazed groove, β — orientation angle of the conical surface K, $(90° - \theta)$ — orientation angle of the working part of the blazed groove, δ — vertical projection of the groove, $\Delta\, x$ — the period of the grooves.

principle of this device is presented in Fig. 3.94. Here we use the Scheimpflug condition [94] adopted to the axiconic system of the light sectioning and described in the Section 2.7.4 [92 – 94]. With Scheimpflug condition we get sharp focus for all points of the working volume of depth h. Thus we provide the angular selectivity and the depth selectivity simultaneously without any scanning operations. The convergent light beam from the source 1 illuminates the working volume. The system 2 of the motor drive and position indicator is attached to the working volume. The meso–optical element with ring response 3 produces two meso–optical images of the line object, the left one L and the right one R. According to the Scheimpflug condition a line object with any z-coordinate produces on the conical surface K well focused meso–optical images. The generating line of the conical surface K is oriented at the angle β to the optical axis and is of the length L. As the required number of the information degrees of freedom along z-coordinate in the working volume of the depth $h = 200\,\mu$m is of the order of $N_z \approx 20 \div 50$, the generating line of the conical surface K is subdivided into N_z short elements of the height δ. The light rays are reflected from the working parts of these short elements so that the light rays are going perpendicular to the generating line of the conical surface K.

The centers of these short elements are located on the Scheimpflug's line which forms an angle β with optical axis of the device.

To observe two meso–optical images in the frame of one common CCD-matrix we cut from the cone K two sectors and dispose them in such a way that they are illuminated by sunset light rays reflected from plane mirrors 4 and 5 (Fig. 3.95). The light rays which produce two meso–optical images are reflected from the conical pieces 6 and 7 and directed into the aperture of the imaging lens 8. The latter produces traditional images of the picture on the conical sectors 6 and 7 on the screen 9.

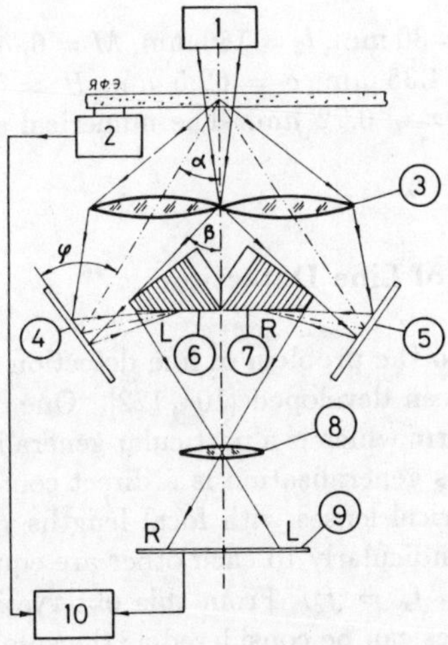

Fig. 3.95. The real design of the MFTM with large depth-of-focus: 1 — source of the convergent light beam, 2 — $(X - Y)$ stage with position indicator, 3 — meso–optical element with ring response, 4 — plane mirror, 5 — plane mirror, 6 — the left part of the conical surface with blazed grooves, 7 — the right part of this conical surface with blazed grooves, 8 — imaging lens, 9 — photodetector of two meso–optical images and, 10 — computer memory.

The angle θ between the generating line of the short element and optical axis of the system is chosen according to equation

$$\theta = 45^\circ + \alpha , \qquad \text{with} \qquad \alpha = \tan^{-1}(D_\circ M/2l_2) , \tag{3.71}$$

where l_1 and l_2 are distances from the line object to the meso–optical element with ring response 3 and from this meso–optical element to the center of the cone K, respectively, and $M = l_2/l_1$. The angle β between the line of the centers of the short elements of the generating figure of the multielement conical mirror K and the optical axis of the system is given by equation

$$\beta = \tan^{-1}[D_\circ(M+1)/2l_2] . \tag{3.72}$$

The distance Δx and the height δ are given by

$$\Delta x = \lambda M / \sin 2\alpha \,, \qquad \delta = \Delta x / \tan(\beta - \alpha) \,, \tag{3.73}$$

where $2D_\circ$ is the external diameter of the meso–optical element with ring response 3. The mirrors 4 and 5 produce an angle φ with marginal optical axis of the system, where

$$\varphi = 45^\circ + \beta/2 \,. \tag{3.74}$$

In the real construction $l_1 = 30$ mm, $l_2 = 180$ mm, $M = 6$, $\alpha = 30^\circ$, $D_\circ = 34.65$ mm, $\theta = 75^\circ$, $\beta = 34^\circ$, $\Delta x = 4.35$ μm, $\delta = 62.5$ μm, $H = 7.2$ mm, $L = 8.32$ mm, $N = 134$, $\Delta z = 1.5$ μm, $\delta x = 0.72$ μm. The numerical aperture of the lens 8 is equal to 1 : 4.

3.10. Another Systems of Line Detection

Some new approaches to the problem of line detection and directional analysis of images in optics have been developed [104, 122]. One of them is based on the anamorphic Fourier transform which is a particular generalization of the traditional $2D$ Fourier transform. This generalization is a direct consequence of the observation [163], that two cylindrical lenses with focal lengths f_x and f_y posed in close contact and oriented perpendicularly to each other are equivalent to one spherical lens with focal length $f = f_x = f_y$. From this observation it follows that two-dimensional process in optics can be considered as the sum or superposition of two independent one-dimensional processes. In the case of $f_x \neq f_y$ we have two different lateral magnifications $M_x \neq M_y$.

The anamorphic Fourier transform [122] is performed by two crossed cylindrical lenses. In the simplest configuration two cylindrical lenses (Fig. 3.96) have unequal focal lengths f_x and f_y, but their back focal planes coincide. A plane object is placed in any plane between (x_1, y_1)- and (x_2, y_2)-planes. In the (x_3, y_3)-plane we observe FT-spectrum with different linear scales along x- and y-direction, the phase factors of this FT-spectrum being different along x- and y-directions.

In the configuration presented in Fig. 3.97, an exact FT-spectrum is observed. Four cylindrical lenses are used, and the plane object is in the (x_1, y_1)-plane. Along x-direction this system acts as a single cylinder L_x of focal length f_x placed in the (x_5, y_5)-plane. Along y-direction there are three cylindrical lenses. Two of them, L_{y1} and L_{y2}, act as an optical telescope which produces an inverted image of the horizontal object component in the (x_6, y_6)-plane. This inverted image is converted by the cylinder L_y. To put into coincidence $1D$ spectra in the (y_8, y_8)-plane, the focal lengths of the cylindrical lenses must satisfy the condition

$$f_x = f_{y1} + f_{y2} + f_y \,. \tag{3.75}$$

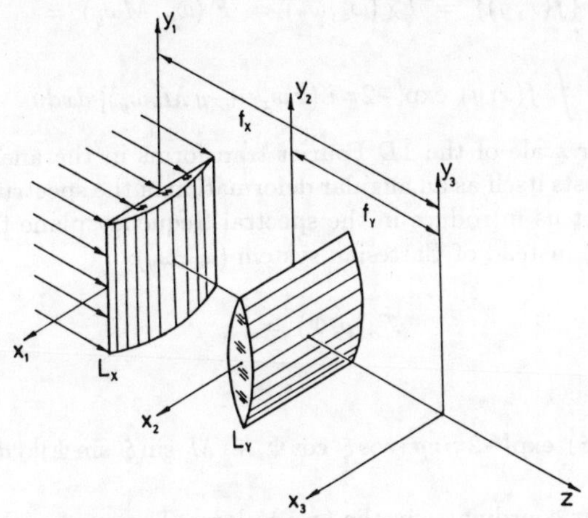

Fig. 3.96. System with two crossed cylindrical lenses, designed for production of the anamorphic Fourier transform.

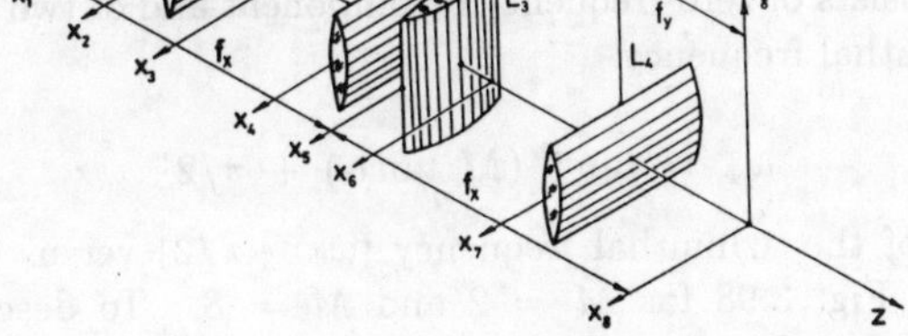

Fig. 3.97. System with four crossed cylindrical lenses, designed for production of the anamorphic Fourier transform.

The coefficient of anamorphosis of the combined transform defined by equation

$$M = \frac{f_x \, f_{yz}}{f_{y1} \, f_{y1}}, \tag{3.76}$$

governs the anamorphic $2D$ Fourier transform

$$\hat{\mathcal{F}}\{f(x,y)\} = F_A(\omega_x,\omega_y) = F(\omega_x, M\omega_y) =$$

$$= \int\int f(x,y)\, \exp[-2\pi\, i\, (x\,\omega_x + y\, M_y\, \omega_y)]\, dxdy\,. \tag{3.77}$$

A different linear scale of the $1D$ Fourier transforms in the anamorphotic $2D$ FT-spectrum manifests itself as an angular deformation of the spectrum. To demonstrate this feature let us introduce in the spectral frequency plane the polar coordinate system (ρ, Φ) instead of Cartesian system (ω_x, ω_y):

$$\hat{\mathcal{F}}_A(\rho, \Phi) =$$

$$= \int_0^{\infty} \int_0^{2\pi} f(r,\xi)\, \exp[-2\pi i r\rho\, (\cos\xi\, \cos\Phi + M\, \sin\xi\, \sin\Phi)]\, r dr d\xi\,, \tag{3.78}$$

where (r, ξ) are polar coordinates in the (x,y)-plane. The anamorphotic spectrum of the cosine diffraction grating

$$g\,(x_1, y_1) = 1 + \cos[2\pi\mu\, (-x_1\, \sin\alpha + y_1\, \cos\alpha)] \tag{3.79}$$

of period $a = 1/\mu$, rotated by the angle α, is given by equation

$$G_A\,(\omega_x,\omega_y) = \frac{1}{M}\,\delta(\omega_y) + \frac{1}{2M}\,[\,\delta(\omega_x + \mu\,\sin\alpha)\,\delta\,(\omega_y - \frac{\mu\,\cos\alpha}{M}) +$$

$$+ \delta\,(\omega_x - \mu\,\sin\alpha)\,\delta\,(\omega_y + \frac{\mu\,\cos\alpha}{M})]\,. \tag{3.80}$$

This spectrum consists of zero-frequency δ-component and of two off-axis δ-components at the azimuthal frequency

$$\omega_\Phi = \tan^{-1}(M\,\tan\alpha) + \pi/2\,. \tag{3.81}$$

The dependence of the azimuthal frequency $(\omega_\Phi - \pi/2)$ versus orientation angle α is presented in Fig. 3.98 for $M = 2$ and $M = 8$. To describe the angular magnification of azimuthal frequencies in an anamorphic spectrum with respect to rotation-invariant frequencies in a traditional $2D$ Fourier spectrum, a function

$$K\,(\alpha, M) = \frac{M}{1 + (M^2 - 1)\,\sin^2\alpha} \tag{3.82}$$

is to be introduced. The larger is the coefficient of anamorphosis M, the narrower is the working range of angles of a spectrum with angular magnification greater than one (Fig. 3.99). This working range is given by the inequality

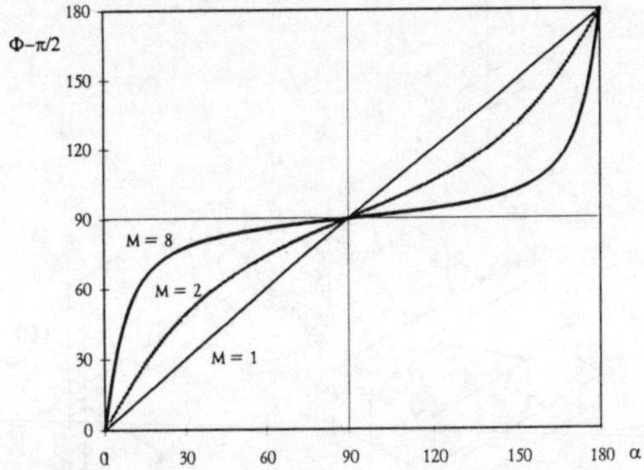

Fig. 3.98. Azimuthal frequency $(\omega_\Phi - \pi/2)$ versus orientation angle α for $M = 2$ and $M = 8$.

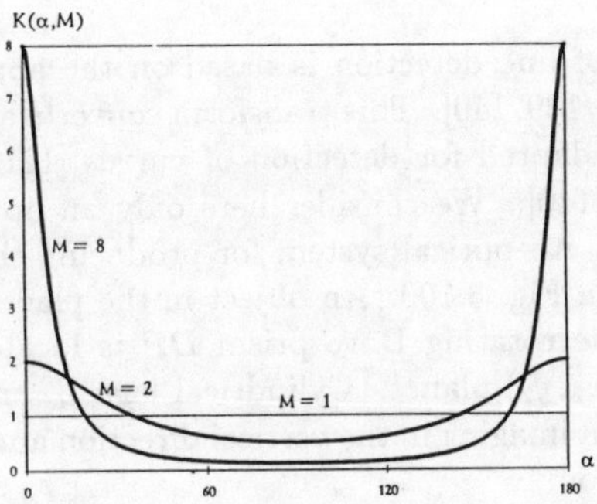

Fig. 3.99. The function $K(\alpha, M)$, given in Eq. (3.82), and tabulated for different M versus orientation angle α.

$$|\alpha| < \sin^{-1}\sqrt{\frac{1}{M+1}}. \tag{3.83}$$

To procedure, recommended in [122] to get accurate angular spectrum, involves the stepwise rotation of the object in the input plane.

The application of this technique for line detection is restricted only to the plane line objects such as we see on seismograms. Here we can use the angular selectivity of the anamorphic Fourier transform which in the system with wedge is $K(\alpha, M)$ times higher than the classical axial symmetric Fourier transform [122]. Anamorphotic Fourier transform can be used as well in an optical correlation for pattern recognition. No on-line compression of the optical information have been proposed.

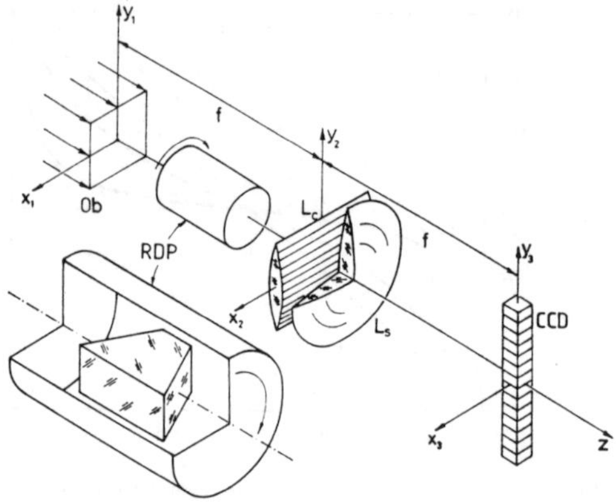

Fig. 3.100. Coherent optical Hough transformation system.

Another technique of line detection is based on the application of the Hough transform [135, 136, 138, 139, 140]. This transform converts a line into a point [137], but it can be as well adapted for detection of curves [121, 138] and for tracking of the moving targets [140]. We consider here only an optical realization of the Hough transform [125]. An optical system for producing the Hough transform in coherent light is given in Fig. 3.100. An object in the plane (x_1, y_1) is illuminated by the plane wave. The rotating Dove prism DP is located on the optical axis between (x_1, y_1)- and (x_2, y_2)-planes. Cylindrical lens L_c and spherical lens L_s of focal length f perform an imaging in the vertical direction and $1D$ Fourier transform in the horizontal direction.

A linear array of photodetectors in the form of the $1D$ CCD-line is oriented vertically. This device records the zero-order component of the Fourier transform, the latter being proportional to the intensity of line of the Hough transform at a given θ. Angular scanning of the object accomplishes a rotating Dove prism. A non-coherent system of the Hough transform was also suggested [125]. Until now no proposals concerning adaptation of the anamorphic Fourier transform or Hough transform for detection of the arbitrary orientation straight line objects in $3D$ space were published.

3.11. Applications

The MFTM has been designed as a special purpose unit for use in the nuclear emulsion technique [161]. Owing to extremely high spatial resolution this technique remained until recently the most precise, convenient and sometimes the only conceivable particle detector. With this technique we can observe in $3D$ space a $3D$ event of interaction of elementary particles and nuclei with nuclear emulsion or the spontaneous decay of these particles in time. The nuclear emulsion is gaining in importance recently in the experimental investigations of rare processes with particles of very short life time ($10^{-13} \div 10^{-16}$ s) [2].

Typically, a nuclear emulsion particle detector has the form of a solid block of volume $50 \div 300$ liters and of a total mass up to 1 ton. The thickness of each element of this block is of the order of 0.6 mm. To distinguish the elements of the particle tracks in the form of silver grains of a dimension $0.5 - 1\,\mu$m, a light microscope of very high numerical aperture is used. The depth-of-focus is of the order of $2\,\mu$m, and the $3D$ volume of the picture element (voxel) is equal to $2\ \mu m^3$. So the number of voxels in the nuclear emulsion block of the volume 200 liters is expressed by the astronomical number of $2 \cdot 10^{17}$(!). If the scanning system is operating with the speed of 10^6 pixels/second we need at least $2 \cdot 10^{11}\ s \approx 6 \cdot 10^3$ years to perform only the scanning of this volume block. To reduce the scanning time, the solid block is made of two parts with different functional purposes: of the emulsion target and of the tracking emulsion plates.

The nuclear emulsion target consists of many pellicles of thickness 0.6 mm and of dimensions $100 \times 200\ mm^2$. Each charge particle which transverses these pellicles of nuclear emulsion, produces in the developed pellicle a track of silver grains with linear density of silver grains as low as 20-30 per 100 μm. The tracking emulsion plate involves a backing of polystyrene foil of thickness 70 μm with two nuclear emulsion layers of thickness 0.5 mm on both sides of the backing. As a rule the particle tracks are oriented perpendicular to the backing plane.

The experimental facility with nuclear emulsion is usually equipped with various external particle detectors such as scintillating fiber tracer, bubble track chamber, streamer gas chamber, streamer tube tracers and high resolution calorimeter. As a rule the scientific collaboration around these experiments is supported by very rich counties. By means of these external devices arranged over the length of the order of 100 m, the location of the nuclear event, which consists of many secondary particle tracks with one vertex, inside the nuclear emulsion target can be estimated with spatial uncertainties of the order of $2 - 3$ mm. The volume to be examined in a such experiment by means of the light microscope is of the order of $0.2\ cm^3 = 2 \cdot 10^{11}\ \mu m^3$ or 10^{11} voxels.

Several attempts have been made to automate the search and the measurements of the events in the nuclear emulsion. The first automatic system in nuclear emulsion technique was the device for measurement of the multiple scattering angle of the particle track [164]. The automatic systems with electromotors, position indicators and computer were built is several scientific laboratories, e.g. ICAMS (Interaction Computer Assisted Measurement System) in the Lawrence Berkeley Laboratory [2] and in the measurement center of Nagoya University [2]. All these apparatuses are suitable for observation and measurement rather than for finding the events. The time required to find an event has been reduced by using the tracking emulsion plates in the E–531 experiment in the Fermi National Accelerator Laboratory [112].

An equipment called CADIM (Computer Aided Digitized Microscope) was built in CERN, Geneva [2], to make the search easier by giving an artificial picture of tracks predicted by the external particle detectors. This picture could be compared with real image of the event. Making decisions, however, remains the operator's task.

A fully automated emulsion analysis system to find and to recognize the particle tracks in the tracking emulsion plate [105] has been built in the Nagoya University. In this system $3D$ motor driven stage with 1 μm read-out linear indicator, and a traditional video image processor with 16 frame memories for each of 16 sublayers have been used. The distance between the adjacent sublayers is equal to 5-30 μm.

However, the problem of the fully automatic search for particle tracks with known orientation in the nuclear emulsion target remains unresolved. To day some cruel problems are facing the scientist: 1) huge volume of the initial information, as a result of very long exposure time of the nuclear emulsion target, generally more than about 1/2 year; 2) the need to accomplish the $3D$ scanning of the working volume in the traditional light microscope; 3) very intense background of the silver grains randomly distributed over the working volume; 4) very complicated computer process or operator's decision is needed to say whether the observed silver grain belongs to real particle track or not.

The first device which perceives a straight line particle track as a single object and does not subdivide the particle track into many separate dots has been built in the Joint Institute for Nuclear Research, Dubna, as MFTM [108, 110, 115]. This nontraditional light microscope sees only straight line particle tracks and accomplishes inside the illumination stage, on-line, a drastically compression of information about particle track in nuclear emulsion. Any other objects are converted by MFTM into an uniformly distributed background or noise. As was explained above, two spots on the TV-screen of the MFTM contain indeed the whole information about the position of the particle track in $3D$ space.

To illustrate the most fundamental performance of the MFTM we consider here the event-searching algorithm. As was shown in [106, 117], the meso–optical images of the particle tracks, which form a real event in the nuclear emulsion and which have a common vertex, are lying on the sinogram in the frame of (θ_{xy}, x)-coordinate system. For small orientation angles θ_{xy} the meso–optical images lie on a straight line [117]. By means of this event-searching algorithm, an operator can draw a conclusion about the existence of the event even in the case, when the vertex of the event is beyond the nuclear emulsion layer to be observed in the MFTM.

As an example we present a multitrack disintegration event shown in Fig. 3.101. Accelerated neon-nucleus is disintegrated into four secondary particle tracks 1, 2, 3, and 4. The results of the measurements which have been performed with the MFTM are presented in Fig. 3.102, in the frame of (θ_{xy}, x) coordinate system for several y-values. We see that all points are grouped around straight lines with one common vertex, with x-coordinate which coincides with x-coordinate of the event. In the course of these measurements no depth scanning have been made.

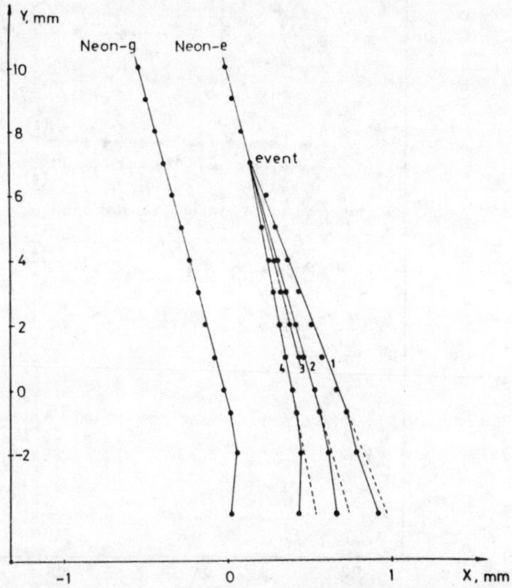

Fig. 3.101. Event with disintegration of accelerated Ne-nucleus into four secondary particles.

If the vertex of the secondary particles is located on the optical axis of the MFTM, all meso–optical images of the secondary particle tracks should lie on a focal circle with its center on the optical axis of the MFTM. The radius of this focal circle is defined by z-coordinate of the event vertex. This technique of the direct event searching has been tested in the simulation experiments [117] in which a system of many narrow concentric transparent circles has been used. This technique of the direct search of the events in the nuclear emulsion does not impose any restrictions on the orientation angles θ_{xy}.

There is no doubt that the feature of the MFTM to concentrate the information about the microscopic line objects into **two** spots on the output display will find a wide application in the microscopic robotics in the near future.

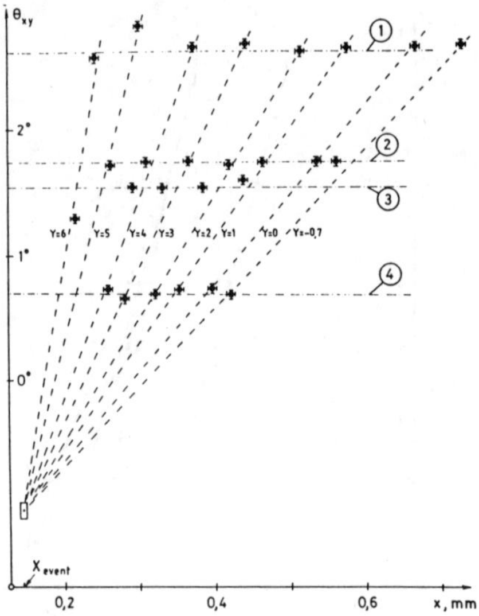

Fig. 3.102. Test of the event searching algorithm [106, 107] in the event presented in Fig. 3.101. The expected zone of the event vertex is shown as rectangular with transverse coordinate $x_{VERTEX} = (0.13 \pm 0.01)\,\text{mm}$.

Chapter 4

DIFFRACTION FREE WAVEFIELDS

4.1. Introduction

We know that a wavefield, localized in the restricted volume of the lateral dimension r, diffracts and at distances $z \gg r^2/\lambda$ assumes the form of the divergent spherical wave with beam angular width of the order of λ/r. Meanwhile it has been shown [72] that there exists an axial symmetrical wavefield in the free space that meets the wave equation

$$\left(\nabla^2 - \frac{1}{c^2}\,\frac{\partial^2}{\partial t^2}\right) E\,(\vec{r},t) \;=\; 0 \tag{4.1}$$

where ∇ denotes an operator nabla, $\vec{r}(x,y,z)$ is the observation vector, and that does not change over the transverse plane at any propagation distances z. At $z > 0$ this wavefield is of the same structure as at $z = 0$. These characteristics exhibits the wavefield of the form [72]

$$E\,(\vec{r},t) \;=\; \exp\,(i\,\beta\,z)\;\exp\,(\,-\,i\,\omega\,t)\,J_\circ\,(\alpha\rho)\,, \tag{4.2}$$

where $\rho^2 = x^2 + y^2$ and where $J_\circ(\bullet)$ denotes the Bessel function of zero order of the first kind. A special limited case of this class "diffraction free" wavefields represents a plane wave, into which the wavefield, Eq.((4.2), goes at $\alpha \to 0$.

The "diffraction free" wavefields postulated in [72, 73, 165] were named "Bessel-form beams" [165]. The restricted Bessel beam can be generated in lasers with ring mirrors and with output beams in the form of a ring. The production of the Bessel beams in the system with narrow ring slit was proposed in [73]. Besides it has been remarked [167] that z-component of the wave vector of the monochromatic diffraction free wavefield is kept constant for all partial components of this beam. This follow from the fact that any phase relations between partial components will be changed along z-axis and will produce spatial variations in the whole wavefield. It was pointed in [167] that an axial non-symmetrical wavefield of the form

$$E\,(\rho,\Phi) \;=\; \cos m\,\Phi\cdot J_m(\alpha\,\rho)\,, \tag{4.3}$$

with Bessel function $J_m(\bullet)$ of the m-order, generates another new class of the diffraction free wavefields. We know from Chapter 2, that the wavefield of the form Eq. (4.2), is a conical one, while the wavefield of the form, Eq. (4.3), has the helical structure.

The properties of the diffraction free wavefields were treated in detail many years ago by J. Dyson [38]. The annular aperture of small width, which creates diffraction free wavefields with $J_\circ^2$-light intensity distribution along the radial coordinate, has been described by G.C. Steward [36]. Thus we may conclude that diffraction free wavefields are indeed very old ones, and only in 1987 J. Durnin has introduced this very useful term [72]. It is evident that these diffraction free wavefields are embraced by the term "MESO–OPTICS" [3].

An ideal diffraction free wavefields of the form, Eq. (4.2), should carry infinite light power because of the divergence of the function $|J_\circ(\alpha\rho)|^2$. To remove the divergence problem, a model of Bessel–Gauss beam has been introduced [168]. Unlike the ideal diffraction free wavefields, the Bessel–Gauss beam, which is a solution of the paraxial wave equation, carries finite light power.

Recall, that classical axial symmetrical diffraction free wavefield can be described by the pictorial model of many plane waves whose wave vectors lie on a conical surface with axis on the z-axis of the system. Therefore the diffraction pattern of this beam should have the same shape at any z-coordinate. The Bessel–Gauss beam of zero order represents a superposition of many Gaussian beams with axes uniformly distributed on a cone of the angular half aperture θ, just the same as in the case of the plane wave superposition [169].

Any Bessel–Gauss beam at $z = 0$ is defined by the Bessel amplitude with Gaussian profile [168]:

$$E(\rho, 0) \;=\; E_\circ\, J_\circ(\alpha\,\rho)\, \exp\left(-\rho^2/w_\circ^2\right), \tag{4.4}$$

where $E_\circ$ is a constant, $w_\circ$ is the effective width of the Gaussian beam (waist parameter), and α is a parameter which defines the angular half aperture of the conical surface, with

$$\alpha \;=\; k\,\sin\theta, \quad \alpha \approx k\,\theta, \quad \theta < 10^\circ, \tag{4.5}$$

and $k = 2\pi/\lambda$ is the absolute value of the wave vector $\vec{k}$. The angular spread of a single Gaussian beam component is given by

$$\theta_G \;=\; \lambda\,/\,\pi\,w_\circ\,. \tag{4.6}$$

The propagation features of the Bessel–Gauss beam are given by the ratio of the angular aperture of the cone θ to the Gaussian beam spread θ_G:

$$\frac{\theta}{\theta_G} \;=\; \frac{\sin^{-1}(\alpha/k)}{(\lambda/\pi w_\circ)} \;\approx\; \frac{\alpha\, w_\circ}{2}, \qquad \theta < 10^\circ\,. \tag{4.7}$$

If $\alpha\, w_\circ \ll 1$, all components of the Gaussian beams are overlapping. An interference diffraction pattern is generated even in the far-zone ($z \to \infty$). The cross-section of this Bessel–Gauss beam resembles the central part of a diffraction free beam. In other words the spot-size $w_\circ$ is smaller than radius of the central spot of the function $J_\circ(\alpha\rho)$. The transverse profile of this beam has the same form at any z-coordinate.

If $\alpha\, w_\circ \gg 1$, the propagation picture is changed severely with z-coordinate. At the distances from the input aperture

$$z > D \cong \frac{w_\circ}{\theta} \cong \frac{w_\circ k}{\alpha}\,, \tag{4.8}$$

even the central Gaussian component is receding from z-axis, and the spot-size is concerted into the annular shape picture with mean radius

$$w\,(z) \,=\, w_\circ\,[1 + (z/Z_R)^2\,]^{1/2}\,, \tag{4.9}$$

where

$$Z_R \,=\, k\,w_\circ^2\,/\,2 \,=\, \frac{\pi w_\circ^2}{\lambda} \tag{4.10}$$

is the so-called Rayleigh distance. For example, for Bessel–Gauss beam with $\theta = 10^{-3}\,\mathrm{rad}$, $w_\circ = 1\,\mathrm{mm}$, $\lambda = 0.63\,\mu\mathrm{m}$, $\theta/\theta_G = 4.96$, $D = 1\,\mathrm{m}$, $Z_R = 5\,\mathrm{m}$.

A confocal resonator with an annular active medium has been proposed for production of Bessel beams [170]. An annular beam generated in the annular active medium, is reflected from the end mirrors and directed into the central region as Bessel beam. After the second reflection from end mirrors this conical beam is converted into the annular beam with waist region inside the annular medium (Fig. 4.1). The generating line of each meso–optical mirror is a circle arc of the radius $R = L/2$, where L is the length of the resonator.

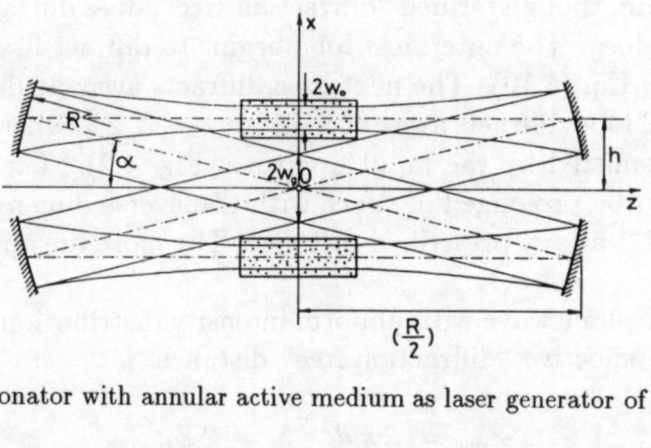

Fig. 4.1. Confocal resonator with annular active medium as laser generator of the Bessel beam.

If the length of the resonator $L \ll Z_R$, the half-width of the Bessel beam zone is equal to

$$w_f \,=\, \lambda\,/\,\pi\,\sin\theta\,, \tag{4.11}$$

where

$$\sin\theta = 2w_\circ / L, \tag{4.12}$$

The angle α between the marginal optical axis and central optical axis defines the diameter of the central spot of the diffraction picture, which is equal to

$$\Lambda_\circ = \frac{3}{4}\frac{\lambda}{\sin\alpha}, \tag{4.13}$$

and the thickness of the next rings, equal to [170]

$$\Lambda = \frac{\lambda}{2\sin\alpha}. \tag{4.14}$$

And now some remarks about power transportation in Bessel- and in Gaussian-beams. In the case of pure Bessel beam each side lobe contains the same energy as does the central spot. Thus, for example, in the Bessel beam with N side lobes in the plane $z = 0$, the central spot will contain only about $1/N$ of the total energy of the beam. The Gaussian-beam contains about 50 per cent of the total light energy within the full width of the half-maximum. The useful relation is taking place [166] for ranges of these two beams at which the peak intensity falls by a factor of 2:

$$\text{Bessel range} \approx \text{N} \cdot \text{Gaussian range}, \tag{4.15}$$

and the demanded total light powers:

$$\text{Total Bessel power} \approx \text{N} \cdot \text{ Total Gaussian power}. \tag{4.16}$$

From these relations it follows that Bessel type beams can be used only for information purposes but not for power transportation.

The Bessel-beam, though termed "diffraction-free", does diffract, but not so fast, as spherical wave does. The outermost lobe begins to diffract first, at a distance of Rayleigh range Z_R Eq. (4.10). The next lobe diffracts away at distance $2\,Z_R$ [171]. Finally, the central lobe diffracts away at a distance $\sim N\,Z_R$, where N is the number of side lobes transmitted by the input aperture (Fig. 4.2). Thus the information carrying signal can be prolonged in space with transverse dimension of the central maximum width α^{-1} at any prescribed distance. The more the number N, the more this distance.

Recall, that for plane wave with uniform intensity distribution over the aperture of diameter α, the effective "diffraction free" distance

$$Z_{pw} = 2\pi\, d^2 / \lambda = 2\,Z_R. \tag{4.17}$$

For Gaussian beam with effective width $w_\circ$ the "diffraction free" distance is equal to

$$Z_G = 2\pi\, w_\circ^2 / \lambda = 2\,Z_R. \tag{4.18}$$

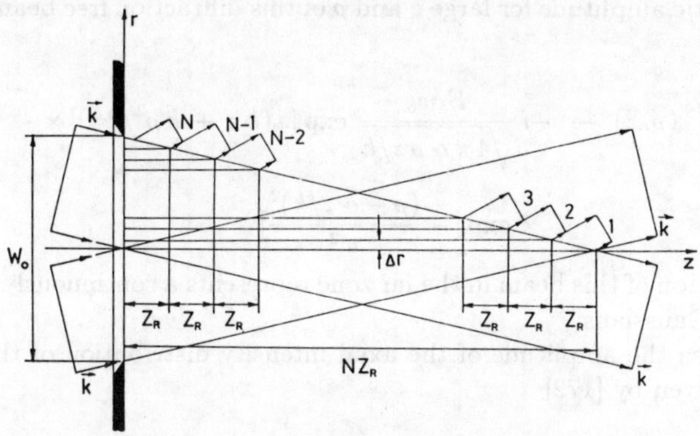

Fig. 4.2. A model of the sequential generation of the Bessel beam side lobes.

Thus, any beam has its proper "diffraction-free" distance. The main advantage of the Bessel beams is that the "diffraction-free" distance can be made as long as you wish. And this feature of the Bessel-beam does not contradict neither the general concept of diffraction, nor the uncertainty principle.

4.2. Bessel–Gauss Beams

In this Section we present analytical expressions for the Bessel–Gauss beams[168] in two extreme cases and give general expressions for axial and transversal light intensity distributions which have been found as solutions of the scalar Kirchhoff theory [172].

If $\alpha \to 0$ in Eq. (4.4), the Bessel–Gauss beam is converted into an ordinary Gaussian-beam with light amplitude

$$E_G(\rho, 0) = E_\circ \exp(-\rho^2/w_\circ^2). \tag{4.19}$$

If $w_\circ \to \infty$, $w_\circ/w \to 1$, we have pure diffraction free beam with light amplitude

$$E_B(\rho, z) = E_\circ J_\circ(\alpha, \rho) \exp(i k_z z) \tag{4.20}$$

with projection k_z of the wave vector $\vec{k}$, equal to

$$k_z = k - \alpha^2/2k. \tag{4.21}$$

It is interesting to note, that for $z \gg Z_R$ the beam width w is equal to

$$w \approx w_\circ \frac{z}{Z_R}, \tag{4.22}$$

and the asymptotic amplitude for large z and ρ of this diffraction free beam assumes the form:

$$E_B^{ASS}(\rho, z) \rightarrow -i \frac{E_\circ w_\circ}{\sqrt{4\pi\alpha\rho z/k}} \exp[i(kz + k\rho^2/2z)] \times$$

$$\times \exp[-\frac{(\rho - \alpha z/k)^2}{w^2}]. \tag{4.23}$$

The cross section of this beam in the far zone represents a continuously divergent annular-shaped Gaussians.

In general case the amplitude of the axial intensity distribution of the Bessel–Gauss beam is given by [172]

$$E(0, z) = E_\circ \frac{i k_\circ \exp(-i k_\circ z)}{z} \int_0^a J_\circ(\alpha\rho') \exp(-q\rho'^2)\rho' d\rho' \tag{4.24}$$

with complex parameter

$$q = \frac{1}{w_\circ^2} + \frac{i k_\circ}{2z}, \tag{4.25}$$

and can be written analytically as

$$E(0, z) = E_\circ \frac{i k_\circ \exp(-i k_\circ z)}{2qz} \left[\exp(-qa^2) \sum_{n=1}^{\infty} \left(\frac{2qa}{\alpha}\right)^n \cdot J_n(\alpha a)\right],$$

$$\text{for} \left|\frac{2qa}{\alpha}\right| < 1, \tag{4.26}$$

and

$$E(o, z) = E_\circ \frac{i k_\circ \exp(-i k_\circ z)}{2qz} \left[\exp(-\alpha^2/4q) - \exp(-qa^2) \cdot \sum_{n=1}^{\infty} \left(-\frac{\alpha}{2qa}\right)^n J_n(\alpha, a)\right],$$

$$\text{for} \left|\frac{2qa}{\alpha}\right| > 1. \tag{4.27}$$

These expressions are valid from the Fresnel zone to infinity.

The axial light intensity distribution of this Bessel–Gauss beam is given by

$$I^{BG}(z) = \left(\frac{k_\circ}{2|q|z}\right)^2 \exp[-2(a/w_\circ)^2] \cdot (S_1^2 + S_2^2),$$

$$\text{for} \left|\frac{2qa}{\alpha}\right| < 1, \tag{4.28}$$

where

$$\left.\begin{aligned} S_1 &= \sum_{n=1}^{\infty} \left(\tfrac{2\,a}{\alpha}\right)^n \, |q|^n \, \cos(n\,\sigma)\, J_n(\alpha\, a)\,, \\ S_2 &= \sum_{n=1}^{\infty} \left(\tfrac{2\,a}{\alpha}\right)^n \, |q|^n \, \sin(n\,\sigma)\, J_n(\alpha\, a)\,, \end{aligned}\right\} \tag{4.29}$$

with

$$\sigma = \tan^{-1}(Z_r\,/\,z)\,. \tag{4.30}$$

For $\left|\dfrac{2\,q\,a}{\alpha}\right| > 1$ we have

$$\left.\begin{aligned} I^{BG}(z) = \left(\tfrac{k_\circ}{2|q|z}\right)^2 \{&\exp[-\alpha^2/(2w_\circ^2|q|^2)] - 2\,S_3\exp(-r_1/w_\circ^2)\cdot\cos\left(\tfrac{k_\circ r_1}{2z}\right) - \\ &-2\,S_4\exp(-r_1/w_\circ^2)\cdot\sin\left(\tfrac{k_\circ r_1}{2z}\right) + \exp[-2(a^2/w_\circ^1)]\;\cdot\;(S_3^2\,+\,S_4^2)\} \end{aligned}\right\}, \tag{4.31}$$

where

$$\left.\begin{aligned} S_3 &= \sum_{n=1}^{\infty} (-1)^n \left(\tfrac{\alpha}{2\,a}\right)^n \tfrac{1}{|q|^n} \cos(n\,\sigma)\, J_n(\alpha\, a)\,, \\ S_4 &= \sum_{n=1}^{\infty} (-1)^n \left(\tfrac{\alpha}{2\,a}\right)^n \tfrac{1}{|q|^n} \sin(n\,\sigma)\, J_n(\alpha\, a)\,, \end{aligned}\right\} \tag{4.32}$$

with

$$r_1 = \frac{\alpha^2}{4\,|q|^2} + a^2\,. \tag{4.33}$$

If $w_\circ \to \infty$, $q \to \dfrac{i k_\circ}{2\,z}$, and if $\alpha \to 0$, we have plane wave incidence. The axial light intensity distribution of a plane wave in the Fresnel region is given by

$$I^{PW}(z) = I_\circ\left[2 - 2\cos\left(\frac{k_\circ a^2}{2\,z}\right)\right], \tag{4.34}$$

with $S_3 \to 1$, $S_4 \to 0$ and for $k_\circ a^4/8z^3 \ll 1$. In the Fresnel region the zero intensity is taking place at points on the optical axis with coordinates

$$z_m = \frac{a^2}{2m\lambda_\circ}\,, \qquad m = 1, 2, \ldots\,. \tag{4.35}$$

The maxima of the axial intensity distribution $I(z)$ are taking place at points on the optical axis with coordinates

$$z_l = \frac{a^2}{l\,\lambda_\circ}\,, \qquad l = 1, 2, \ldots\,. \tag{4.36}$$

The rightmost first maximum of the axial intensity occurs for $l = 1$ at $z_1 = a^2/\lambda_\circ$.

For pure Gaussian incidence with $\alpha \to 0$ we have

$$I^G(z) = I_\circ \left(\frac{k_\circ}{2|q|z}\right)^2 \{1 - 2\exp[-(a_\circ/w_\circ)^2]\cos\left(\frac{k_\circ a^2}{2z}\right) + \exp[-2(a/w_\circ)^2]\}. \quad (4.37)$$

For Bessel beam incidence with $w_\circ \to \infty$ and $0 < \alpha < k_\circ$ the axial light intensity is equal to

$$I^B(z) = I_\circ \left[1 - 2S_3\cos\left(\frac{k_\circ r_1}{2z}\right) - 2S_4\sin\left(\frac{k_\circ r_1}{2z}\right) + S_3^2 + S_4^2\right],$$

$$\text{for} \quad \left|\frac{k_\circ a}{\alpha z}\right| > 1. \quad (4.38)$$

where

$$\left.\begin{aligned} S_3 &= \sum_{n=0}^{\infty}(-1)^n\left(\tfrac{\alpha z}{k_\circ a}\right)^n \cos\tfrac{n\pi}{2}\, J_n(\alpha a), \\ S_4 &= \sum_{n=0}^{\infty}(-1)^n\left(\tfrac{\alpha z}{k_\circ a}\right)^n \sin\tfrac{n\pi}{2}\, J_n(\alpha a). \end{aligned}\right\} \quad (4.39)$$

From Eq. (4.34) it follows that plane-wave axial intensity oscillates between 0 and $4I_\circ$ and this oscillatory region is more far away from the aperture than for other types of incidence.

There is a conical shadow boundary with

$$z_{max} = \left| k_\circ a \Big/ \left[\alpha^2 - \left(\frac{2a}{w_\circ^2}\right)^2\right]^{1/2} \right| \quad (4.40)$$

for Bessel–Gauss beam incidences, when $\left|\frac{2qa}{\alpha}\right| = 1$, and

$$z_{max} = \frac{k_\circ a}{\alpha} \quad (4.41)$$

for conical shadow zone in the case of the Bessel-beam.

The transverse intensity distribution in the Bessel–Gauss beam incidence can be expressed analytically in the Fresnel region [172] as:

$$I(\rho, z) = I_\circ \left(\frac{k_\circ a^2}{r}\right)^2 \sum_{n=0}^{\infty} \frac{(-1)^n (q\,a^2)^n}{n!} \sum_{k=0}^{\infty} B_k, \quad (4.42)$$

where

$$B_k = \frac{(-1)^k \left(\frac{\alpha a}{r}\right)^{2k} {}_2F_1\left[-k, -k; 1; \left(\frac{\gamma}{\alpha}\right)^2\right]}{(k!)^2 \, 2\,(n+k+1)}, \tag{4.43}$$

$$\text{for} \quad \left|\frac{\gamma}{\alpha}\right| < 1,$$

or

$$B_k = \frac{(-1)^k \left(\frac{\gamma a}{r}\right)^{2k} {}_2F_1\left[-k, k; 1; \left(\frac{\alpha}{\gamma}\right)^2\right]}{(k!)^2 \, 2\,(n+k+1)}, \tag{4.44}$$

$$\text{for} \quad \left|\frac{\alpha}{\gamma}\right| < 1,$$

where $\gamma = k_\circ \rho / r$, ${}_2F_1$ — hypergeometric function. The point, at which the hypergeometrical function changes argument, is the point, at which transverse shadow zones appear. Here $\gamma = \alpha$ and

$$r_{max} = \frac{k_\circ \rho}{\alpha}. \tag{4.45}$$

When $w_\circ \gg a$, the Bessel–Gauss beam transverse intensity does not differ significantly from the pure Bessel-beam intensity distribution. For $w_\circ \approx a$ the Bessel–Gauss intensity is diminished, the number of side lobes is reduced, and the intensity of the side lobes is decreased. For $w_\circ \ll a$ the intensity is very small, and the profile is getting the same, as pure Gaussian, in the Fresnel region [172].

Now we present the Fraunhoffer diffraction pattern of the pure Bessel beam:

$$I(\rho, z) = 4\, I_\circ \left[\frac{t\, J_\circ(\eta)\, J_1(t) - \eta\, J_1(\eta)\, J_\circ(t)}{t^2 - \eta^2}\right]^2, \tag{4.46}$$

with $t = \gamma\, a$, $\eta = \alpha\, a$, $t^2 \neq \eta^2$. As was pointed out in [172], Eq. (4.46) can be considered as a generalization of the classical Airy function, into which Eq. (4.46) goes for $\alpha \to 0$. The extrema positions in the classical Airy function are fixed by a parameter γ. In the case of the generalization Airy function, Eq. (4.46), the extrema positions are defined by the parameter α of the Bessel beam. The amount of light energy within one ring of the Airy picture is governed by the parameter α as well.

4.3. Darkness in Optics

It is well known that in the near-field diffraction region of the monochromatic plane wave, transmitted by circular aperture, the points of zero light intensity are observed on the optical axis (Fig. 4.3). The positions of dark points are governed

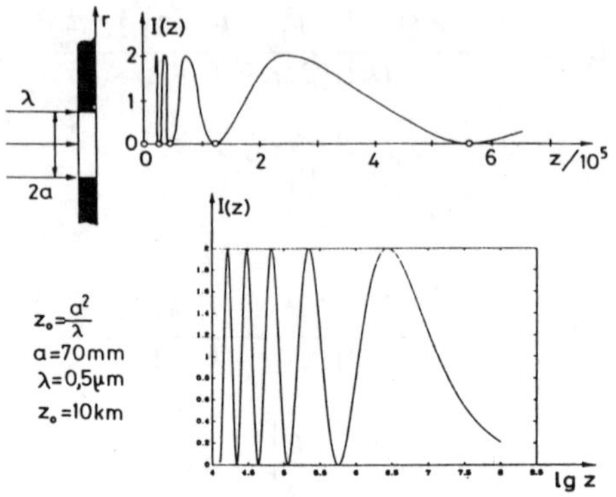

Fig. 4.3. The on-axis points of zero intensity in the near field diffraction of the truncated plane wave.

by Fresnel's condition: the open aperture of diameter 2a must include even number of Fresnel's zones. This phenomenon can be explained by Fresnel's theory of diffraction, in which the phase term in the diffraction integral has the form of a binomial series.

But recently it has been shown [173] that positions of these dark points, predicted by the Fresnel approximation theory, do not coincide with positions defined by the rigorous Rayleigh–Sommerfeld theory. This contradiction is especially severe if the optical system has high numerical aperture and if the Fresnel's number is small. A new approximate generalization of Fresnel diffraction theory has been suggested [173], by means of which the validity region of the classical Fresnel approximation can be expanded. The light field amplitude is presented in the analytical form in term of the Lommel functions used in optics [141].

The idea of this generalization is to approximate the phase term in the diffraction integral by a paraboloid terms instead of binomial expansion accepted by the old theory. In our case the light field amplitude at point (ρ, z) is given by Rayleigh–Sommerfeld diffraction integral

$$u(\rho, z) = \frac{a^2}{2\pi} \int_0^1 \int_0^{2\pi} \frac{\exp(ikr)}{r} \left(\frac{z}{r}\right) \left(ik + \frac{1}{r}\right) d\Phi \, \rho' \, d\rho' \tag{4.47}$$

where

$$r = (z^2 + a^2\rho^2 + a^2\rho'^2 - 2a\rho\rho' \cos\Phi)^{1/2} \tag{4.48}$$

with aperture plane at $z = 0$. In the conventional Fresnel approximation [141]

$$r \approx z(1 + \frac{a^2\rho^2}{2z^2} + \frac{a^2\rho'^2}{2z^2} - \frac{a^2\rho\rho'}{z^2} \cos\Phi). \tag{4.49}$$

By introducing

$$R^2 = z^2 + a^2\,\rho^2\,, \tag{4.50}$$

we have

$$r = R\left[1 + \frac{a^2\rho'^2}{2r^2} - \frac{a^2\rho\rho'}{R^2}\cos\Phi\right]. \tag{4.51}$$

The exponent phase term $(i\,k\,r)$ can be written in terms of the optical coordinates

$$\left.\begin{aligned} u &= \frac{k\,a^2}{R}\,,\\ v &= \frac{k\,a^2\,\rho}{R}\,,\\ w &= k\,R\,, \end{aligned}\right\} \tag{4.52}$$

in the form:

$$i\,k\,r = i\left(\frac{u\,\rho'^2}{2} - v'\rho'\cos\Phi + w\right). \tag{4.53}$$

The light field amplitude

$$u\,(\rho, R) = a^2\,(\frac{z}{r})\,(ik + \tfrac{1}{R})\,\frac{\exp\,(i\,\omega\,t)}{R}\int_0^1 \exp\,(i\,\tfrac{1}{2}\,u\,\rho'^2)\;J_{\circ}\,(v\rho')\,\rho'\,d\rho' \tag{4.54}$$

has been estimated by the generalized Fresnel's method [173]. For $z^2 \gg a^2$ and $u^2 \ll 4\,k\,a^2$ the on-axis light intensity is given by equation:

$$I^G\,(z) = \frac{1}{4}\,\{[1 + (1 + \frac{a^2}{z^2})^{1/2}]^2\,/\,(1 + \frac{a^2}{2z^2})\}\cdot\sin^2\{\frac{ka^2}{2z}\,/\,[(1 + \frac{a^2}{z^2})^{1/2} + 1]\}\,, \tag{4.55}$$

whereas the exact solution has the form

$$I^E(z) = \frac{1}{4}\,(1 + [1/(1 + \frac{a^2}{z^2})] - [2/(1 + \frac{a^2}{z^2})^{1/2}]\cdot\cos\{\frac{ka^2}{z}/[(1 + \frac{a^2}{z^2})^{1/2} + 1]\})\,. \tag{4.56}$$

The conventional Fresnel approximation

$$I^C\,(z) = \sin^2\left(\frac{ka^2}{2z}\right) = \frac{1}{2}\left[1 - \cos\left(\frac{ka^2}{z}\right)\right] \tag{4.57}$$

yields incorrect positions of the zero intensity points on the z-axis and predicts infinite number of zero intensity points. In the reality the number of these points is equal to a/λ. The generalized Fresnel technique, Eq. (4.55), predicts correct positions of zero intensity points down to a distance of the order of the aperture's diameter.

4.4. Darkness in Meso–optics

The mathematical support of the zero intensity points on the z-axis in traditional optics is of zero measure. A completely different situation occurs in meso–optics. To produce arbitrary long segment of zero intensity points a wavefront in the form of helical structure must be used [58]. An Archimedean spiral of constant pitch d produce on-axis light intensity distribution

$$I_1(\rho, z) = |J_1(\rho)|^2 \tag{4.58}$$

for one-start circular grating, and

$$I_2(\rho, z) = |J_2(\rho)|^2 \tag{4.59}$$

for two-starts circular grating. The length of this zero intensity segment does not depend on light diffraction and can be made in practice as long as 20 – 50 meters. The effective diameter of the central light tube with dark line on its axis is equal to $m\,\lambda/\theta$, where m is the number of threads of the spiral grating and θ is the angle between two wave vectors near z-axis.

The structure of the macroscopic model of the Archimedean circular grating of the diameter 120 mm with pitch 0,5 mm and with $m = 1$ is shown in Fig. 4.4. The zero intensity point on z-axis is converted into a dark spot on the photographic negative film surrounded by opaque ring of width $\lambda/m\theta$. At strong overexposure the transparent spot is getting smaller. This method yields a superresolution in the positioning of the zero intensity spot up to (0.07 – 0.1) $\lambda/m\theta$, as compared with λ/θ value in system without helical wavefront. This method works well with CCD-photodetector.

To produce long darkness segment on the z-axis another method can be recommended. It is based on the proprieties of the phase filter with amplitude transparency [174]

$$t(\rho, \Phi) = \exp(-i\,m\,\Phi)\,\mathrm{sgn}\,J_m(\rho/\rho_\circ), \tag{4.60}$$

where

$$\rho_\circ = \frac{\lambda f}{2\pi R_\circ} \tag{4.61}$$

and $R_\circ$ is the radius of the narrow annular zone in the focal plane , "sgn" is the sign-function and $J_m(\bullet)$ is the Bessel function of m-order of the first class. This filter

Fig. 4.4. View of the macroscopic model of the Archimedean diffraction grating.

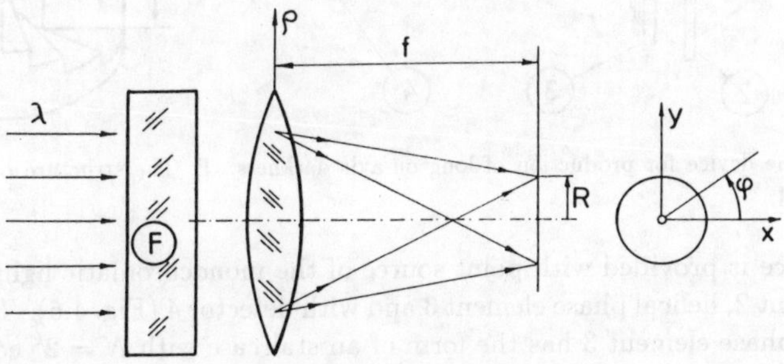

Fig. 4.5. The helical spatial filter with amplitude transparency $F(\rho, \varphi) = \exp[i\,(-\alpha\,\rho + m\,\varphi)]$, $\alpha > 0, m = 1, 2, \ldots$. The radius of the focal circle is equal to $R = \alpha\, f\, \lambda / 2\,\pi$.

with m-fold axial symmetry, named "helical spatial filter" [174] provides continuous linear azimuthal phase variations and induces radial phase π-jumps (Fig. 4.5).

The simplest axial symmetric binary phase Bessel filter of the type

$$t_\circ\,(\rho, \Phi) \;=\; \operatorname{sgn} J_\circ\,(\rho/\rho_\circ) \tag{4.62}$$

approximates well the binary circular phase grating with amplitude transparency

$$t_\circ^A \;=\; \operatorname{sgn}\, \cos\,(\frac{\rho}{\rho_\circ} - \frac{\pi}{4})\,. \tag{4.63}$$

This filter provides π-phase jumps at radii:

$$\rho_\nu = (\nu - 1/4)\,\frac{\lambda f}{2R_\circ}, \qquad \nu = 1, 2, \ldots . \tag{4.64}$$

These circles separate adjasent zones of constant phase of the width $\Delta\rho = \lambda f/2R_\circ$.

A filter with amplitude transparency

$$T(\rho, \Phi) = \exp(-i\,b\,\rho)\,\exp(i\,n\,\Phi), \tag{4.65}$$

focuses all the energy into a ring of radius $R = bf/k$, and for $n \neq 0$ produces zero light intensity on the optical axis at any propagating distances. A device for laser sighting of this type has been proposed in [175].

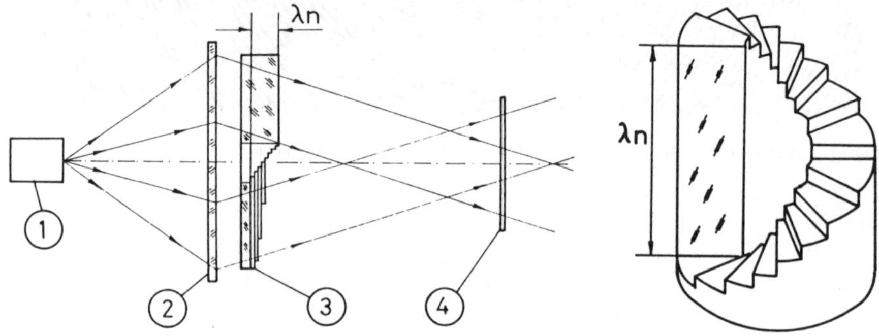

Fig. 4.6. A. The device for production of long on-axis darkness. B. The structure of the helical phase element 3.

This device is provided with point source of the monochromatic light 1, meso–optical element 2, helical phase element 3 and with detector 4 (Fig. 4.6). The optical relief of the phase element 3 has the form of an staircase with $N = 2^k$ equal steps, where $k \geq 4$. The height jump between the first and the second stages of this phase element is chosen equal to λn, where λ is the wavelength and n is the refractive index of the bulk material of the staircase. The position of the dark spot is fixed on the photographic film exposed with strong overexposure. The diameter of central bright spot on this photonegative is $\sim 10 \div 20$ times smaller than the effective width of the dark spot in the system with kinoform [52]. Thus the diameter of the meso–optical element 2 can be diminished as much as 10 times. A method of the manufacturing of the helical phase element 3 which goes back to J. Dyson [58] was proposed in [175].

In the experiments [176] an interference of two helical waves [176] has been observed. The light beam from laser, operating on many longitudinal modes, was split with amplitude division, lateral inverted and spatially separated. The helical waves are generated at high power of certain laser blue green lines. An unusual fringe pattern was detected, which can be explained by adopting that phase of the helical wave is given by

$$\Phi_R(\Psi, z) = 2\pi z / a + \Psi_1 \tag{4.66}$$

for right-handed helix and

$$\Phi_L(\Psi, z) = 2\pi z / a - \Psi_1 \tag{4.67}$$

for left-handed helix.

Two types of interference patterns observed in two experiments can be explained as follows. In the first experiment the mixing of two non-inverted, separated helical wavefronts was made, and in the second experiment one of the wavefront was inverted. In the first experiment the phase difference has been produced by helical like waves, whereas in the second one these helical waves were unlike ones. In both cases a points or zero intensity are detected.

The dimensions of points where the field amplitude is zero are equal to $\Delta\rho \approx (k\theta)^{-1}$ in transverse plane and $\Delta z \approx (k\theta^2)^{-1}$ in longitudinal direction, where θ is the angular divergence of the beam. The zero field points represent dislocations of the wavefronts of the helical structure. In the vicinity of dislocation point a parabolic wave equation is valid for monochromatic polarization beam with inhomogeneities [177]. The phase increment $\Delta\varphi = \pm 2\pi$ acquired along the closed loop around the dislocation point is stable in topological sense. The difference between the zero numbers $N_+ - N_-$ with positive and negative sign of $\Delta\varphi$ is constant in the beam propagation, and the singularities points are created and annihilated in average in pairs. The mean number of dislocations can be estimated as θ^2/λ^2. In the experiments [178] an interference picture, which demonstrates the phase structure of dislocations, is observed by adding a reference plane wave to the speckle inhomogeneity field. The titled angle exceeded the angular divergence of the beam under study. The experimental results reveal that the number of dislocations $N \approx (kd)^{1/2}$, where d is the diameter of the pinhole diaphragm of the point-like source.

A phase singularity is defined as a point, where the field is zero and where the closed loop integral of the phase gradient along the path surrounding this point is integer multiple of 2π: [179]

$$\Delta\Phi = \oint_l \nabla\Phi \cdot dl = 2\pi \cdot m, \qquad m = \pm 1, \pm 2, \dots . \tag{4.68}$$

The integer m is called the topological charge of the singularity and its sign is chirality of the helical wave. This definition of the phase singularity differs from that of phase discontinuity which appears as a dark region. However the topological charge of the phase discontinuity is zero. These two darkness phenomena have unlike physical nature. The technique of the astigmatic imaging enables one to identity these two phase phenomena [185] in a laser beam. A "doughnut" transversal pattern with zero point in the center, typical for phase singularity, can be explained as a multimode cooperative frequency locked region of the laser. It has been proved [186] that equiphase surface of the doughnut pattern has helical structure. In the frequency-locked multimode regime [180], two transverse modes are locked to a common frequency, at which they oscillate in a synchronized manner. The relative phase of these modes is locked as well, and the output pattern assumes a stationary transverse configuration, which resembles a single-mode regime. The radial p and the angular l indexes of these modes with axis symmetric wavefront are met to the condition

$$2p + l = q\,, \tag{4.69}$$

where q is a fixed number. If $q = 2$ the competition regime in ring laser involves three different modes of the empty cavity, in general case $(q+1)$ different modes. For $q = 0$ there are no singularities.

In the experiments [179] the two lower-order modal patterns have been used as active ones in the cooperatively frequency locked modal regime. For example, the simplest singularities are generated by TEM_{01} and TEM_{10} modes locked in frequency with phase difference of $\pi/2$, which are propagating along the z-direction. The sketch of the interference pattern observed with plane reference wave is shown in Fig. 4.7 in the form of boundaries between bright and dark interference fringes. These boundaries are defined by equation [179]

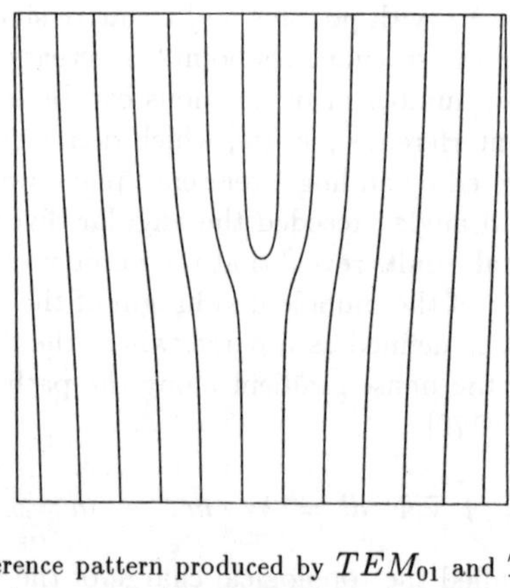

Fig. 4.7. Sketch of the interference pattern produced by TEM_{01} and TEM_{10} modes which are locked in frequency and which have phase difference $\pi/2$.

$$y = \pm x \cdot \tan(k_x\, x)\,. \tag{4.70}$$

The singularity point is seen as origin point of a new fringe. The sketch of the interference pattern observed in the interferometer with two beam-splitters and two

Fig. 4.8. Sketch of the interference pattern for input doughnut structure of topological charge $m = 1$: "a" — perpendicular to the interference fringes, "b" — parallel to the interference fringes.

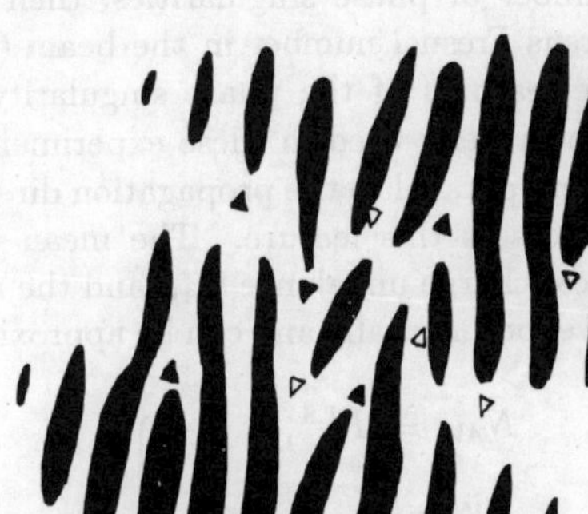

Fig. 4.9. Sketch of the expected interferogram observed for stationary input wave with five spots.

plane mirrors is shown in Fig. 4.8 for input wave having doughnut structure with topological charge equals to one ($m = 1$). The sketch of the expected interferogram for five-spot stationary solution is presented in Fig. 4.9 [179]. The interferometric method of the phase singularities detection [179] is well suited to provide information both on the position and on the relative topological charge of the singularities, as well as to distinguish between phase discontinuities and phase singularities.

The topological effects induced by the nonlinear defects have been investigated [181]. In contrast to the experiments [177], in which a random phase plate has been used to induce linear defects, the defect pattern in [181] evolves in time. There have

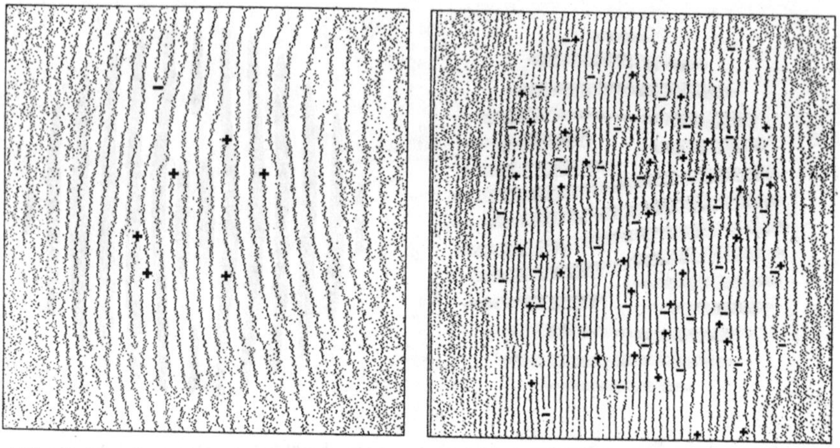

Fig. 4.10. Sketch of two destructive features of the phase singularity configurations.

been measured the mean number of phase singularities, their mean separation, and the charge unbalance versus Fresnel number in the beam ("aspect ratio" F). The sketch of two distinctive features of the phase singularity configurations is given in Fig. 4.10. The nonlinear field, used in these experiments, was truly two-dimensional and defect lines were parallel to the propagation direction. The speckle field used in [177] does not possess this feature. The mean number of defects N_{AV}, the normalized topological charge unbalance U_{AV} and the average interdefect distance D_{AV} have been found experimentally and can be approximated as following function of F: [181]

$$\left.\begin{aligned} N_{AV} &\cong F^{1,8}\,, \\ U_{AV} &= \tfrac{|N_+ - N_-|}{|N_+ + N_-|} \cong F^{-1,2}\,, \\ D_{AV} &\cong F^{-0,5}\,. \end{aligned}\right\} \tag{4.71}$$

To gain an insight into the nature of the phase singularities, a computer-generated hologram technique has been used [182]. It was proved that boundaries between a bright fringe and a dark fringe of the interference pattern, produced by a plane reference wave and the lowest-order doughnut mode

$$u\,(\rho,\theta) \;=\; E_\circ\,\frac{\rho}{\rho_\circ}\,\exp\,(-\,\frac{\rho^2}{\rho_\circ^2})\cdot\exp\,[-1\,(\frac{k\rho^2}{2R}\,\pm\,\theta)\,]\,, \tag{4.72}$$

occur when [182]

$$\left.\begin{aligned} \pm m\,\theta &= (n\,+\,1/2)\,\pi\,+\,\tfrac{k\rho^2}{2R} \\ \text{with}\quad & n=0,\,\pm1,\,\pm2,\ldots\,. \end{aligned}\right\} \tag{4.73}$$

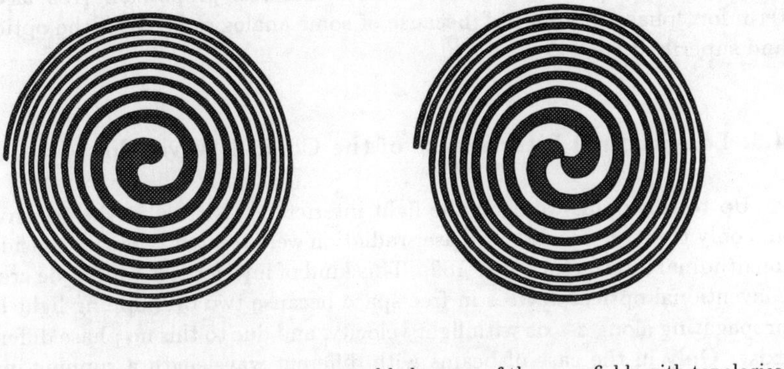

Fig. 4.11. The optical relief of the computer-generated holograms of the wave fields with topological charges $m = 1$ and $m = 2$.

Fig. 4.12. Computer-generated hologram, which produces the helical wave beam on the stage of reconstruction.

The sketch of two binary computer-generated holograms of the wave fields with topological charges $m = 1$ and $m = 2$ are shown in Fig. 4.11. The difference between a meso–optical device shown in Fig. 4.6 and the computer-generated holograms in Fig. 4.11 is that the latter gives the focused point only at definite z-coordinate. The sketch of the interferometric pattern for Gaussian Hermite TEM_{01} hybrid mode is given in Fig. 4.12. Such a hologram can be used both for helical wave production and for detection of these waves in the complex fields.

The analysis of the wavefront folding interferometer as a device for observation of coherence properties of the doughnut beams with and without phase singularities was given in [184]. A term "optical vortice" has been proposed in [183] as doublicate term for "phase singularity" because of some analogies between the optical vortice and superfluid vortex.

4.5. Longitudinal Interference of the Conical Wavefields

Up to this point we dealt the light interference pattern in the transverse plane and only transversal modes of laser radiation were treated. Now we consider on-axis longitudinal interference [187-189]. This kind of interference cannot be created with conventional optical devices in free space because two overlapping light beams are propagating along z-axis with light velocity, and due to this no phase differences can exist. Only in the case of beams with different wavelength a running interference pattern can be arisen. As it will be explained here, the on-axis longitudinal interference can be produced only with conical or helical waves when wave vectors $\vec{k}_1$ and $\vec{k}_2$ of two coaxial conical waves form a mutual angle $\theta \neq 0$. The on-axis velocities of wavefronts, both above the light velocity, are different. To produce the longitudinal marks on the central lobe of these two coaxial beams, Bessel beams are to be used with $\vec{k}_1$ and $\vec{k}_2$ and with $|\vec{k}_1| = |\vec{k}_2|$ (Fig.4.13). In the meridional cross section in the region of the mutual overlapping of these Bessel beams the interference fringes are oriented parallel to the bisectrix of the mutual angle θ_{12} between two wavevectors $\vec{k}_1$ and $\vec{k}_2$. The period of these fringes is equal to

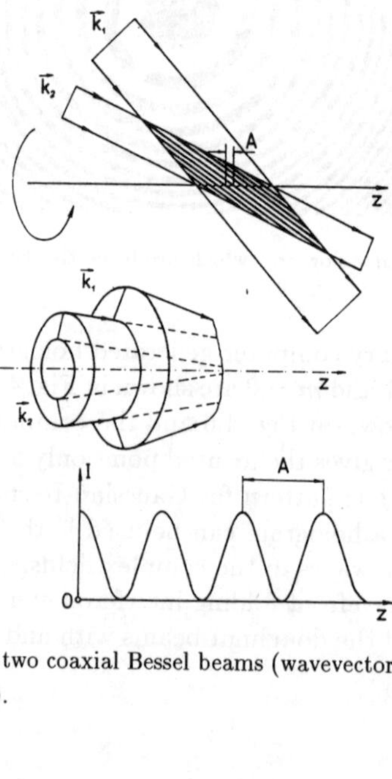

Fig. 4.13. Overlapped region of two coaxial Bessel beams (wavevectors $\vec{k}_1$ and $\vec{k}_2$), and the on-axis light intensity distribution $I(z)$.

$$a = \frac{\lambda}{\sin(\theta_2 - \theta_1)}, \tag{4.74}$$

where θ_1 and θ_2 are the orientation angles of the wavevectors $\vec{k}_1$ and $\vec{k}_2$ with respect to the optical axis of the system. The orientation angle θ_o of these interference fringes is equal to

$$\theta_o = \frac{1}{2}(\theta_1 + \theta_2). \tag{4.75}$$

In the real axial symmetry system, each point of this meridional cross section corresponds to a circle. Only the points lying on the optical axis of the system are remained fixed. The period A of the on-axis longitudinal interference of the Bessel beams is equal to

$$A = \frac{a}{\sin\left(\frac{\theta_1+\theta_2}{2}\right)} = \frac{\lambda}{\sin(\theta_2 - \theta_1)\cdot \sin\frac{\theta_2+\theta_1}{2}}. \tag{4.76}$$

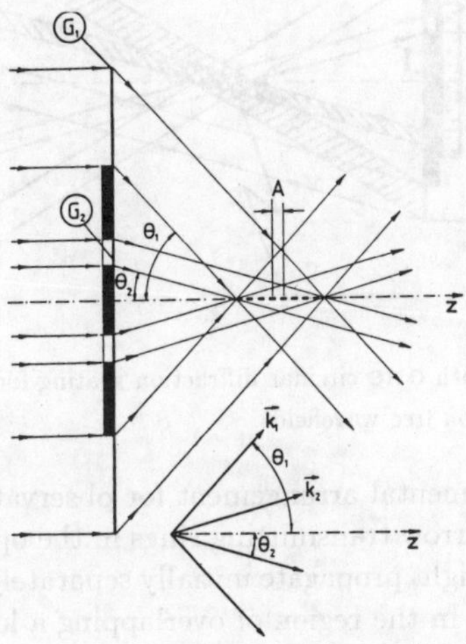

Fig. 4.14. Experimental setup with two coaxial circular diffraction gratings G_1 and G_2 for observation of the longitudinal interference of the diffraction free wavefields.

The most straightforward experiment for observation of this longitudinal interference is shown in Fig. 4.14 with two coaxial circular diffraction gratings G_1 and G_2. The working parts of these two coaxial circular diffraction gratings are chosen as two annuals, and the region of the mutual overlapping of two conical wavefields is located near the optical axis of the system. One circular diffraction grating G can be used, if the later produces at least two diffraction orders (Fig. 4.15). To match the light intensities of these diffraction orders, a rotating polarizer P is inserted in front of the circular diffraction grating G. The external diameter is chosen smaller than the diameter of the external working part of the grating G. In such a configuration we have $\theta_2 = 2\theta_1$ and the period A_1 of the on-axis longitudinal interference is equal to

$$A_1 = \frac{\lambda}{\sin\theta_1 \cdot \sin\frac{3}{2}\theta_1}, \tag{4.77}$$

where θ_1 is the diffraction angle of the first order.

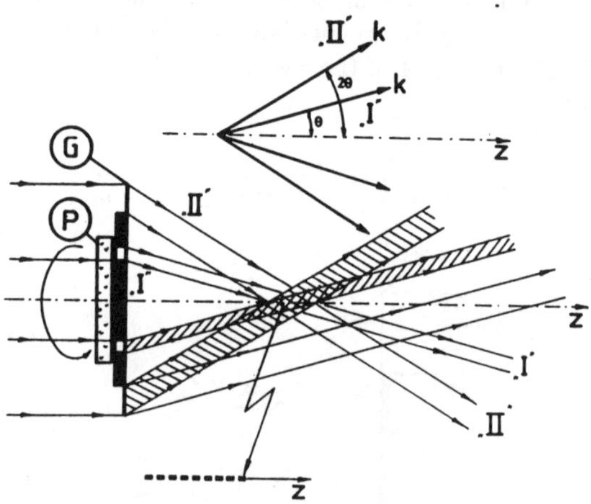

Fig. 4.15. Experimental setup with one circular diffraction grating for observation of the longitudinal interference of the diffraction free wavefields.

The most simple experimental arrangement for observation of the longitudinal interference involves two narrow transmitting rings in the opaque screen. The components of the transmitted light propagate initially separately and at some distances they cross the optical axis. In the region of overlapping a longitudinal interference is arisen.

If this phenomenon is undesirable we may suppress longitudinal on-axis light intensity modulation by means of the circular Hilbert phase jump filter. This phase filter consists of two halves each of 180° azimuthal angular interval (Fig. 4.16). One half of this filter is a glass plate. The second half represents a ring-shaped phase jump with phase difference $\Delta\Phi$ for given wavelength λ. The radius of the ring currying phase jump is just equal to the intermediate value of the radii of two coaxial narrow rings. Under the influence of this ring-shaped Hilbert phase filter the interference picture produced by upper part (L) of this filter is in antiphase with respect to the interference picture produced by the bottom part (R) of this filter. The sum of these mutually antiphase interference pictures is the constant function

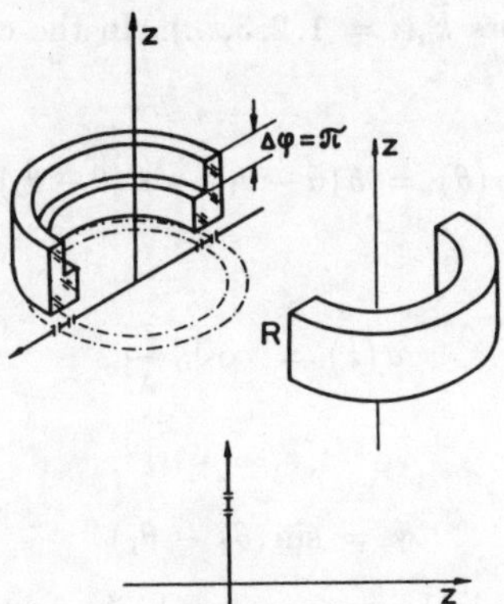

Fig. 4.16. Circular Hilbert filter with radial phase jump, dedicated for suppression of the longitudinal on-axis interference: L — the left half and R — the right half of this filter.

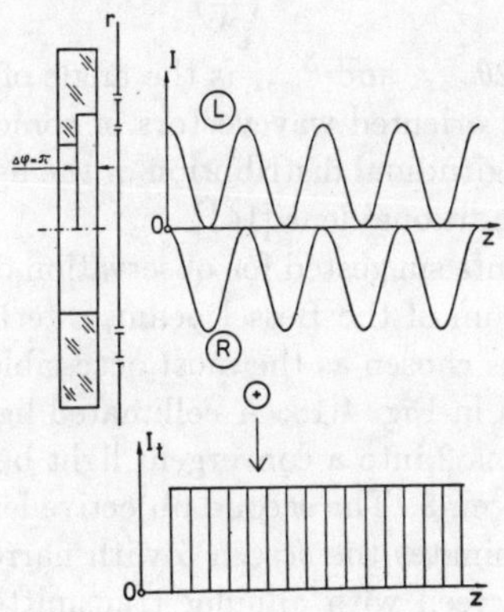

Fig. 4.17. The mutual arrangement of the circular Hilbert filter with radial phase jump and of the diaphragm with two coaxial narrow ring slits (to the left) and on-axis light intensity distribution (to the right).

of z-coordinate. As this cancellation is taking place in each meridional cross section we observe the complete smoothing of the initial light intensity modulation on the optical axis of the system. The construction of two halves of this filter is shown in Fig. 4.17.

It is easy to prove that the amplitude of the longitudinal interference of many coaxial conical wavefronts $a(z)$ is equal to Fourier transform of the angular field

$g(\theta)$ with many wavevectors $\vec{k}_i (i = 1, 2, 3, ...)$. In the case of two wavevectors $\vec{k}_1$ and $\vec{k}_2$ we have

$$g(\theta) = \delta(\theta - \theta_1) + \delta(\theta - \theta_2) \tag{4.78}$$

and

$$a(z) = \cos[\gamma \frac{z}{\lambda}], \tag{4.79}$$

where

$$\gamma = \sin(\theta_2 - \theta_1). \tag{4.80}$$

When the number of the conical components is high and the corresponding wavevectors are distributed equidistantly along the axis $\sin\theta_i$, the periodic structure of the longitudinal interference is conserved but with small contrast. In the case when the number of the conical wavefronts is very high and the angular distribution of the wavevectors is irregular, we have

$$a(z) \to \frac{\sin \frac{\mu z}{\lambda}}{(\frac{\mu z}{\lambda})}, \tag{4.81}$$

where $\mu = \sin\theta_{AV}$, $\theta_{AV} = 1/2\theta_{max}$, and θ_{max} is the angle of the cone which is filled by the irregular but compact oriented wavevectors of conical components. We see that Eq. (4.81) is just the longitudinal distribution of the light in the vicinity of the crossover produced by the traditional lens [141].

Among possible experiments suggested for observation of the longitudinal interference of the central maximum of the Bessel beam, a technique with two narrow coaxial transmitting rings was chosen as the most accessible one [188]. The experimental arrangement is shown in Fig. 4.18. A collimated light beam from laser 1 is converted by the objective lens 2 into a convergent light beam which goes through the pinhole in the opaque screen 3. The second objective lens 4 produces the single mode light beam which illuminates the screen 5 with narrow coaxial transmitting rings. The second opaque screen with annular transmitting zone 6 restricts the number of the narrow coaxial transmitting rings which contribute to the interference pattern. Each narrow transmitting ring of width 20 μm on the screen 5 was fabricated by the laser photosynthezator [53, 54]. The diameter of the focused laser spot was equal to 1 μm. It must be note that the groves on the glass backing of the screen 5, produced on the stage of the fabrication, are working as a circular diffraction grating, and the observer sees the rainbow spectra by placing the screen 5 between the glow lamp and eye. The system of 10 narrow coaxial transmitting rings with spacing 200 μm has been fabricated [54, 119].

The components of the diffracted light from each transmitting ring are propagating initially separately, then they are mutually overlapping near the optical

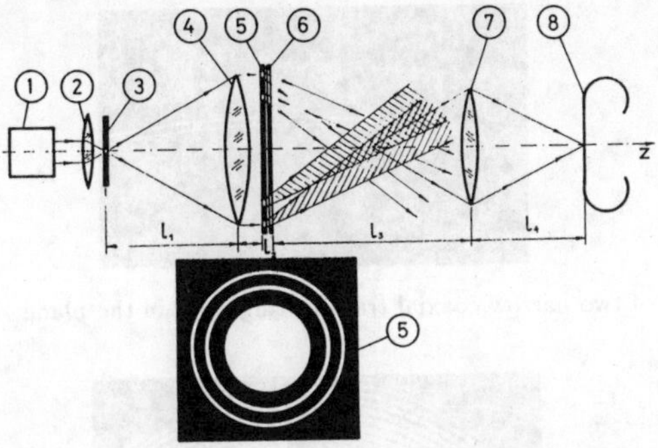

Fig. 4.18. The experimental setup with two narrow coaxial transmitting rings for observation of the longitudinal interference of the diffraction free waves.

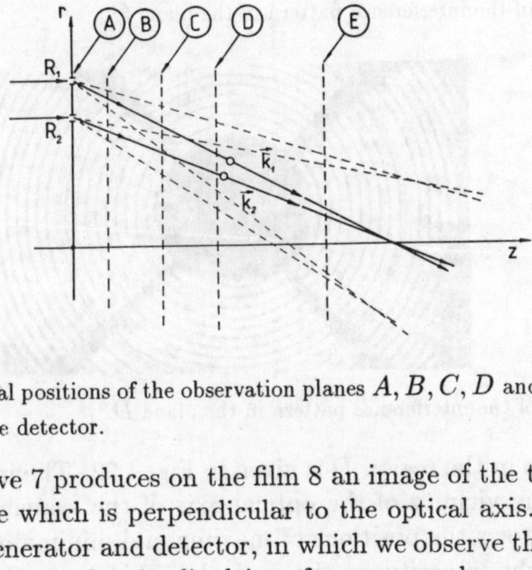

Fig. 4.19. Geometrical positions of the observation planes A, B, C, D and E in the region between the generator and the detector.

axis. The objective 7 produces on the film 8 an image of the transversal diffraction picture in a plane which is perpendicular to the optical axis. Several parts of the region between generator and detector, in which we observe the structure of various evolution stages of the longitudinal interference are shown in Fig. 4.19. In the plane A we see narrow coaxial transmitting rings in focus of the objective 7. In the plane B some kind of the Talbot self-imaging is observed. In the plane C only partial overlapping of the light components going from adjacent coaxial rings is taking place. In the region D we see the complete overlapping of these components. This region is at some distance from the optical axis. Only inside the region E the interference fringes approach the optical axis from all azimuthal directions.

The structure of the fragment of two coaxial transmitting rings in the plane A is shown in Fig. 4.20. The interference picture in the plane C is presented in

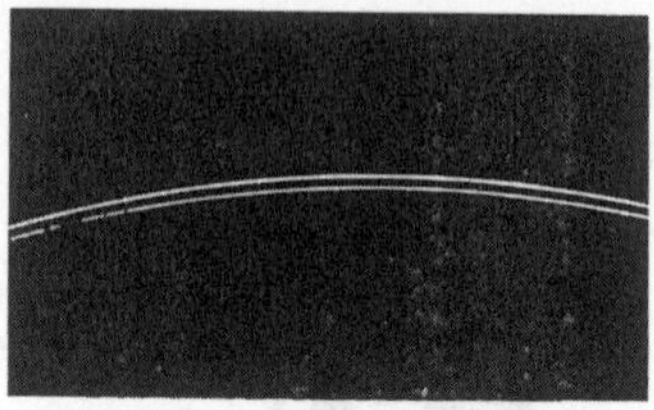

Fig. 4.20. The structure of two narrow coaxial transmitting rings in the plane A.

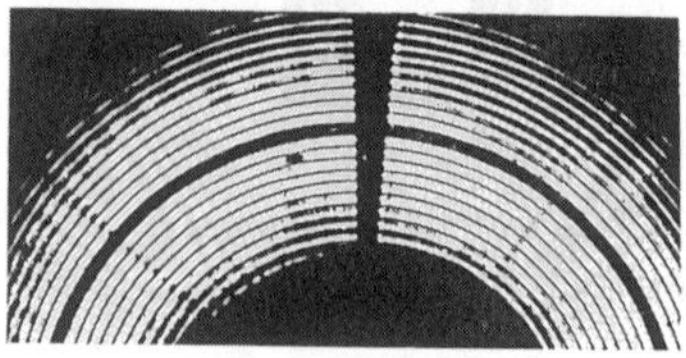

Fig. 4.21. The structure of the interference pattern in the plane C.

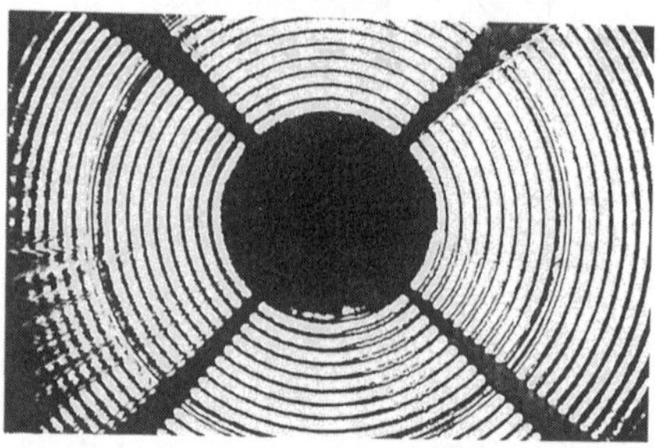

Fig. 4.22. The structure of the interference pattern in the plane D.

Fig. 4.21. The picture in the region D is given in Fig. 4.22. The interference fringes in the region E in the vicinity of the optical axis of the system are presented in Fig. 4.23. Fig. 4.24 shows the positions of maxima and minima of the longitudinal modulation of the light intensity on the optical axis of the system. As an angle between two wavevectors $\vec{k}_1$ and $\vec{k}_2$ is very small, $\sim 10^{-4}$, the fringe period A of this modulation is of the order of 10 mm. There were estimated 12 maxima and minima on the optical axis. From Fig. 4.25, we may conclude that fringe period A is a slow varying function of z-coordinate. This is a consequence of the fact that each narrow transmitting ring contributes to the interference picture as a point-like light source and therefore the angle between the wavevectors $\vec{k}_1$ and $\vec{k}_2$ is varying inversely proportional to the distance from the screen 5 to the observation point.

The on-axis longitudinal interference of the light is the result of overlapping of two interference pictures propagating to z-axis from two opposite directions. Each

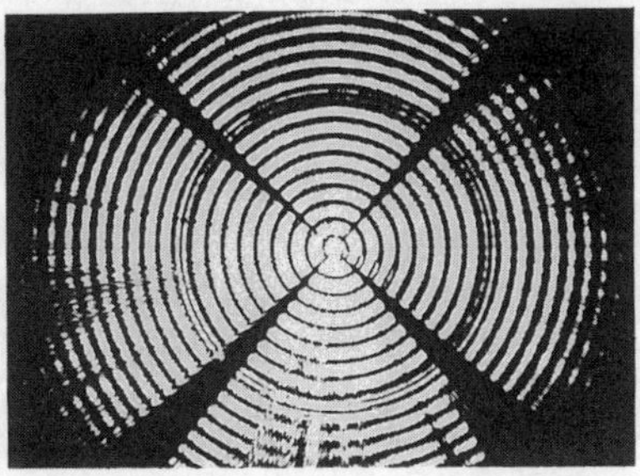

Fig. 4.23. The structure of the interference pattern in the plane E.

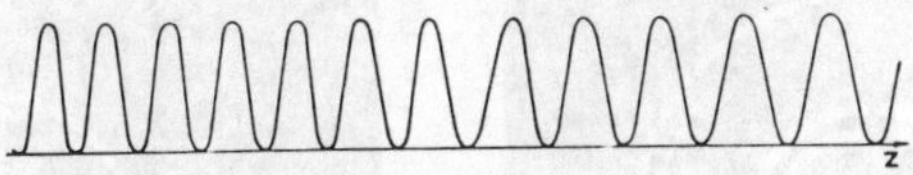

Fig. 4.24. The measured intensity distribution of the longitudinal interference of two coaxial conical wavefronts.

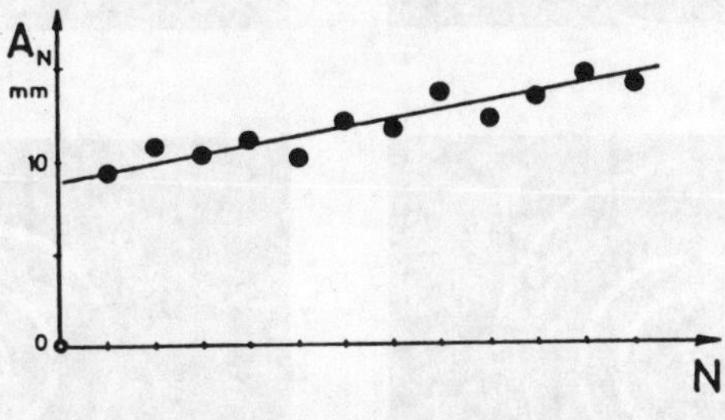

Fig. 4.25. The dependence of the observed fringe period A versus z-coordinate.

contributing picture moves to the optical axis, then it crosses this axis and finally propagates from the optical axis. To demonstrate this process pictorially, we have chosen the experimental arrangement, in which the number of the narrow coaxial transmitting rings was equal to 10. In this condition the width of the main maxima of the interference pictures was small with respect to the fringe period A and this facilitates the analysis of the above mentioned dynamics of the mutual overlapping of two contributing components of the whole interference pattern.

The picture shown in Fig. 4.26"a", was detected in the position, in which two contributing interference components, propagating from opposite directions, mutually overlap and the light intensity on the optical axis is maximal. The picture presented in Fig. 4.26"b", was detected at z-coordinate, at which these two contributing interference components are separated. A typical doublet is seen. At the

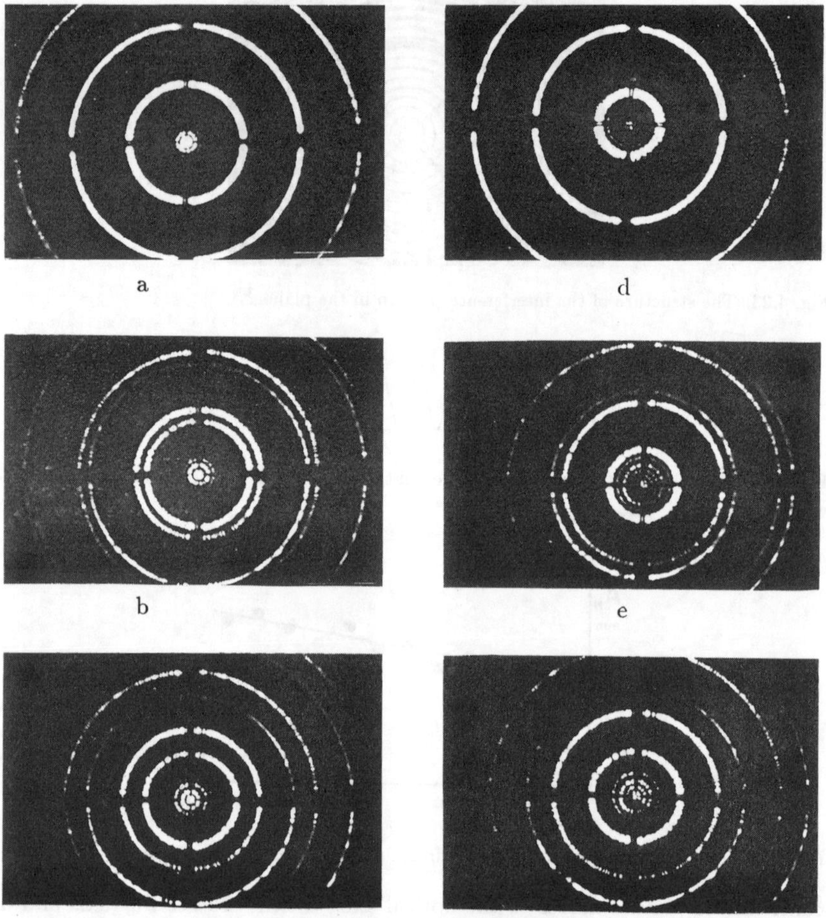

Fig. 4.26. The sequence of six interference patterns observed at different z-coordinates for diaphragm of 10 transmitting rings.

next stage shown in Fig. 4.26"c" the gap between two components of this doublet is getting larger than in Fig. 4.26"b". In some position these two contributing interference components are in antiphase (Fig. 4.26"d"). The total on-axis light intensity is minimal (zero). At the next stage these components are separating again (Fig. 4.26"e"). In Fig. 4.26"f" we see the secondary maximum when the on-axis light intensity is maximal though the contributing interference components do not coincide completely in this plane.

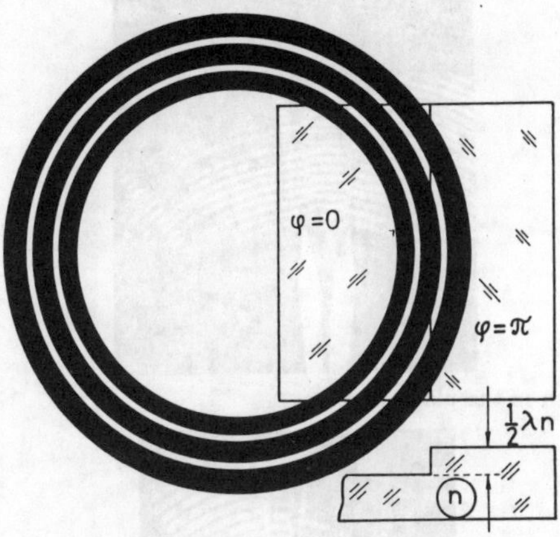

Fig. 4.27. Mutual arrangement of the opaque screen with two coaxial transmitting rings and of the straight line 180° phase jump.

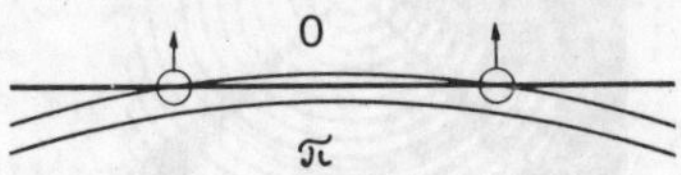

Fig. 4.28. Straight line 180° phase jump crosses one (external) transmitting ring in the diaphragm of two transmitting rings.

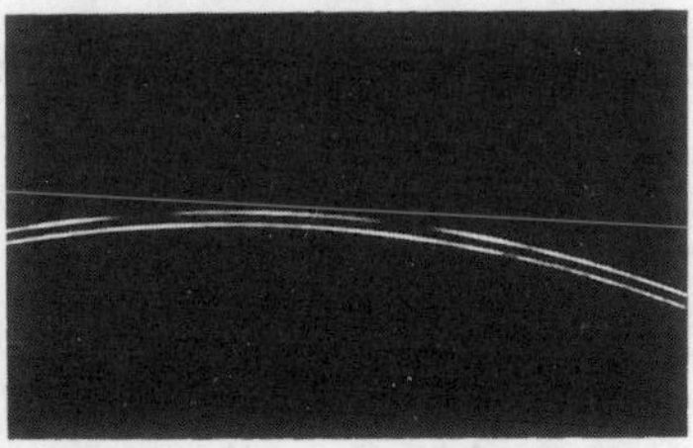

Fig. 4.29. The picture in the plane A.

To demonstrate the feasibility of the technique for suppressing the longitudinal light modulation exactly on the optical axis of the system with two coaxial narrow transmitting circles, an experiment with straight line of 180°-phase jump has been performed [188]. This straight line of 180°-phase jump is shown in Fig. 4.27 together with two coaxial annular slits on the opaque screen. The straight line of 180°-phase jump crosses at least one of two annular slits. One such arrangement is shown in Fig. 4.28. The structure of the pattern in the plane A is given in Fig. 4.29. The points of zero light intensity are marked by the white arrows. The structure of

Fig. 4.30. Interference pattern in the plane D.

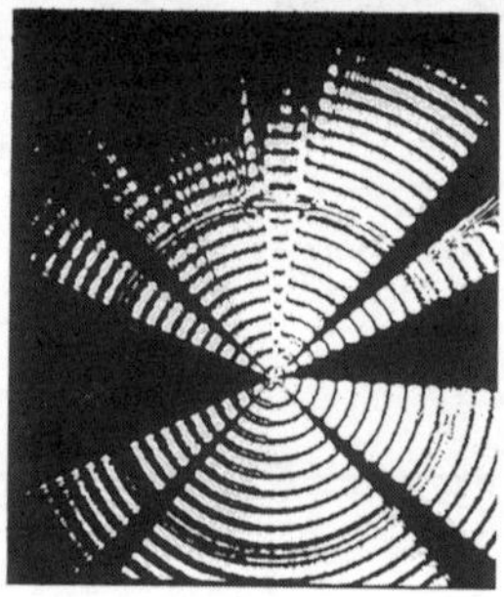

Fig. 4.31. Interference pattern in the plane E.

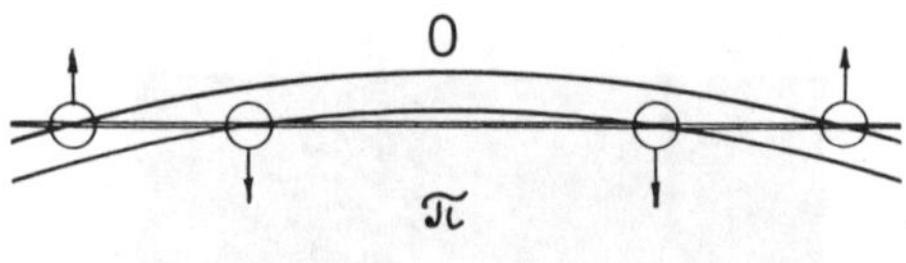

Fig. 4.32. Straight line $180°$ phase jump crosses two transmitting rings.

the interference picture in the plane D is given in Fig. 4.30. The radial lines with phase jumps of the interference picture are marked by the white and black arrows. These phase jumps are in the points where two adjacent interference pictures are in antiphase. The dark azimuthal ring arc goes into the bright ring arc and vice versa, the bright ring arc goes into the dark ring arc. Fig. 4.31 shows the interference picture taken in the position, when it approaches the optical axis of the system. As the length of the phase jumped arc is small with respect to the length of the whole circle, these local phase jumps cannot change the light intensity on the optical axis.

In the second experiment the straight line with phase jump crosses two coaxial transmitting rings (Fig. 4.32). The photo of this arrangement in the plane A is given in Fig. 4.33. Now two points of zero intensity marked by the white arrows are on both rings. The structure of the interference picture in the region D is presented

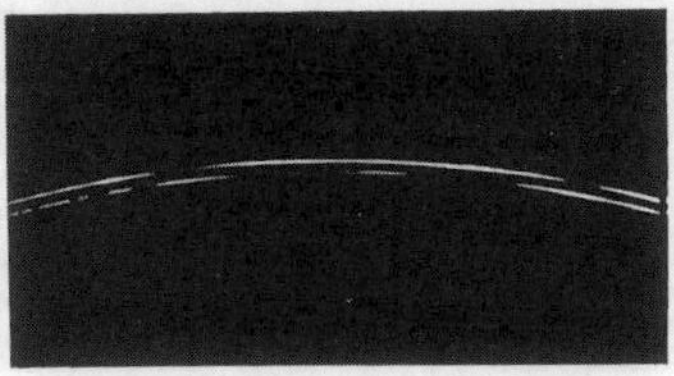

Fig. 4.33. The picture in the plane A.

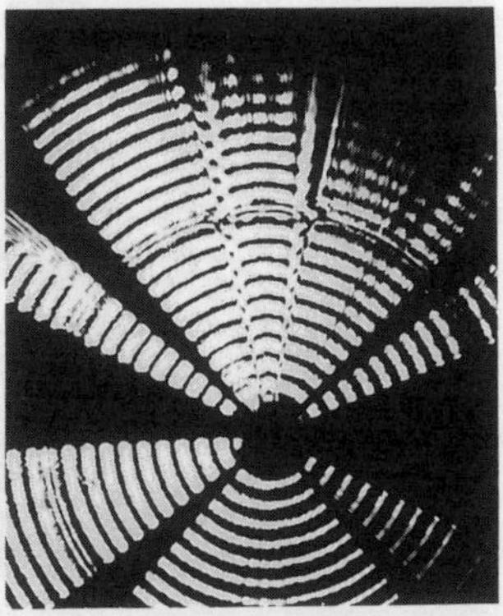

Fig. 4.34. Interference pattern in the plane D.

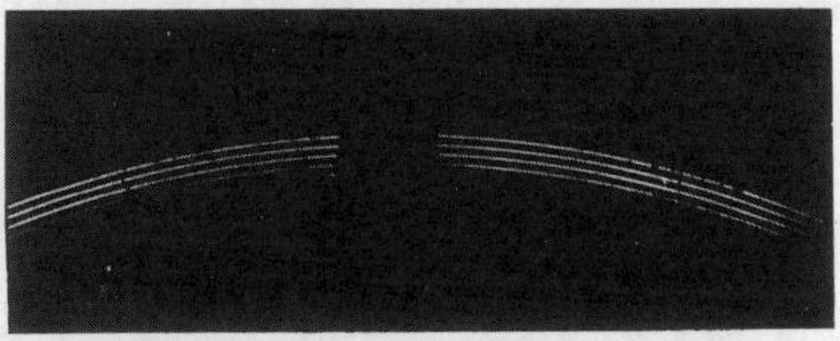

Fig. 4.35. The diaphragm of 4 transmitting rings.

in Fig. 4.34. Four radial lines with phase jumps are marked by the black arrows. We see that the central part of the ring arc is in phase with the external parts of this ring arc.

If the number of the working narrow coaxial transmitting rings in the opaque screen 5 is more than two, the interference fringes assume more complicated structure. We have tested two arrangements with 4 and 10 rings (Fig. 4.35 and Fig. 4.36). The dark gap in the central part of these photos is a shadow made by one of four supporting streeps of the opaque screen 6 (see Fig. 4.18). The interference pattern in the region E, taken for aperture of four transmitting rings, is shown in Fig. 4.37. It incorporates three intermediate minima and two secondary maxima. This pattern can be explained by the theorem: the axial amplitude $a(z)$ is equal to the Fourier transform of the angular field $g(\theta)$, generated by the contributing wave vectors.

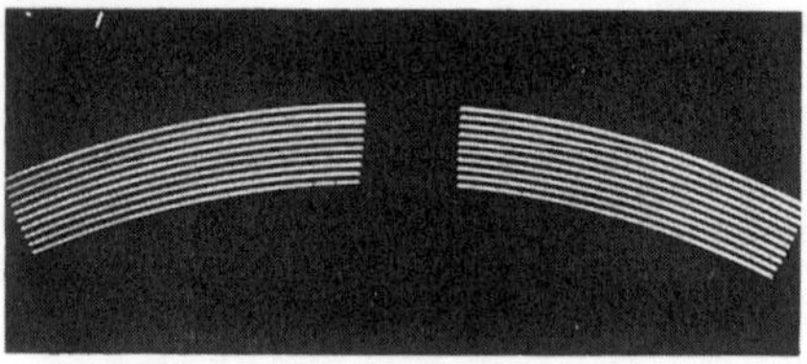

Fig. 4.36. The diaphragm of 10 transmitting rings.

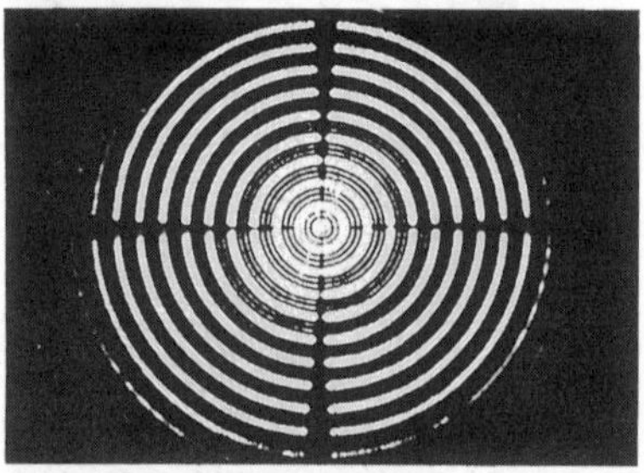

Fig. 4.37. Interference pattern produced by the diaphragm of 4 transmitting rings in the plane E.

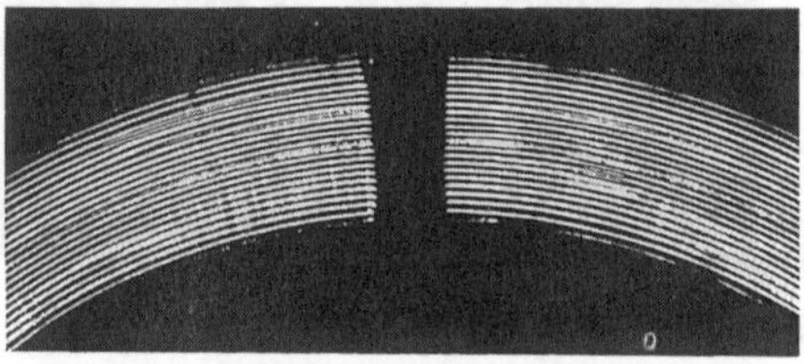

Fig. 4.38. Radial self-imaging Talbot effect observed in the plane B.

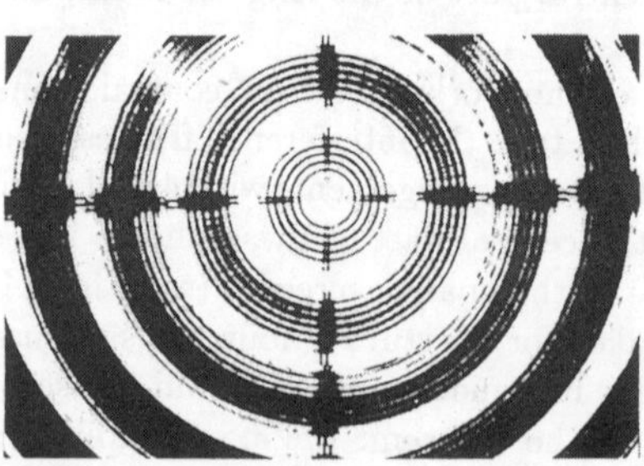

Fig. 4.39. Interference pattern produced by the diaphragm of 10 transmitting rings in the plane E.

The most pronounced phenomena, which have been used above, are observed for aperture of 10 transmitting rings. A circular self-imaging Talbot effect manifests itself in the region B (Fig. 4.38). The interference pattern in the region D, shown in Fig. 4.39, consists of one main maximum and weak secondary maxima.

4.6. Production of Diffraction-free Waves

In his first paper, in which J. Durnin has introduced the term "diffraction free wavefields" [72], the Bessel beam generated by narrow circular transmitting ring (annular slit) placed in the focal plane of an imaging lens. The same technique has been used for observation of the longitudinal interference of diffraction free waves [188], but without imaging lens. The deflection of the light rays towards z-axis has been produced by circular diffraction grating, generated in each open ring by virtue of the accepted manufacturing technology. It is evident that the annular ring (annular slit) can be considered as the most simple method of diffraction free waves generation, but the output light intensity is very small. The intensity of the conical waves, produced by axicon (or conical lens) will be higher than for annular slit. Each element of the focal range on the z-axis is illuminated by the light which is going through the proper annular ring of the axicon. As the length of the annular ring is proportional to the radius of the corresponding annular, the on-axis light intensity is a linear increasing function of z-coordinate. To eliminate this drawback the axicons with logarithmic generating line have been proposed [25, 88]. The same problem has been effectively resolved by means of focusators [62] and computer-generated meso–optical hologram [194].

It is worth noting that spherical lens with great spherical aberration can be used to produce diffraction free beams [190, 191] with high light efficiency. An on-axis stop is used to isolate the marginal light rays from the central ones [190]. The scale of the transverse structure, the length of the focal range and recommendations to optimize the system are presented in [190].

To increase the intensity of the Bessel beams a Fabry–Perot etalon has been used in the generator of the Bessel beam [192] (Fig. 4.40). This device consists of point-like source of the monochromatic radiation 1, Fabry–Perot etalon mirrors 2, a lens with focal length f_1 3, annular spatial filter (annular slit) 4 and a lens with focal length f_2. The fixed cavity length was $d = 0.635$ cm. The length of the focal range is equal to

$$L = \frac{F\,d}{\pi\sqrt{R}}\left(\frac{f_2}{f_1}\right)^2 \tag{4.82}$$

where

$$F = \pi\sqrt{R}/(1-R) \tag{4.83}$$

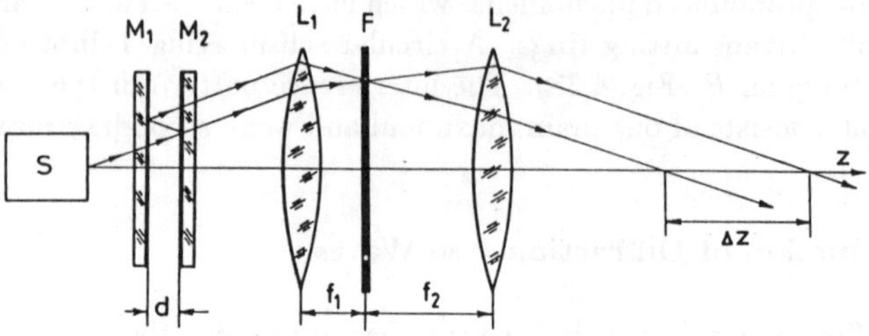

Fig. 4.40. Generator of the Bessel beam which consists of Fabry–Perot etalon and annular transmitting ring.

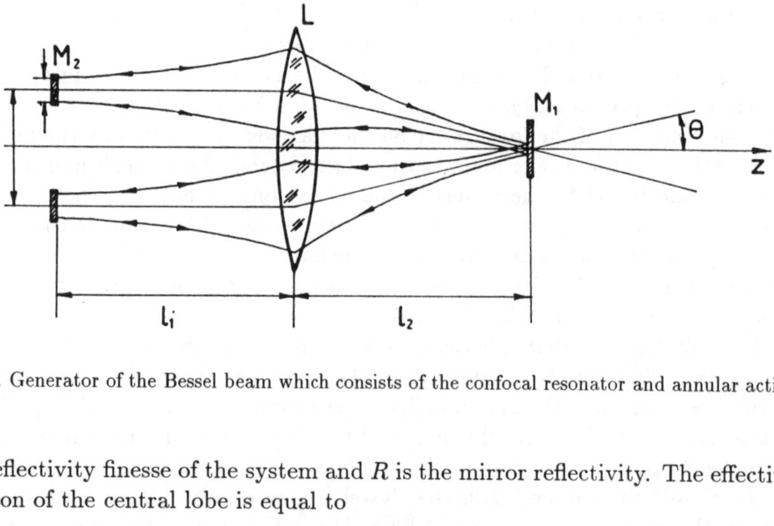

Fig. 4.41. Generator of the Bessel beam which consists of the confocal resonator and annular active medium.

is the reflectivity finesse of the system and R is the mirror reflectivity. The effective dimension of the central lobe is equal to

$$\rho_\circ = 0.383 \left(\frac{f_2}{f_1}\right) (\lambda\, d)^{1/2}\,, \tag{4.84}$$

where d is the cavity length. The on-axis light intensity distribution has no oscillations which are present in the system with annular slit without Fabry–Perot etalon.

A system with active confocal resonator and with annular active medium was treated in [170]. The idea of this method is based on the fact that Bessel beam can be produced by coherent source in the form of annular located in the front focal plane of a lens. The confocal resonator of the length R is provided with toroidal mirrors with generating line as circle arc of the radius R (Fig. 4.41). After two reflections the annular shaped beam of the waist $2w_\circ$ and diameter 2 h is retransformed into the initial annular beam. The performance of this system is defined by the parameter

$$n = [n'] \qquad \text{with} \qquad n' = (\frac{h}{w_\circ})^2 . \tag{4.85}$$

The field amplitude at $z = 0$ is equal to

$$u(x, z=0) = (x/w_\circ)^{2n} \exp[-(x/w_\circ)^2], \tag{4.86}$$

The field amplitude $u_1(x)$ after passing through an optical system is defined by

$$u_1(x) = N_1 (1 - i N_1 A)^{-n-1} L(\xi) \exp(-\xi), \tag{4.87}$$

where

$$\xi = \frac{(\alpha/w_1)}{1 - iN_1 A}, \quad w_1 = \lambda B/\pi w_\circ, \quad N_1 = \pi w_\circ^2 / \lambda B, \tag{4.88}$$

and $L_n(\bullet)$ is Laguerre's polynomial.

The field inside the resonator at $z \neq 0$ consists of two components, one before the reflection and another after the reflection at mirrors. The relation between these two field components is defined by parameter

$$\varepsilon = z / Z_R, \tag{4.89}$$

with $Z_R = \pi w_\circ^2/\lambda$ the diffraction Rayleigh range. If the resonator length is small and $R \ll Z_R$, the component before reflection dominates. The half-width of the diffraction-free zone is equal to

$$w_f = \lambda / (\pi \sin\theta), \tag{4.90}$$

with $\sin\theta = 2w_\circ/R$. The half-width over which the light intensity decreases m times, is defined by

$$\left. \begin{aligned} L_f &= Z_\alpha n \sqrt{m^{1/(n+1)} - 1}, \\ Z_\alpha &= \lambda / (\pi \sin^2\alpha), \qquad \sin\alpha = 2n/R. \end{aligned} \right\} \tag{4.91}$$

The relative length of the diffraction free zone is equal to L_f/Z_α. The thickness of the diffraction-free ring is equal to the central spot diameter

$$\Lambda_\circ = \frac{3}{2}\Lambda, \tag{4.92}$$

where Λ is the thickness of the next ring,

$$\Lambda = \lambda / (2\sin\alpha). \tag{4.93}$$

Holographic technique has been adapted for generation and reproduction of the diffraction free beams [193 – 196]. An annular slit, used originally in [72], has been replaced by holographic optical element, which was an equivalent of the axicon [193]. The performance of this system has been verified theoretically by using the Fresnel diffraction integral and experimentally with binary-amplitude form of the overall phase function. The mock-hologram of diameter 18 cm was fabricated with 300-pixel in Xerox laser printer. The real hologram was 10 mm of diameter. The diffraction efficiency was only 10% in comparison with $\sim$ 60% of the kinoform-equivalent of this hologram.

The physical hologram of the diffraction free beam has been manufactured by exposing the photographic emulsion with Gaussian reference beam [195]. The diameter of the central lobe was $\approx 60\mu$m over the focal range from $z = 20$ cm up to $z = 80$ cm, but the on-axis light intensity was abnormally small for $z < 40$ cm. The diameter of the hologram was 1 cm.

A holographic axilens with relatively small focal range was described in [196]. A nontraditional structure of the light rays, intermediate between conventional lens and axicon was used. Therefore this element was called "aspheric holographic optical element". Its computer-generated equivalent of the diameter 12.5 mm provided the lateral resolution 80 μm at average focal length $f = 1250$ mm over the focal range 30 mm. This was only three times larger than that of the aberration free lens with the same parameters.

In conclusion we present some fresh ideas concerning the necessary conditions for diffraction free fields to be exist [197]. The reasoning behind these ideas was induced by the theory of the generalized aperture [198]. According to this theory, the $3D$ field amplitude $u(\vec{r})$ of a monochromatic scalar field, propagating along z-axis, is equal to $3D$ Fourier transform of a generalized pupil $P(\vec{q})$ as a planar aperture stretched over a sphere of unit radius:

$$u(\vec{r}) = \iiint P(\vec{q}) \exp(i\, k\, \vec{q} \cdot \vec{r})\, d^3 q\,, \tag{4.94}$$

where

$$P(\vec{q}) = A(k\, \vec{q}_\perp)\, \delta(\vec{q}_\perp^{\,2} + q_z - 1)\,. \tag{4.95}$$

with q_z as a longitudinal spatial frequency and $\vec{q}_1$ as a $2D$ transverse spatial frequency in the aperture. To be independent from z, it is necessary that

$$q_z = \beta = const\,. \tag{4.96}$$

The plane $q_z = \beta$ cuts a ring of radius

$$q_\perp = (1 - \vec{q}_\perp^{\,2})^{1/2} = \rho \tag{4.97}$$

on the unit sphere. Hence a diffraction free field must have the spatial-frequency spectrum located on a single ring of spatial frequency $k\rho$. This theorem explains analytically why a narrow annular aperture plays a key role for diffraction free beams and for meso–optics as a whole.

In general case the field amplitude over this annular aperture has an azimuthal distribution

$$A(k\,\vec{q}_\perp) = P(\theta)\,\delta(q_\perp - \rho), \tag{4.98}$$

where the function $P(\theta)$ can be expanded into azimuthal harmonics:

$$P(\theta) = \sum_n p_n \exp(i\,n\,\theta). \tag{4.99}$$

Thus, the general expression for a diffraction free field amplitude has the form:

$$u(\vec{r}) = \exp(i\,k\,\beta\,z) \cdot \sum_n p_n \exp(-i\,n\,\theta)\,J_n(k\,\rho\,\rho_\perp) \tag{4.100}$$

with circular coordinates (ρ, θ) on the plane perpendicular to the z-axis.

The selection of only one **longitudinal** spatial frequency β is equivalent to selecting a ring of **lateral** frequency ρ. This selection can be accomplished with Fabry–Perot etalon as in [199] and [192]. An achromatic diffraction free field can be prepared [197].

Hence we may recognize two principally different methods of generating of diffraction free fields. In the first one we select a **radial** spectral frequency by diffraction on a narrow annular aperture. In the second one we select a **longitudinal** spatial frequency by means of resonance cavity. The latter operation is an achromatic one, but the first one is not.

The theory of the generalized aperture [198] offers an alternative approach to this problem. The main point is that diffraction free fields can be treated as a subgroup of the class of self-imaging fields [200].

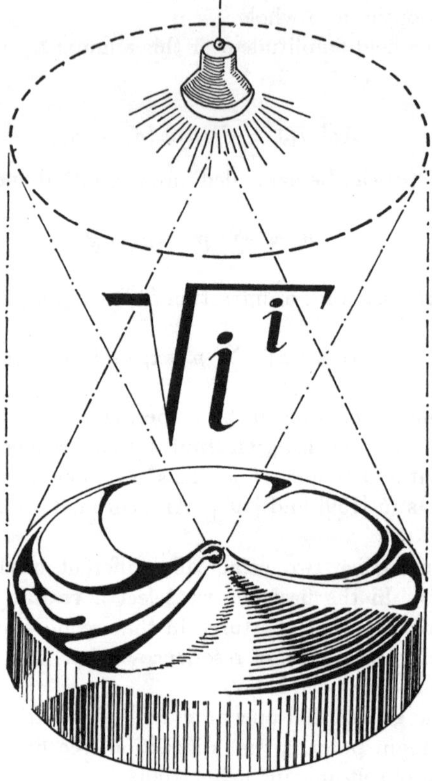

Artists understanding of the meso-optical element with ring response.

Chapter 5

LOCALIZED ELECTROMAGNETIC WAVES

5.1. Introduction

Unlike the monochromatic diffraction free beams treated in Chapter 4, we consider here the light pulse, that has $3D$ localized structure and that is propagating with velocity of light along z-axis without dispersion. The remarkable feature of these localized light pulse trains is that they involve conical beam components, described by Bessel functions of the radial coordinate. These localized light pulse trains enter into **nonseparable** class of solutions and represent indeed exact solutions to the wave equation for the scalar fields or to Maxwell's equations for the vector fields.

Recall that the solution to the scalar wave equation in the $1D$ case,

$$\left(\frac{\partial^2}{\partial z^2} - \frac{\partial^2}{c^2 \partial t^2}\right) u\,(z,\, t) = 0. \tag{5.1}$$

can be presented in the d'Alembert form

$$u\,(z,t) \;=\; f_1\,(z \,-\, c\,t) \,+\, f_2\,(z \,+\, c\,t)\,, \tag{5.2}$$

which involves the sum of two plane waves moving in the opposite directions.

To find a solution to the $3D$ wave equation, we use the product method, proposed by Bernoulli, in the form

$$u\,(\vec{r},t) = u_1\,(\vec{r}) \cdot u_2\,(t)\,, \tag{5.3}$$

or in more general form

$$u\,(\vec{r},t) \;=\; u_1\,(\rho) \cdot u_2\,(z) \cdot u_3\,(t)\,. \tag{5.4}$$

It is evident that by this method we may construct only separable solutions.

This Chapter is devoted to the nontraditional wave compositions which involve conical wave fronts and which are called "localized electromagnetic pulse trains" or "localized wave fields". With these localized electromagnetic beams we can transport the electromagnetic energy over the distances, which are many times longer than those in the case of the traditional monochromatic steady state beams.

We begin with a new description of the light beams which involve the complex source point. Then it will be shown that a solution to the wave equation in the form of the forward propagating plane wave

$$u\,(\vec{r},t) \;=\; \exp\left[\,i\,k\,(z - c\,t)\,\right] \cdot G\,(\rho)\,, \tag{5.5}$$

where $G(\rho)$ is the radial envelope function, cannot give the localized solution. Next we explain why the bidirectional representation, as new decomposition form of the exact solution to the wave equation, helps to construct localized solutions with finite total energy. The bidirectional representation involves both forward **AND** backward travelling plane waves, which appear as **PRODUCT** unlike the **SUM**, used up to now. With this method we can construct the particular and also the general solutions to the wave equation. The bidirectional representation has its proper Fourier interpretation. With this approach we can find the localized waves with smooth transversal envelope, which recovers itself periodically along z-axis over large propagating distances. All these solutions have nonseparable form.

The convincing verifications of this new approach have been demonstrated experimentally in the equivalent conditions with acoustical waves. The theory of the localized wave solutions can be extended to the Klein–Gordon equation, which describes the massive elementary particles moving with group velocity without any contradiction with causality principle. In all these localized solutions there are involved universally the transversal Bessel conical components in the form $J_n(k\rho)$.

In conclusion to this Chapter there will be given a short overview of several other attempts to solve the problem of energy transmission over large distances, such as electromagnetic "missiles" and electromagnetic "bullets".

5.2. Light Beams and Complex-source Point Method

Let us fix our attention only on the light beams propagating along z-axis. To describe these beams a coordinate system shown in Fig. 5.1 will be used. As an angle θ between the wave vector $\vec{k}$ and z-axis of the beam is small, it is worthwhile to introduce transverse component $\vec{k}_\perp$ and longitudinal component $\vec{k}_\parallel$ with $k_z = \xi$ of the wave vector $\vec{k}$. In this configuration [201]

$$\xi = k\cos\theta \cong k(1-\sin^2\theta/2) = k - \frac{1}{2k}(k_\perp)^2, \tag{5.6}$$

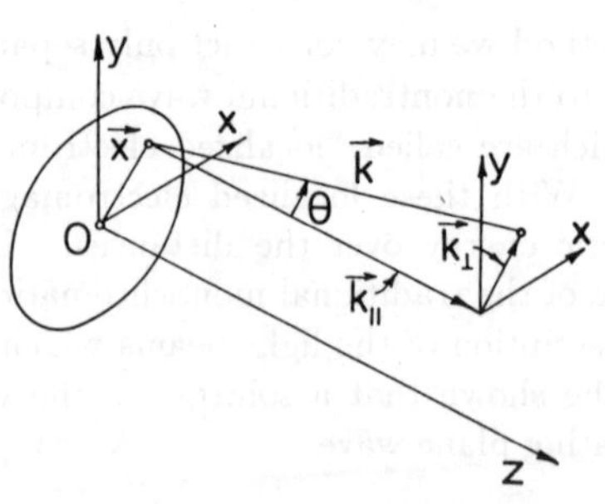

Fig. 5.1. Parameters of the light beam propagating along z-axis.

where $k_\perp = k \sin\theta$. Eq. (5.1) shows that the spherical surface of radius k can be approximated in the vicinity of z-axis by the paraboloid of the same curvature.

As an example we consider a Gaussian beam specified by the wave field over the plane $z = 0$ by the function

$$u(\vec{x}, 0) = \exp\left(-\frac{1}{2}\frac{\vec{x}^{\,2}}{\sigma_\circ^2}\right) = \exp\left(-\frac{k\vec{x}^{\,2}}{2a}\right), \tag{5.7}$$

with scaled variance $a = k\,\sigma_\circ^2$.

In paraxial region, for small θ and small x, the wave equation

$$(\Delta + k^2)\,u = 0, \tag{5.8}$$

for the light beams is converted into the parabolic equation

$$\frac{1}{i}\frac{\partial}{\partial z} = k - \frac{1}{2k}(\nabla_\perp^2), \tag{5.9}$$

with transverse Laplasian

$$\nabla_\perp^2 = \frac{\partial^2}{\partial x^2} + \frac{\partial^2}{\partial y^2} \tag{5.10}$$

and with solution for $z \geq 0$ in the form:

$$u(\vec{x}, z) = (1 + i\,z/a)^{-1} \exp(i\,k\,z) \cdot \exp\left(-\frac{k\vec{x}^{\,2}}{2a}\right) \cdot \exp\left(-i\,\frac{k\vec{x}^{\,2}}{2R}\right). \tag{5.11}$$

Here $R(z)$ is the wavefront curvature. If we introduce the scaled variance $A(z) = k\,\sigma^2$, then we have

$$\left.\begin{aligned} A(z) &= a + z^2/a, \\ R(z) &= z + a^2/z, \end{aligned}\right\} \tag{5.12}$$

The structure of the light beam in the cross-section at any z is defined by two real parameters (a, R), which can be combined into a single complex number Z,

$$Z = z - i\,a. \tag{5.13}$$

Then we have

$$Z^{-1} = R^{-1} + i\,A^{-1}, \tag{5.14}$$

and

$$u(\vec{x}, z) = -\frac{i\,a}{Z}\exp(i\,k\,Z)\exp\left(i\frac{k}{2}\frac{\vec{x}^{\,2}}{Z}\right). \tag{5.15}$$

Eq. (5.15) represents a complex form of the Gaussian beam [201], which is valid only near z-axis. The complex number $Z = z - i\,a$ is the distance from the complex center C to the observation point.

From Eq. (5.15) it follows that Gaussian beam is equivalent paraxially to the spherical wave with center C at a complex location and with spatial structure as a bundle of complex rays originating from this center C. To get a real function describing the amplitude of the beam, the complex conjugate must be added to the final expression in the Eq. (5.15).

The same description can be expressed in terms of the Green function

$$G(\vec{x}, z) = \frac{\exp(i\,k\,r)}{r}, \tag{5.16}$$

if we place the complex source in the point $\vec{r}\,' = \vec{r}\,'_b(0, 0, i\,b)$ and if we introduce the complex distance

$$r_b = [x^2 + y^2 + (z - ib)^2]^{1/2}. \tag{5.17}$$

In the far zone the complex distance is equal to

$$r_b \approx r - i\frac{b\,z}{r}, \tag{5.18}$$

with $r = (x^2 + y^2 + z^2)^{1/2}$ and $r \ll k\,b^2$. In the paraxial region we have

$$r_b = (z - i\,b) + \frac{(z + ib)\,\rho^2}{2(z^2 + b^2)}, \tag{5.19}$$

with $\rho^2 = (x^2 + y^2) \ll z^2 + b^2$. In this domain the Green function assumes the form

$$G(\vec{r}, \vec{r}\,'_b) = \frac{1}{4\pi(z - ib)} \exp[\,i\,k\,(z - ib)\,] \cdot \exp[\,i\,k\,\rho^2/(2z - 2ib)\,]. \tag{5.20}$$

The distance "b" to the complex-source point defines the beam width at any z.

This complex-source point method explained above is used throughout this Chapter for description of the localized electromagnetic pulse trains. This class of the nonseparable solutions is a part of the huge reservoir of the nontraditional solutions to the wave equation. The method allows to avoid some difficulties encountered in the classical calculations and manifests itself as an elegant approach valuable also in the case of the inhomogeneous medium [202].

5.3. Focus Wave Modes

The explicit expression of the electromagnetic modes that represents the nondispersive wave packets have been constructed by J.N. Brittingham [203]. These modes, referred to as "focus wave modes" (FWM), are remaining localized as $3D$ structure for all times as they are propagating in free space. The heuristic approach gives for the FWM the product form

$$FWM = F_1 \cdot F_2 \cdot F_3 , \tag{5.21}$$

where F_1 describes a $3D$ light pulse, propagating with light velocity c, F_2 is a plane wave moving with velocity $c_1 < c$ in the same direction, and F_3 is a function of the angular variables. Electric flux density $\vec{D}$ and magnetic field $\vec{H}$ of this beam are defined by five wave functions Ψ_q, $q = 1, 2, ..., 5$ as follows:

$$\left.\begin{aligned} D_\rho(\rho, \Phi, z, t) &= \Psi_1 + \Psi_1^*, \quad D_\Phi(\rho, \Phi, z, t) = \Psi_2 + \Psi_2^*, \\ H_\rho(\rho, \Phi, z, t) &= \Psi_3 + \Psi_3^*, \quad H_\Phi(\rho, \Phi, z, t) = \Psi_4 + \Psi_4^*, \\ H_z(\rho, \Phi, z, t) &= \Psi_5 + \Psi_5^*, \end{aligned}\right\} \tag{5.22}$$

where

$$\Psi_q = I_q\left[\rho, (z - c\,t)\right] \cdot \exp\left[i\,k\,z\,(z - c_1\,t)\right] \cdot \Phi_q , \tag{5.23}$$

with

$$I_q = A_q\left[\rho, (z - c\,t)\right] \cdot G_1\left[\rho, (z - c\,t)\right] \cdot G_2\,(z - c\,t) . \tag{5.24}$$

The functions I_q represent a fixed $3D$ pulse, moving along z-axis at light velocity, the complex exponent in Eq. (5.23) describes the plane wave, moving along z-axis with a velocity less than light velocity c. The functions Φ_q describe the angular dependencies and their absolute value is constant, for example, $|\Phi_q| = 1$. The functions A_q and G_1 define the asymptotic behavior of the Ψ_q functions.

In the subsequent papers [204, 205] a rigorous mathematical formulation of the FWM has been constructed. There have been generated more extended family of such packetlike beams. In particular, the packetlike solution to the wave equation is written as [204]:

$$G(\vec{r}, t) = F\,(x, y, z - c\,t) \cdot \exp\left[-i\,\frac{K}{2}\,(z + c\,t)\right] . \tag{5.25}$$

This form includes two plane waves moving in the opposite directions. The function $F(x,\ y,\ z - c\,t)$ is a solution to the paraxial wave equation:

$$\left.\begin{array}{r}\nabla_\perp^2 F \;-\; 2i\cdot\frac{K}{2}\left(\frac{\partial F}{\partial z} \;-\; \frac{1}{c}\frac{\partial F}{\partial t}\right) = 0\,, \\ \text{or}\;\; \nabla_\perp^2 F \;-\; 2iK\,\frac{\partial F}{\partial \xi}(x,y,\xi) \;=\; 0\,. \end{array}\right\} \tag{5.26}$$

with

$$\xi \;=\; z \;-\; ct\,. \tag{5.27}$$

Notice that until now we have made no approximations, and therefore the function $F(x,\, y,\, \xi)$ represents an exact solution to the paraxial wave equation (5.26).

As an illustration we consider the zero mode solution which can be written in the form:

$$F_{\circ,\circ} \;=\; \left(\frac{z_\circ}{\xi + iz_\circ}\right)\cdot\exp\left(-i\,\frac{K}{2}\cdot\frac{\rho^2}{\xi + iz_\circ}\right), \tag{5.28}$$

where

$$z_\circ \;=\; \frac{K\,a^2}{2} \tag{5.29}$$

and a is the beam waist at $\xi = 0$. Higher-order mode solutions are expressed by Laguerre polynomials $L_m(x)$ of the order $m = 0, 1, 2, \ldots$, as

$$F_{\circ,m}(\rho,\xi) \;=\; \left(\frac{z_\circ}{\xi + iz_\circ}\right)^m L_m\left(i\,\frac{K}{2}\,\frac{\rho^2}{\xi + iz_\circ}\right). \tag{5.30}$$

Starting with these Gaussian–Laguerre packetlike solutions, the Brittingham's solutions, which have been found before heuristically [204], can be constructed analytically.

Another more direct derivation of the FWM solutions has been developed in [205], where the difference between $1D$ and $3D$ localized solutions has been analyzed more carefully. The $1D$ solution, written in the d'Alembert form, Eq. (5.2), can be made localized without problems, if we chose properly the functions f_1 and f_2. In the case of $3D$ solution with $\vec{x}_\perp$ as transversal component of the position vector $\vec{x}$ we have

$$(\Delta \,+\, k^2)\,\Psi\,(\xi,\vec{x}_\perp) \;\equiv\; (\nabla_\perp^2 \,+\, 4\,\partial_\xi\,\partial_\eta)\,\Psi\,(\xi,\,\vec{x}_\perp) \;=\; \nabla_\perp^2\,\Psi\,(\xi,\vec{x}_\perp) \;=\; 0\,. \tag{5.31}$$

We know that any function which satisfies the $2D$ Laplace equation, as Eq.(5.31), cannot attain a maximum or a minimum within restricted domain of the transversal $\vec{x}_\perp$ plane. Hence, $\Psi\,(\xi,\,\vec{x}_\perp)$ cannot be localized function in this domain, and FWM, written in the form of $\Psi\,(\xi,\,\vec{x}_\perp)$, cannot be pure travelling wave.

One of the correct algorithm of FWM construction consists in adopting explicitly that the solution, to be searched for, does depend on both ξ and η, where

$$\eta = z + ct. \tag{5.32}$$

For example, for this approach it follows that Ψ function can be written in the form:

$$\left.\begin{aligned} \Psi &= \psi(\xi, \vec{x}_\perp) \cdot \Phi(\eta), \\ \text{with} \quad |\Phi(\eta)| &= const. \end{aligned}\right\} \tag{5.33}$$

or

$$\Psi = \exp(ik\eta) \cdot \psi(\xi, \vec{x}_1). \tag{5.34}$$

Here ψ-function is a solution of the Schrödinger equation

$$(\nabla_\perp^2 + 4ik\partial\xi)\,\psi(\xi, \vec{x}_\perp) = 0, \tag{5.35}$$

which allows localized solutions, as was shown in [205]. The envelope $|\psi|$ is a function only of $\xi = z - ct$ and $\vec{x}_\perp$, and this envelope is propagating without deformation along z-axis with light velocity. The structure of the function ψ in the transversal plane at any z can be expressed via Schrödinger propagator for function ψ specified initially at the plane with $\xi = z - ct = 0$ as follows [205]

$$\psi(z, \vec{x}_\perp) = \exp[ik(z+ct)] \int d\vec{x}\,'_\perp \cdot \psi_\circ(\vec{x}\,'_\perp) \cdot \frac{k_\perp}{i\pi(z-ct)} \cdot \exp\left[\frac{ik_\perp(\vec{x}_\perp - \vec{x}\,'_\perp)}{(z-ct)}\right] =$$

$$= \exp[ik(z+ct)] \cdot \int d\vec{q}_\perp \cdot F(\vec{q}_\perp) \cdot \exp[i\,\vec{q}_\perp(\vec{x}_\perp - \vec{q}_\perp \cdot \frac{z-ct}{4k_\perp})], \tag{5.36}$$

where

$$\psi_\circ(\vec{x}_\perp) = \psi(\xi = 0, \vec{x}_\perp), \tag{5.37}$$

and

$$F(\vec{q}_\perp) = (2\pi)^{-2} \int d\vec{x}_\perp \cdot \psi_\circ(\vec{x}_\perp) \exp[-i\,\vec{q}_\perp \cdot \vec{x}_\perp], \tag{5.38}$$

is the spatial Fourier transform of $\psi_\circ(\vec{x}_\perp)$.

It must be emphasized that there is a substantial difference between solution to the paraxial wave equation Ψ_{PAR} and exact solution to the wave equation Ψ_{FWM}. We have

$$\left.\begin{aligned} (\Delta + k^2)\,\Psi_{FWM} &= 0, \\ \text{but} \quad (\Delta + k^2)\,\Psi_{PAR} &\neq 0. \end{aligned}\right\} \tag{5.39}$$

The focus wave modes, constructed by J.N. Brittingham [203], can be generated now by means of the Eqs. (5.36)–(5.38). For the initial waist profile of the structure

$$\Psi(\xi = 0, \vec{x}_\perp) = A(\frac{\rho}{a})^n \exp[\pm i n \Phi - (\rho/a)^2], \tag{5.40}$$

the FWM-solutions have been constructed in [205].

The principle problem for FWM solution was its infinite total energy. As has been proved in [206], there exist two different kinds of FWM solutions, both with infinite total energy. The solution of the second kind can be constructed from the FWM solution of the first kind by multiplying with function $\exp(-i\xi w/2)$, where w is a scalar parameter. By proper superposition of the conventional focus wave modes, the localized solution with finite total energy can be constructed [209].

Of a particular interest is the paper [207] in which a relativistic theory of scalar diffraction has been used to resolve the problems in the field of the localized solutions to the wave equation. Namely, the plane waves diffracted by the pinhole aperture, which is moving with light velocity, do generate some localized FWM solutions. To prove this conclusion, let us start with the traditional scalar theory of light diffraction. In the frame of this theory, the diffracted field Ψ_D can be written as

$$\Psi_D(\vec{r}, ct) = \int_s \Psi_{IN}(\vec{r}, ct')\, h(\vec{r}, ct; \vec{r}', ct')\, d\vec{r}\, dt, \tag{5.41}$$

where $\Psi_{IN}(\vec{r}, ct)$ is the incident field, h is the diffracting kernel and s is the surface of the aperture. If the aperture and the observer are at the fixed planes z' and z, the diffracting kernel in the Fresnel approximation can be written as:

$$\begin{aligned} h(\vec{r}, ct; \vec{r}', ct') &= (z - z')^{-1} \exp[i k (\eta - \eta')] \cdot \\ &\cdot \exp[i k \frac{(x - x')^2 + (y - y')^2}{2(z - z')}], \end{aligned} \tag{5.42}$$

$$\text{where} \quad \eta = z - ct \quad \text{and} \quad \eta' = z' - ct. \tag{5.43}$$

As was proved in [207], the field Ψ_D represents a solution of the paraxial wave equation

$$(\partial_x^2 + \partial_y^2)\Psi_D + 2\partial_z\partial_\eta\Psi_D = 0, \tag{5.44}$$

which in turn must be considered as a nonrelativistic approximation of the wave equation. The planes z and z' are the characteristic surfaces of the Eq. (5.44). There are solutions of the characteristic differential equation

$$(\partial_x \tilde{u})^2 + (\partial_y \tilde{u})^2 + 2\partial_z \tilde{u}\, \partial_\eta \tilde{u} = 0, \tag{5.45}$$

associated to the paraxial wave equations, Eq. (5.44). Hence, Eq. (5.45) represents the nonrelativistic approximation of the characteristic equation, associated to the wave equation, and, as was shown in [207], the Fresnel theory of diffraction is actually an exact nonrelativistic theory and its diffracting kernel is a solution of the paraxial equation, the aperture plane and the observer being at points which are characteristic for this equation.

The next topics of [207] is the relativistic theory of diffraction, the kernel of which is equal to the nondispersive solution of the wave equation, called above as the focus wave mode (FWM). Namely,

$$h(\vec{r},t;\vec{r}',t') = (\xi - \xi') \cdot \exp[ik(\eta - \eta'')] \cdot \cdot \exp[-ik\frac{(x-x')^2 + (y-y')^2}{(\xi - \xi')}], \tag{5.46}$$

where

$$\left.\begin{array}{ll} \xi = z + ct, & \xi' = z' + ct', \\ \eta = z - ct, & \eta'' = z' - ct'. \end{array}\right\} \tag{5.47}$$

The diffracting kernel in Eq. (5.46) is a solution of the wave equation

$$(\partial_x^2 + \partial_\eta^2)\Psi + 4\,\partial_\xi\,\partial_\eta\,\Psi = 0. \tag{5.48}$$

The aperture and the observer are located at hyperplanes with $4D$ coordinates ξ' and ξ, the latter being the solutions of the associated characteristic equation

$$(\partial_x u)^2 + (\partial_y u)^2 + 4\,\partial_\xi u\,\partial_\eta u = 0. \tag{5.49}$$

To get back to the nonrelativistic theory it is sufficient to change ξ and ξ' by $2z$ and $2z'$, respectively. It is evident, that when the aperture and the observer are fixed in space, both theories give the same solutions.

The relativistic theory of diffraction, outlined above, has been applied to a plane wave

$$\Psi_{IN}(\vec{r}', ct') = \exp(ik\eta'), \tag{5.50}$$

incident on a circular aperture, located at the characteristic point with $\xi' = 0$. The diffracted field is given by the following equation [207]:

$$\Psi_D(\vec{r}, ct) = \frac{ik}{\xi}\exp(ik\eta)\int_s \exp\{-i\frac{k}{\xi}[(x-x')^2 + (y-y')^2]\,dx'\,dy' = \\ = \frac{ik}{\pi\xi}\exp[-ik(\eta + \frac{x^2+y^2}{\xi})] \cdot F(a), \tag{5.51}$$

where

$$F(a) = \int_0^a r'dr'\int_{-\pi}^{+\pi} d\theta' \cdot \exp[-i\frac{k}{\xi}(r'^2 - 2r'x\cdot\cos\theta' - 2r'y\sin\theta')] = \\ = 2\pi\int_0^a r'dr' \cdot J_\circ(\frac{2krr'}{\xi})\exp[-i\frac{kr^2}{\xi}], \tag{5.52}$$

with $r'^2 = x'^2 + y'^2$, $r^2 = x^2 + y^2$. In the limit $a \to \infty$

$$F(a) \;=\; \pi\, a^2 \;+\; \mathcal{O}(a^3)\,. \qquad (5.53)$$

The field, diffracted by a pinhole aperture, is equal to

$$\Psi_D \;=\; i\,\frac{k\,a^2}{\xi}\,\cdot\,\exp\left[\,i\,k\,(\eta\,+\,\frac{x^2+y^2}{\xi})\right]. \qquad (5.54)$$

We see that the structure of the Eq.(5.54) is analogous to that of the FWM. On the other hand, the wave field at a fixed distance from the input plane is expressed as Gaussian beam of the form:

$$\Psi_G \;=\; i\,\frac{k\,a^2}{z}\,\cdot\,\exp\left[\,i\,k\,(\eta\,+\,\frac{x^2+y^2}{\xi})\right]. \qquad (5.55)$$

Hence we arrive at the very interesting conclusion: to generate the focus wave modes, we must bring into moving the pinhole aperture with velocity equal to the light velocity. This conclusion can be verified only for acoustic waves.

5.4. Bidirectional Representation

The bidirectional representation, which offers new decomposition forms of the exact solutions to the wave equation, involves the **product** of the forward and backward propagating plane waves [208]. To explain this new approach, let us start with the elementary solution of the scalar wave equation, written in the coordinate system (ρ, ξ, η) as:

$$\Psi_l\,(\vec{r}, t) \;=\; \Psi_l\,(\rho, \xi, \eta) \;=\; [C_n \cdot J_n\,(k_\perp \rho) \,+\, D_n \cdot N_n\,(k_\perp \rho)]\,\cdot$$
$$\cdot\,\exp\,(\pm\, i\, n\,\Phi)\cdot\exp\,(-\,i\,\alpha\,\xi)\cdot\exp\,(i\,\beta\,\eta)\,, \qquad (5.56)$$

where $\xi = z - t$, $\eta = z + t$, $\alpha\beta = k_\perp^2/4$, $J_n(\bullet)$ — Bessel function, $N_n(\bullet)$ — Neumann function, and the light velocity $c = 1$. The main feature of this solution is that forward, $\exp\,(-\,i\,\alpha\,\xi)$, and backward, $\exp\,(i\,\beta\,\eta)$, travelling plane waves appear as a product.

A general solution, which is chosen finite at $\rho = 0$, assumes the form in the same coordinate system

$$\Psi\,(\rho, \xi, \eta) \;=\; (2\pi)^{-2}\,\sum_{l=-1}^{+1}\,\sum_{n=0}^{\infty}\,\int_0^{\infty} dk_\perp \int_0^{\infty} d(l\,\alpha) \int_0^{\infty} d(l\,\beta)\,\cdot\, C_n\,(l\,\alpha,\, l\,\beta,\, k_\perp)\,\cdot$$
$$\cdot\, k_\perp\, J_n\,(k_\perp \rho)\cdot\exp\,(\pm\, i\, n\,\Phi)\,\cdot\exp\,(-\,i\,l\,\alpha\,\xi)\,\cdot\,(\,i\,l\,\beta\,\eta)\,\cdot\,\delta\,(\alpha\beta - k_\perp^2/4)\,, \qquad (5.57)$$

where functions $C_n(\bullet)$ in this bidirectional representation have the meaning of the spectral coefficients. It is evident that Eq. (5.57) represents a superposition of properly weighted elementary Bessel beams. In the frame of the conventional coordinate system $(\vec{r}, t)$ the bidirectional representation can be translated into the Fourier interpretation form [208]. A general solution, defined by Eq. (5.57), will be rewritten as

$$\Psi(\vec{r},t) = (2\pi)^{-2} \sum_{n=0}^{\infty} \int_0^{\infty} d\,k_\perp \int_{-\infty}^{+\infty} d\omega \int_{-\infty}^{+\infty} d\,k_z \cdot A_n(\omega, k_{z_1} k_\perp) \cdot$$

$$\cdot\, k_\perp\, J_n(k_\perp \rho) \cdot \exp(\pm i\, n\, \Phi) \cdot \exp(-i\, k_x\, x) \cdot \exp(i\, \omega\, t) \cdot \delta(\omega^2 - k_\perp^2 - k_z^2), \quad (5.58)$$

where $A_n(\omega, k_z, k_\perp)$ is the spectrum of the function $\Psi(\vec{r}, t)$ in the conventional representation. The spectra $C_n(\bullet)$ and $A_n(\bullet)$ are coupled by the relation:

$$\left.\begin{aligned} A_n(\omega, k_z, k_\perp) &= 2\, C_n[\tfrac{1}{2}(\omega + k_z),\, \tfrac{1}{2}(\omega - k_z),\, k_\perp], \\ k_z &= \alpha - \beta, \quad \omega = \alpha + \beta. \end{aligned}\right\} \quad (5.59)$$

After integration over time frequency ω, we arrive at

$$\Psi(\vec{r},t) = (2\pi)^{-2} \sum_{n=0}^{\infty} \int_0^{\infty} dk_\perp \int_{-\infty}^{+\infty} dk_z \cdot \{ A_n[\sqrt{k_\perp^2 + k_z^2}, k_z, k_\perp] \cdot$$

$$\cdot \exp[-i(k_z z - \sqrt{k_\perp^2 + k_z^2} \cdot t)] + A_n[-\sqrt{k_\perp^2 + k_z^2}, k_z, k_\perp] \cdot$$

$$\cdot \exp[i(k_z z + \sqrt{k_\perp^2 + k_z^2} \cdot t)]\} \cdot \frac{k_\perp}{2\sqrt{k_\perp^2 + k_z^2}} \cdot J_n(k_\perp \rho) \exp(\pm i\, n\, \Phi). \quad (5.60)$$

We see once again that the radial dependence of the solution $\Psi(\vec{r}, t)$ is described here by the Bessel functions $J_n(k_\perp \rho)$.

After integration over $(l\,\alpha)$ the bidirectional representation, Eq. (5.57), assumes the form:

$$\Psi(\rho, \xi, \eta) = (2\pi)^{-2} \sum_{n=0}^{\infty} \int_0^{\infty} dk_\perp \int_0^{\infty} d\beta \cdot \Big\{ C_n(\frac{k_\perp^2}{4\beta}, \beta, k_\perp) \exp(-i\,\frac{k_\perp^2}{4\beta}\,\xi) \cdot \exp(i\,\beta\,\eta) -$$

$$- C_n(-\frac{k_\perp^2}{4\beta}, -\beta, k_\perp) \exp(i\,\frac{k_\perp^2}{4\beta}\,\xi) \cdot \exp(-i\,\beta\,\eta) \Big\} \cdot \frac{k_\perp}{\beta}\, J_n(k_\perp \rho) \exp(\pm i\, n\, \Phi). \quad (5.61)$$

For focus wave modes the bidirectional representation is:

$$\Psi_\beta^{FWM}(\vec{r}, t) = \exp[i\beta(z+t)] \cdot \exp[-\beta \rho^2 V] \cdot \frac{1}{4\pi V}, \quad (5.62)$$

where V is the complex spatial variance, with

$$\frac{1}{V} = \frac{1}{A} - \frac{i}{R}, \tag{5.63}$$

and β is the parameter, which defines the behavior of the pulse. The beam spread parameter A, the wavefront curvature R and the beam waist w are defined by

$$\left.\begin{aligned} A &= a_1 + \xi^2 / a_1 , \\ R &= \xi + a_1^2 / \xi , \\ w &= (A / \beta)^{1/2} . \end{aligned}\right\} \tag{5.64}$$

We present the bidirectional zero-order spectrum $C_\circ(\bullet)$ of the FWM and the corresponding bidirectional representation FWM function:

$$C_\circ(\alpha, \beta, k_\perp) = (\sqrt{2}/2)\sigma \cdot \exp[-\sigma^2(\beta - \beta')^2] \cdot \exp(-\alpha a_1), \tag{5.65}$$

and

$$\Psi^{FWM}(\rho, \xi, \eta) = (2\pi)^{-2} \int_0^\infty dk_\perp \int_0^\infty d\beta \int_0^\infty d\alpha \cdot C_\circ(\alpha, \beta, k_\perp) \cdot$$

$$\cdot\, k_\perp J_\circ(k_\perp \rho) \cdot \exp(-i\alpha\xi) \cdot \exp(i\beta\eta) \cdot \delta(\alpha\beta - k_\perp^2/4). \tag{5.66}$$

The analogous relations exist for monochromatic Bessel beams [72, 73], for localized Gaussian pulse trains [209, 211] and for moving modified Bessel–Gauss pulses [210].

The bidirectional approach offers certain advantages over the conventional one. The first one is that, owing to its simple structure, the bidirectional spectrum $C_n(\bullet)$ can be evaluated with relative ease, whereas the equivalent Fourier spectrum $A_n(\bullet)$ might be very complicated and due to this can be overlooked. This is one of the reasons why the localized wavefields have not been discovered earlier. The second positive feature of the bidirectional representation is that here there are only algebraic singularities rather than branch-cut problems typical for the Fourier transform representation.

5.5. Localized Gaussian Fields

Several efforts have been made to construct the localized solution with finite total energy, but only in the papers [209, 210] this problem was resolved finally. In this section we consider the moving modified Gaussian pulses [209, 211].

The main idea of the approach of R.W. Ziolkowski et al. is that fundamental Gaussian pulse fields, described by Eq. (5.62), can be used as basic functions for construction, with proper weighting of these functions in their superposition a solution with finite total energy. As was shown in [208], an exact solution has a form:

$$\Psi(\vec{r},t) = \int_0^\infty d\beta \cdot \Psi_\rho(\vec{r},t) \cdot F(\beta) =$$

$$= \frac{1}{4\pi[a_1 + i(z-t)]} \int_0^\infty d\beta \cdot F(\beta) \cdot \exp[-\beta\, s(\rho,z,t)], \qquad (5.67)$$

where the spectrum $F(\beta)$ is the weighting function, and in accordance with Eq. (5.62)

$$s(\rho,z,t) = \frac{\rho^2}{[a_1 + i(z-t)]} - i(z+t). \qquad (5.68)$$

The parameter a_1 is a beam-source location term in the presentation with complex-point source, and the product $F(\beta)\,\beta^{-1/2}$ must be square integrable [211].

The inverse functions of the decomposition, presented in Eq. (5.67), are given by

$$\Phi_\beta(\rho,\xi,\eta) = 8\sqrt{\pi} \cdot \exp[-(\xi/4\,\beta\, a_1)^2] \cdot \Psi_\beta(\rho,\xi,\eta). \qquad (5.69)$$

To arrive at various solutions the spectrum $F(\beta)$ is to be changed. The structure of an exact solution, which corresponds to the "modified power spectrum" chosen by R.W. Ziolkowski, has the form

$$\Psi(\vec{r},t) = \frac{1}{4\pi\,[a_1 + i\xi]\,[a_2 + s/p]^q} \cdot \exp[-b\,s/p], \qquad (5.70)$$

where parameters b, a_1, a_2, p and q define concrete view of the modified power spectrum. The function $\Psi(\vec{r},t)$, given by Eq. (5.70), is localized near the beam axis and its amplitude is recovering repeatedly at very large distances. This class of solutions is called **"electromagnetic directed-energy pulse train"** (EDEPT) [211].

5.6. Localized Bessel–Gauss Fields

The new family of exact solutions, called "moving modified Bessel–Gauss pulses," has been discovered [210]. This family encloses fundamental Gaussian pulses, presented in Section 5.5, Bessel beams, and also zeroth-order Bessel–Gauss pulse solutions that are more localized than the localized Gaussian pulse trains.

The fundamental functions of this new family were chosen as:

$$\Phi_n(\vec{r},t) = \frac{a_1}{V}\, J_n(\frac{k_\perp a_1 \rho}{V})\, \exp(\pm i\, n\, \Phi)\, \exp(-\frac{\beta\rho^2}{V})\cdot$$
$$\cdot \exp(-i\,\frac{k_\perp^2 a_1 \xi}{4\beta V})\, \exp(i\,\beta\,\eta)\,, \tag{5.71}$$

where $V = a_1 + i\,\xi$ is the complex variance, and the spectral parameters $k_\perp$ and β are subjected to the constraint

$$k_\perp^2 = 4\,\alpha\,\beta\,. \tag{5.72}$$

The zeroth-order solution

$$\Phi_\circ(\vec{r},t) = \frac{a_1}{V}\, J_\circ(\frac{k_\perp a_1 \rho}{V})\, \exp(-\frac{\beta\rho^2}{V})\cdot \exp(i\,\beta\,\eta)\, \exp(-i\,\frac{k_\perp^2 a_1 \xi}{4\,\beta\, V}) \tag{5.73}$$

has been called "modified moving Bessel–Gauss pulse".

For $k_\perp \to 0$, we have

$$\Psi_G(\vec{r},t) = \lim_{k_\perp\to 0}\, \Phi_\circ(\vec{r},t) = \frac{a_1}{V}\, \exp(-\frac{\beta\rho^2}{V})\, \exp(i\,\beta\,\eta)\,, \tag{5.74}$$

which is just the fundamental Gaussian pulse [209]. In another extreme case, for $a_1 \to \infty$, when $\lim_{a_1\to\infty}(a_1/V) = 1$, we arrive at the solution

$$\Psi_B(\vec{r},t) = \lim_{a_1\to\infty}\, \Phi_n(\vec{r},t) = J_\circ(k_\perp\rho)\, \exp(-\,i\,\alpha\,\xi)\, \exp(i\,\beta\,\eta)\,, \tag{5.75}$$

which is diffraction free Bessel beam [72, 73], written in the bidirectional representation with $\alpha = k_\perp^2\,/\,4\beta$.

Now let us compare the propagation features of various localized pulses. For $\rho = 0$, $z = c\,t$ ($\xi = 0$), we have

$$Re\,[\Psi_G(\vec{r},t)] = Re\,[\Phi_\circ(\vec{r},t)] = \cos 2\,\beta\, z\,. \tag{5.76}$$

Eq. (5.76) shows that the amplitudes of the Ψ_G and $\Phi_\circ$ pulses are recovering periodically with a period $\mathcal{L} = \pi/\beta$. The comparison of these two solutions shows that for $k_\perp > \beta$ the Bessel–Gaussian pulse is localized both transversally and along z-axis better than the fundamental Gaussian pulse [210]. However, for $k_\perp < \beta$ both pulses are alike. It is interesting to note that as $k_\perp/\beta$ increases and a_1 decreases, the backward travelling pulse wave, $\exp(i\,\beta\,\eta)$, gives little contribution to the solution in comparison to the forward propagating plane wave.

The fundamental functions in the Bessel–Gauss family can be used for construction another general localized solutions. The corresponding superposition has the form

$$\Psi(\vec{r},t) = (2\pi)^{-2}\,\frac{a_1}{V}\int_0^\infty dk_\perp \int_0^\infty d\alpha \int_0^\infty d\beta\,\cdot\,C_\circ(k_\perp,\beta,\alpha)\,\cdot$$

$$\cdot\,k_\perp\,J_\circ\Big(\frac{k_\perp a_1 \rho}{V}\Big)\exp\Big(-\frac{\beta\rho^2}{V}\Big)\exp(i\,\beta\,\eta)\,\cdot\exp\Big(-i\,\frac{\alpha\,\xi\,a_1}{V}\Big)\,\delta(\alpha\beta - k_\perp^2/4)\,. \qquad (5.77)$$

After integration over α we have

$$\Psi(\vec{r},t) = (2\pi)^{-2}\,\frac{a_1}{V}\int_0^\infty dk_\perp \int_0^\infty d\beta\,\cdot\,C_\circ\Big(k_\perp,\beta,\frac{k_\perp^2}{4\beta}\Big)\cdot\frac{k_\perp}{\beta}\cdot J_\circ\Big(\frac{k_\perp a_1 \rho}{V}\Big)\,\cdot$$

$$\cdot\,\exp\Big(-\frac{\beta\rho^2}{V}\Big)\exp(i\,\beta\,\eta)\,\exp\Big(-i\,\frac{k_\perp^2 q_1}{4\beta V}\Big)\,. \qquad (5.78)$$

5.7. Experimental Verification

To investigate experimentally the end effect of the localized transmission of the electromagnetic energy over the great distances, we must generate the fields of ultra-wide bandwidth [212 – 215]. As the used solutions are indeed nonseparable ones in space-time, the spatially components of the solution, on the one hand, and the broad bandwidth spectra, on the other hand, have strong correlation. Due to this property, which is completely absent in the monochromatic fields, the moving interference patterns, resulting from constructive interference of the contributions from various wave forms, are forced by this correlation to arrive at correct times at the observation point. To achieve this new effect, a transmitter must be chosen as $2D$ array of transmitting elements, which are driven **independently**. Each element has his proper time driven function. Only in this manner the nonseparable character of the space-time driving functions can be attained. It must be recalled that in the conventional transmitter all elements of the array are driven by the same monochromatic or narrow bandwidth time signal with proper amplitude and phase control for angular scanning and side lobes suppression. In the case of the localized wavefields transmitter, we control not only the amplitude or phase, but also the frequency spectra of the signals delivered to each individual element of the $2D$ array.

In these conditions it is very difficult to compare correctly the signals, having the broad bandwidth spectra, with monochromatic signals in the traditional transmitters. First of all, it is principally difficult to give the diffraction length $\mathcal{L}_d$ for these ultra-broad bandwidth signals [212]. Also the problem of the detected system is not so simple as it may appear at first glance.

Meanwhile it has been proved that the fundamental feature of the localized beams do depend on the higher order spectra moments of the driven signals. The main point of this theorem is that the inevitable process of diffraction degrades first only lower orders of the spectra moments and only later involve moments of higher orders. It is evident that very high degree of correlation in the higher order moments of their spectra is extremely desired, if a localized beams of very long propagating range is needed. The localized beams, treated in the last two Sections, provide these features with high efficiency.

So far the experiments have been accomplished with equivalent acoustic waves [213, 214]. At the first stage the computer simulation of these experiments and of the proposed criteria of the correct comparisons of the localized waves with traditional monochromatic beams has been made. The comprehensive analysis of the both transmitting **and** detecting systems has led to the nontrivial recommendations concerning the justified interpretation of the experimental results.

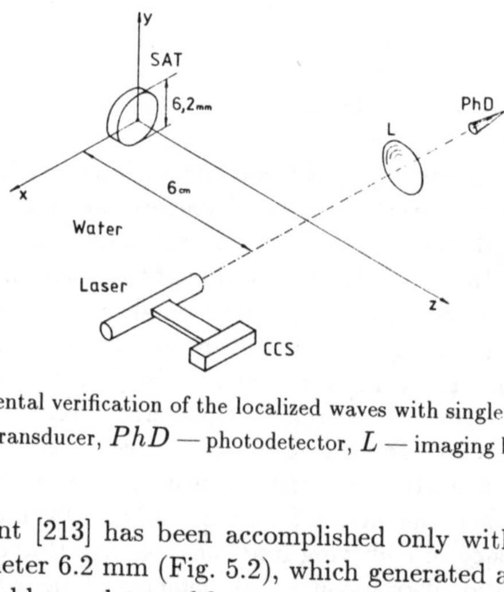

Fig. 5.2. The first experimental verification of the localized waves with single acoustical transducer: SAT — single acoustical transducer, PhD — photodetector, L — imaging lens, CCS - computer controlled scanning.

The first experiment [213] has been accomplished only with single acoustical transducer of the diameter 6.2 mm (Fig. 5.2), which generated an acoustical beam in water. The sound field was detected by means of the laser beam, which crossed the far-field region of the sound field at distance 60 mm from the transducer. The transmitted light beam has been focused by the positive lens onto one-channel photodetector. To interpret the measurements of amplitude and phase of the pressure field in time, a reciprocity theorem has been used. The resonance frequency of the acoustical transducer was well above the spectra of the driving signal. The virtual image of the photodetector was near or inside the sound field. The object of the measurements was the time variation of the optical refractive index of water. Only phase information has been detected.

First, it has been shown, that the signal, detected in this experiment, was equal to the third time derivative of the driving function. The optical system induced the second derivative of the length integral of the acoustical pressure, and the acoustical transducer added the first time derivative.

The synthetic linear array, which did not exist in the reality, consisted of 21 elements with period 3.0 mm and with total array width 60 mm. The duration of the time record was 10.24 μs, there were detected 176 wave forms.

In this experiment the modified-power-spectrum pulse, described by Eqs. (5.70) and (5.68), has been used. The oscillating recovery length was equal to $\Delta z = \pi\beta/b$, in the intermediate zone $z_1 < z < z_2$, where

$$z_1 = \frac{\beta}{2b}, \qquad z_2 = \frac{\beta a}{2}. \tag{5.79}$$

For $z > z_2$ the field amplitude is decaying as $z^{-\alpha}$. The longitudinal localized decay length was $[z_\circ^2 + (z - ct)^2]^{-1}$.

The pulse wave form used in this experiment was defined by the parameters: $a = 1$ m; $\alpha = 1.0$; $b = 600$ m^{-1}; $\beta = 300$; $z_\circ = 4.5 \cdot 10^{-4}$ m, $z_1 = 0.25$ m, $z_2 = 150$ m, and $\Delta z = 1.571$ m. Pulse dimension was equal to $\rho = \sqrt{\beta z_\circ/b} \approx 15$ mm. The sound velocity in water was $c = 1.5 \cdot 10^4$ m/s, the frequency $\nu = 2$ MHz, the effective wavelength $\lambda = 0.75$ mm. The Rayleigh length $L_R = \pi w_\circ^2/\lambda \cong 94$ cm ~ 1 m.

The results of the experiment with $1D$ array and with different driving function in each element demonstrate that 1) the spreading of the conventional Gaussian beam was from 1.5 cm to 240 cm with intensity decay $6.28 \cdot 10^{-3}$; 2) the recovery length for modified-power-spectrum pulse of the form, given by Eq. (5.70), was $z = 150$ cm, and the spreading from 1.5 cm to 1.5 m (!).

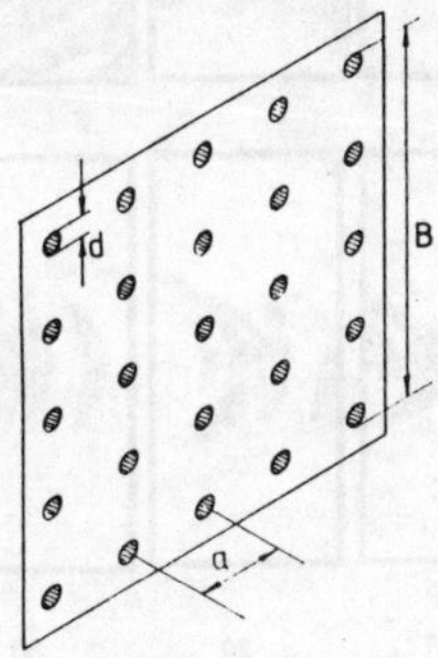

Fig. 5.3. The second experiment with 5×5 diameter d and of the period a.

In the second experiment [214] a $2D$ array of 5×5 transmitting acoustical transducers has been used (Fig. 5.3). The diameter of each transducer was 0.5 mm with separation 2.5 mm. The area of the $2D$ array was $A = 1.0$ cm^2. Six different wave forms have been generated. The time sample interval was 100 ns.

The receiver was 3mm-radius transducer. There were averaged 1000 wave forms. The scanning spatial interval was 0.5 mm over the length 10 cm perpendicular to the beam axis, with 200 samplings. The received energy was $= \int dt\, |voltage|^2$. The measurements has been repeated at distances $z = 5, 10, 15, 20, 30, 40, 50$ cm.

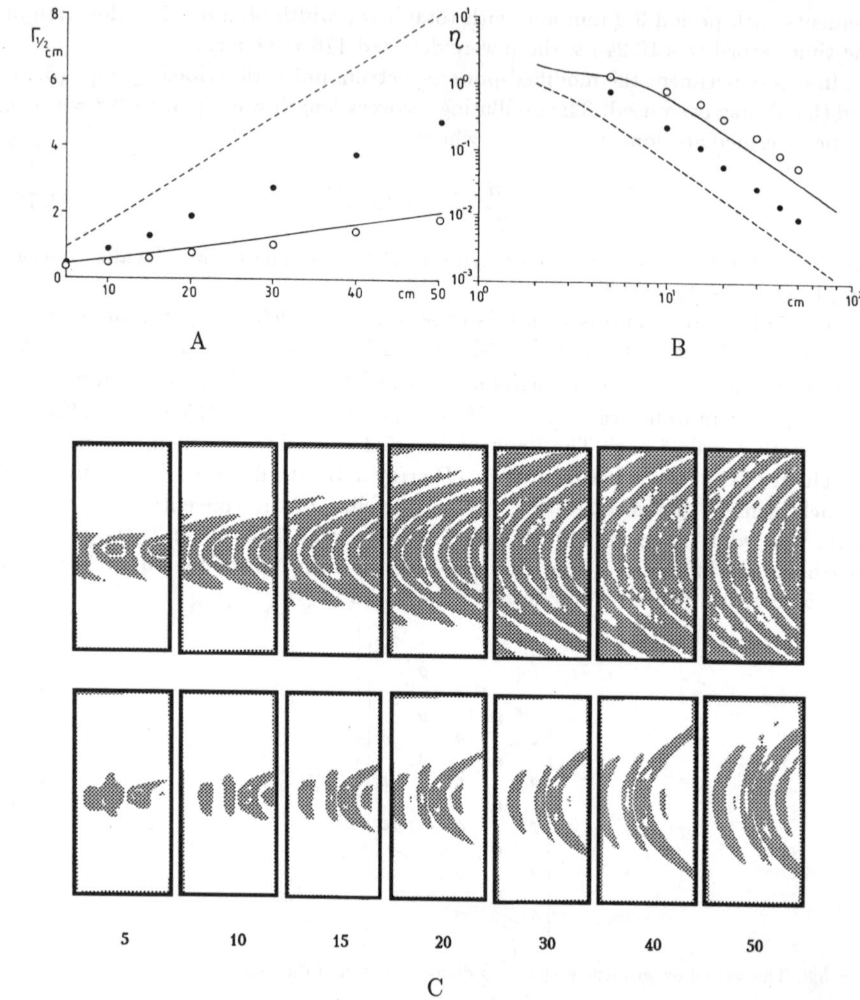

Fig. 5.4. The experimental results and the theoretical calculations:

"A", the half-width $\Gamma_{1/2}$ of the beam at various distances Z from the source, ○ — LW experiment, ● — CW experiment, ——— LW theoretical simulation, - - - - - CW theoretical simulation,

"B", the energy efficiency η at various distances Z from the source, ○ — LW experiment, ● — CW experiment, ——— LW theoretical simulation, - - - - - CW theoretical simulation,

"C", the spatial structure of the CW field (top) and of the LW field (bottom) at various distances from the source.

The experimental results are sketched in Fig. 5.4. The first diagram shows the half width of the beam at various distances z. The second one represents the decay of energy with distance. The outstanding result of the second experiment was that the extended near-field range for localized waves was $> 10\ L_R$ with

$$L_R = \frac{A}{\lambda_R}, \tag{5.80}$$

where A is the area of the transmitting $2D$ array, and α_R is the effective wavelength of the localized waves.

The results of these two experiments [213, 214] provoked some remarks in the literature [216, 217].

5.8. Localized Waves for the Klein–Gordon Equation

The Klein–Gordon equation for particles of mass $m \neq 0$ and with spin zero, has the form

$$(\Delta - \partial^2_{ct} - \mu^2)\,\Psi = 0, \tag{5.81}$$

where $\mu = mc/\hbar$. The bidirectional representation of the localized solution of the Klein–Gordon equation can be presented in the form [218 – 220]

$$\Psi = G(\rho, z, t)\exp(i\,\beta\,\eta), \tag{5.82}$$

where the variable

$$\eta = z + (\frac{c^2}{\nu_g})\,t, \tag{5.83}$$

and the group velocity of the localized wave is

$$\nu_g = \frac{|\beta|c}{(\beta^2 + \mu^2)^{1/2}}. \tag{5.84}$$

The function $G(\rho, z, t)$ must met to the equation

$$[i\,2\,\beta\,(\partial_z - \partial_t/\nu_g) + (\partial_z^2 - \partial_t^2/c^2) + \nabla_\perp^2]\,G(\rho, z, t) = 0. \tag{5.85}$$

If we introduce the distance parameter

$$\tau = \gamma(z - \nu_q\,t) \tag{5.86}$$

with relativistic factor $\gamma = [1 - (v_g/c)^2]^{1/2}$, then we come to the hyperbolic Schrödinger equation

$$(i\,4\,\beta\,\gamma\,\partial_\tau + \partial_\tau^2 + \nabla_\perp^2)\,G(\rho, \tau) = 0. \tag{5.87}$$

With substitution

$$G(\rho,\,\tau) \;=\; g(\rho,\,\tau)\exp(-i\,2\,\beta\,\gamma\,\tau), \qquad (5.88)$$

we come to the Helmholz equation for $g(\rho,\,\tau)$ [218]:

$$[\nabla_\perp^2 + \partial_\tau^2 + 4\,\beta^2\gamma^2]\,g(\rho,\,\tau) \;=\; 0. \qquad (5.89)$$

Its general solution can be written as

$$g_{l\,m}(\rho,\,\tau) \;=\; J_l(2\,\beta\,\gamma\,\xi)\,P_l^m(\tau/\xi)\cos(m\theta) \qquad (5.90)$$

with

$$\xi \;=\; [\rho^2 + \gamma^2(z - \nu_q\,t)^2]^{1/2}. \qquad (5.91)$$

The general solution in the bidirectional representation has the form:

$$\Psi_{l\,m}(\vec{r},\,t) \;=\; J_l(2\,\beta\,\gamma\,\xi)\,P_l^m(\tau/\xi)\cos m\,\theta\cdot\exp(-i\,2\,\beta\,\gamma\,\tau)\cdot\exp(i\,\beta\,\eta). \qquad (5.92)$$

The parameter β, having the sense of the wavevector, is given by

$$\beta \;=\; \gamma\,(\frac{\nu_g}{c})\,\mu. \qquad (5.93)$$

The analogous approach can be applied to the localized waves in the cold plasma [221].

5.9. Approximate Focus Wave Modes

The controversy arises as to whether focus wave modes are causal or not. This not yet unresolved dilemma has two conflicting approaches: analytical one [222] and a physical one [223]. An argument in favour of the first approach is based on the fact that an exact solution of the source-free, time-dependent wave equation in the infinite homogeneous space can be written as [203 – 205, 209]

$$\Phi(\vec{r},\,t) \;=\; Re\,\{\Phi_+(\vec{r},\,t)\},$$
$$\Phi_+(\vec{r},\,t) \;=\; \frac{\exp[-ik\,(\eta + \rho^2/(\xi + iz_\circ)]}{4\pi\,(\xi + iz_\circ)}, \qquad (5.94)$$

where $\Phi_+(\vec{r},\,t)$ represents the analytical continuation of a real signal into the lower half of the complex t-plane. For $|\xi| \ll z_\circ$, we have

$$\Phi_+(\vec{r},\,t) \;\approx\; (4\,\pi\,i\,z_\circ)^{-1}\cdot\exp[-\,i\,k\,(z + \nu t) \;-\; k\rho^2/z_\circ]. \qquad (5.95)$$

This analytical signal represents the Fresnel region field of **backward-propagating time-harmonic Gaussian beam**. The forward moving envelope of the focus wave modes suggests unambiguously [222], that focus wave modes actually represents a superposition of such backward-propagating time-harmonic Gaussian beams, in which the waist location and phase reference are chosen in order to get the forward propagation of the whole localized wave. This interpretation follows also from the results of the paper [224] in which it was succeeded to construct focus wave modes as a superposition of **backward propagating** Gaussian beams, generated by a **forward** moving complex source.

A physical approach to the dilemma with causality emerged from the analysis of the "physical realizability" of the focal wave modes [223]. The first main point is that the focus wave modes must be considered as solutions of the homogeneous wave equation **with** the boundary condition $\exp(-\rho^2/w_\circ^2)$ given on the characteristics $\xi = 0$, $\eta = 0$, Eq. (5.31). This is actually a Goursat problem [225] typical for any well posed physical problems. The second main point is that focus wave modes are **inhomogeneous waves** with frequency and wave vector depending on space **and** time. The notions of local wave vector, local frequency, local phase velocity, local group velocity and local velocity of energy propagation are typical for these waves. The "physical" analysis shows that focus wave modes behave globally as a plane wave, which propagates at global group velocity $= c$. Just the same as plane wave has infinite total energy, the focus wave modes possess this property as well. Therefore the physical realizability of the focus wave modes must be understood as an approximate realizability in the same fashion as ideal plane wave is approximated by its restricted part with finite total energy. As a penalty of this property a some angular and spatial spreading of the beam will arise. Hence it will be naturally to consider some possibilities for approximation of the focus wave modes. Two such approximations have been treated in [222]. The first one is the construction of a packet of focus wave modes as properly weighted sum. The second one is the embedding of the focus wave modes into a class of approximate solutions depending on a parameter. In this approach focus wave modes can be treated as an approximate limit of this parameter [226].

On comparing the focus wave modes with the conventional Gaussian beams we make out two differences between them: 1) focus wave modes are generated by a source located on the characteristic with $\xi = 0$, which represents a plane moving with the light velocity c, whereas this source in the case of the Gaussian beams is placed on a **fixed** plane, 2) as was stated in [222], "the focal wave modes are nondispersive but not the Gaussian beams".

To construct the approximate focus wave modes, the characteristic $\xi = z - x_\circ = 0$, with $x = c\,t$, is to be replaced by the variable $\xi_\beta = z - \beta\, x_\circ = 0$ with parameter β, $0 \leq \beta \leq 1$, or with parameter $\mu = (1-\beta)/(1+\beta)$ [226]. For $\beta = 1$ we have the relativistic description of the focus wave modes, but for $\beta = 0$, which corresponds to the fixed plane with $\xi = z$, we have Gaussian beams. The wave equation

$$\partial_x^2 \Psi + \partial_y^2 \Psi + 4\, \partial_\xi \, \partial_\eta \, \Psi = 0\,, \tag{5.96}$$

valid for $\beta = 1$, in the general case with $\beta \neq 1$, assumes the form

$$\partial_x^2 \Psi + \partial_y^2 \Psi + \frac{4}{(1+\beta)^2} \left[\beta\, \partial_z^2 \Psi - \partial_{z_o}^2 \Psi + (1-\beta)\, \partial_{x_o} \partial_z \Psi \right] = 0\,. \tag{5.97}$$

Its solution, written as

$$\Psi_\beta = \frac{1}{a + i\,\xi_\beta} \exp\left[i\,\chi\,\eta - \frac{\chi \rho^2}{a + i\,\xi_\beta} \right], \tag{5.98}$$

with "a" and χ as some real parameters, has finite energy if $\beta \neq 1$.

The question at issue for causality remains unresolved. It is evident that the causality principle is violated if the causality is meant in its narrow analytical sense as directivity of processes in time. Meanwhile we may interpret the causality principle as requirement that no energy flow must outrun the wave front, and that no energy flow can be detected by the observer earlier than the wavefront has passed over him.

5.10. Electromagnetic "Missiles" and "Bullets"

The aim of these systems was to generate an electromagnetic waves, which are decaying at very large distances R slower than R^{-2}. These pulses can be generated by the electrical currents in antenna of finite size with rising edge in time as $f(t) \sim t^{\varepsilon/2}$, with $\varepsilon < 1$ [227, 228]. In [229] it was shown that for this purpose a spherical lens can be used.

In the system with electromagnetic "bullets", an inverse initial problem is formulated at first [230, 231]. The desired wavefield is taken as an desired function of space and time. The initial values of the field, which generate the desired field in the far zone, is to be found. The methods suitable for choosing these initial conditions, have been treated [230, 231]. It was proved that the initial conditions are unique, and it was shown that there are asymptotic fields, which propagate between two coaxial cones from negative hemisphere to positive hemisphere in the angular range $\theta_1 < \theta < \theta_2$, $0 < \Phi < 2\,\pi$. When $\theta_2 \to 0$, the asymptotic field assumes the "bullets" structure, which gives very high spatial resolution. No experimental evidences and no comprehensive computer simulations have been done untill now.

Chapter 6

GRAVITATIONAL LENS

6.1. Introduction

The deflection of the light rays in the gravitational field of the celestial objects follows from the Newtonian gravitational theory and from the hypothesis that beam composes of particles. The same effect was predicted by special and general relativistic theories. However only the general relativistic theory predicts the correct value of the deflection angle. The term "gravitational lens" has been introduced to describe a phenomenon which we observe in our universe. However we know well that a "gravitational lens" is indeed not at all a lens and represents an example of the meso–optical element of the cosmic scale in nature. The gravitational lens never produces a point-like image and also does not accomplish the central projection of the object space onto the projected plane, what is typical for traditional optical lens. The gravitational lens accomplishes the transformation of light, issued from a point-like source, into a semi-infinite linear "image". As will be shown here the gravitational lens can be considered as a nonlinear transformer of the velocity vector of the source star. There are other features of the gravitational lens which will be explained in this Chapter.

6.2. General Theory

The deformation of the initial wavefront of any radiation is produced by the gradient of the gravitational field in the vicinity of the gravitational lensing object (deflector). The velocities of the light rays at various two points A and B on the wavefront are different due to the difference of the space metric in the gravitational field [232] (Fig. 6.1).

Let us evaluate the resulted deflection of the light rays in the frame of general relativistic theory. The acceleration of the object induced by the gravitational field is equal to

$$w = \frac{GM}{r}, \tag{6.1}$$

where M is the mass of the gravitating object, G is the gravitational constant and r is the distance between the gravitating object and test particle. For small acceleration, the acquired velocity β of the test particle, such as a light photon, is defined by the equation:

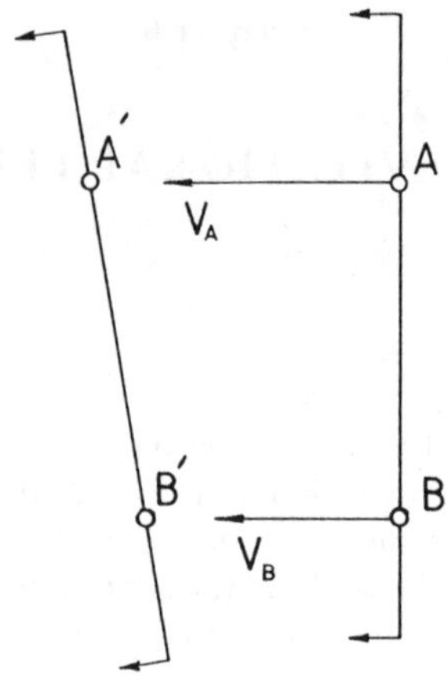

Fig. 6.1. Refraction of the light rays in the gravitational field: the light velocities at the points A and B are different.

$$\beta^2 = \frac{2GM}{r\,c^2} = 2\,\frac{r_g}{r}, \tag{6.2}$$

where r_g is the gravitational radius of the gravitating object:

$$r_g = \frac{GM}{c^2}\,. \tag{6.3}$$

For our Sun $r_{g\odot} = 1.4\,\text{km}$, and for our Earth $r_{gE} = 0.4\,\text{cm}$.

The gravitational field changes the space metric at each point with distance r from the gravitating object. Thus we have:

$$\left.\begin{aligned} dr &= \tfrac{dx}{\gamma} = \sqrt{1-\beta^2}dx \approx (1 - r_g/r)dx\,, \\ dt' &= \gamma dt = \tfrac{1}{\sqrt{1-\beta^2}}dt \approx (1 + r_g/r)dt\,, \\ ds_\theta &= dy\,, \\ ds_\varphi &= dz\,, \end{aligned}\right\} \tag{6.4}$$

where the relativistic factor $\gamma = (1-\beta^2)^{-1/2}$, and the differential distances ds_θ and ds_φ are oriented perpendicular to x and to r [232] (Fig. 6.2).

We may consider two separate effects which change the light velocity in the gravitational field: radial and tangential ones. The light velocity in the radial direction is equal to

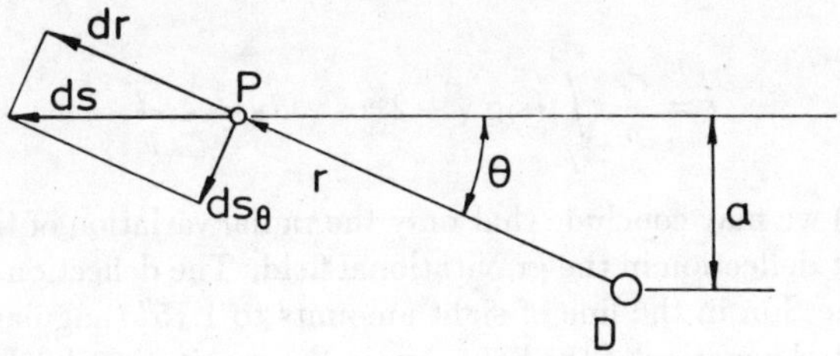

Fig. 6.2. Geometry of the light ray in the vicinity of the deflecting lens D in the coordinate system (θ, r), a is the impact parameter.

$$v_r = \frac{dr}{dt} = \frac{dx}{dt}\frac{(1 - r_g/r)}{(1 + r_g/r)} = c_\circ(1 - 2\,r_g/r), \tag{6.5}$$

where $c_\circ$ is the light velocity at infinity from the gravitating object. The light velocity in the tangential direction is defined by

$$v_{tang} = \frac{dx_\theta}{dt} = \frac{dx_\varphi}{dt} = c_\circ\,(1 - r_g/r)\,. \tag{6.6}$$

The light deflection angle α is equal to

$$\alpha = \int_{-\infty}^{\infty} \frac{\partial v}{\partial y}\frac{1}{v}\,ds \approx \int_{-\infty}^{\infty} \frac{\partial v}{\partial y}\frac{ds}{c_\circ}\,. \tag{6.7}$$

From Eqs. (6.5) and (6.6) we find:

$$\frac{\partial v_r}{\partial r} = \frac{2\,r_g\,c_\circ}{r^2}\,, \qquad \frac{\partial v_\theta}{\partial s_\theta} = \frac{r_g\,c_\circ}{r^2}\,, \tag{6.8}$$

and

$$\frac{\partial v}{\partial y} = \frac{\partial v_\theta}{\partial s_\theta}\sin\chi + \frac{\partial v_r}{\partial r}\cos\chi = \frac{r_g\,c_\circ}{r^2}\,(\sin\chi + 2\,\cos\chi)\,, \tag{6.9}$$

where the angle $\chi = \chi(r,p)$ is a function of the impact parameter p and of the distance r. The total light deflection angle

$$\epsilon = \int_{-\infty}^{\infty} \frac{r_g}{r^2}(\sin\chi + 2\cos\chi)ds. \tag{6.10}$$

As

$$\left.\begin{aligned} s = p\tan\chi, \quad ds = p\sec^2\chi d\chi\,, \\ \text{and} \qquad\qquad \\ r^2 = p^2\sec^2\chi\,, \end{aligned}\right\} \tag{6.11}$$

we find

$$\epsilon = \frac{r_g}{p} \int_{-\pi/2}^{\pi/2} (\sin\chi + 2\cos\chi) d\chi = 4\,\frac{r_g}{p}. \tag{6.12}$$

From Eq. (6.12) we may conclude that only the radial variation of the light velocity induces the light deflection in the gravitational field. The deflection of the star light which is near the Sun in the line of sight amounts to 1.75″ (angular second).

The differential equation of the light rays in the gravitational field can be written in the polar coordinate system (r, ϕ) [233, 234]:

$$(\frac{dr}{d\phi})^2 = \frac{1}{p^2}(r + r_g)^4 - (r + r_g)(r - r_g). \tag{6.13}$$

In the approximation of small gravitational radius $r_g \ll p$ we have

$$\frac{1}{r} = 2\,\frac{r_g}{p^2} + \frac{1}{p}\cos\varphi. \tag{6.14}$$

The trajectory of the light rays in the weak gravitational field represents a hyperbola with asymptote orientations defined by the equation:

$$\cos\varphi = -2\frac{r_g}{p}. \tag{6.15}$$

The asymtotes of this hyperbola are defined by the angles $\phi_{1,2}$ equal to

$$\phi_{1,2} = \pm(\frac{\pi}{2} + \delta), \tag{6.16}$$

where

$$\delta = 2\,\frac{r_g}{p}. \tag{6.17}$$

The total bending angle of the light ray is defined by the equation

$$\epsilon_{tot} = 4\,\frac{r_g}{p}, \tag{6.18}$$

and coincides with Eq. (6.12).

The geometry of the light deflection by the gravitational lens is presented in Fig. 6.3 for the case of point-like source star S and point-like gravitating lens D (deflector star). We explain the meaning of the involved angles. The left splitting angle β_1 separates the image of the source star S, observed in the absence of the gravitating lens D, from the left apparent meso–optical image I_1. The left angular displacement θ_1 separates the left meso–optical image I_1 from the projected image of the gravitating lens D. The left bend angle θ_1 represents an angle between two asymptotes of the left hyperbolic light ray from the source star S to observer O. The analogous definitions of the right splitting angle β_2, of the right angular displacement θ_2 and of the right bend angle θ_2 are evident counterparts of the definitions given above. The off-axis α (the left eccentricity angle) defines the angular displacement of the gravitating lens D with respect to the central line SO connecting the source star S with the observer O.

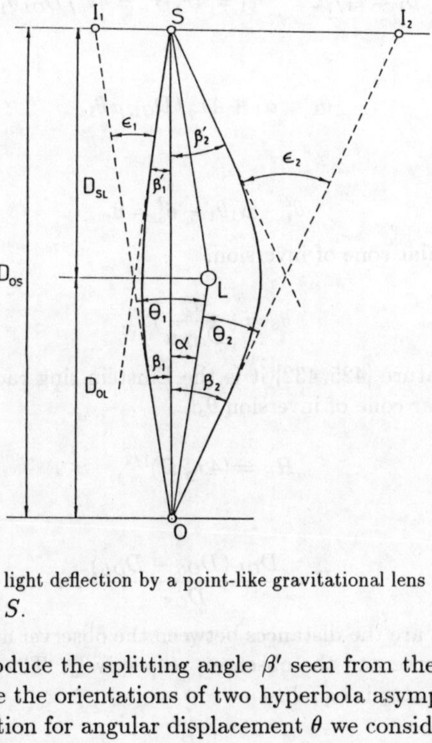

Fig. 6.3. Geometry of the light deflection by a point-like gravitational lens L (deflector) in the case of a point-like source star S.

Also we may introduce the splitting angle β' seen from the source star S. The angles β and β' define the orientations of two hyperbola asymptotes.

To derive an equation for angular displacement θ we consider Fig. 6.3 anew. As was shown in [235], we have

$$\left.\begin{aligned} \theta_1 &= \alpha_1 + \beta_1\,, \\ \epsilon_1 &= \beta_1 + \beta_1'\,, \\ \beta_1' &= \beta_1\,\tfrac{D_{OL}}{D_{SL}}\,, \end{aligned}\right\} \tag{6.19}$$

where D_{OL} and D_{SL} are the distances between the observer and the lens, and between the source and the lens, respectively. If we introduce an auxiliary parameter

$$\mu = 1 + \frac{D_{OL}}{D_{SL}} \tag{6.20}$$

and take into account that

$$r_1 = \beta_1 \cdot D_{OL} = \theta_1 \cdot D_{SL}, \tag{6.21}$$

then we have

$$\beta_1 = \epsilon_1/\mu, \qquad \epsilon_1 = 4r_g/r_1 = 4r_g/D_{OL}\theta_1, \tag{6.22}$$

and

$$\theta_1 = \alpha_1 + 4\,r_g/D_{OL}\,\mu\,\theta_1, \tag{6.23}$$

or

$$\theta_1^2 - \alpha_1\theta_1 - \theta_\circ^2 = 0, \tag{6.24}$$

where $\theta_\circ$ is the angular cone of inversion,

$$\theta_\circ = (\frac{4r_g}{D_{OL}\mu})^{1/2}. \tag{6.25}$$

In the modern literature [425, 432] it is the Einstein ring radius R_E which is used instead of the angular cone of inversion $\theta_\circ$:

$$R_E = (4\,r_g\,D)^{1/2}, \tag{6.26}$$

with

$$D = \frac{D_{OL}\,(D_{OS} - D_{OL})}{D_{OS}}, \tag{6.27}$$

where D_{OL} and D_{OS} are the distances between the observer and the lens (deflector) and between the source and the observer, respectively. The angular Einstein ring radius of the lensing star is equal to

$$\theta_* = \frac{R_E}{D_{OL}}. \tag{6.28}$$

Eq. (6.24), considered with respect to unknown θ_1, has two solutions,

$$\theta_{1,2} = [(\frac{\alpha}{2})^2 + \theta_\circ^2]^{1/2} \pm \frac{\alpha}{2}, \tag{6.29}$$

and gives two relations:

$$\left.\begin{array}{c} \theta_1 - \theta_2 = \alpha \\ \theta_1 \cdot \theta_2 = \theta_\circ^2 \end{array}\right\}. \tag{6.30}$$

The sequence of the image configurations which can be observed in the case, when the object star S is moving to the lens star D, is shown in Fig. 6.4. These stars have finite angular radii $\phi_\circ$ and ϕ_L, respectively [236]. Both meso–optical images of the source star, the primary and the secondary ones, can be observed in the configuration when the source star is inside the Einstein ring radius R_E, Eq. (6.26). In the on-axis configuration we see annular ring image of the source star with radius equal to the Einstein ring radius R_E. It is interesting to note that the angular Einstein ring radius $\theta_\circ$ decreases with distance D_{OL} as $(D_{OL})^{-1/2}$.

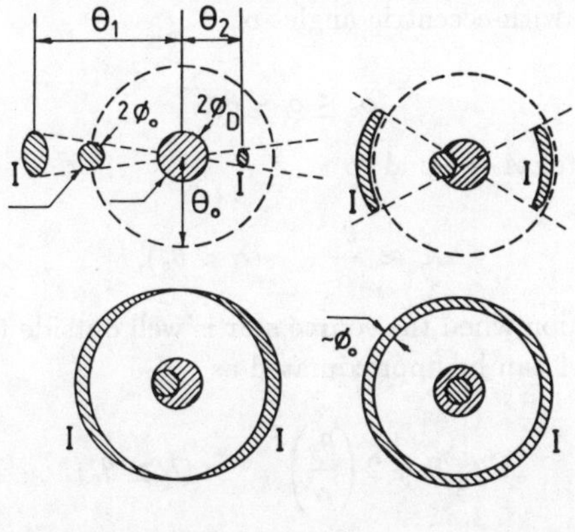

Fig. 6.4. The sequence of the image configurations produced by the lensing star D of finite dimensions at various eccentricities.

The elliptical gravitational lens is defined by the gravitational potential with three parameters: velocity dispersion, ellipticity, and core radius [441].

The gravitational lens is in fact a positive meso–optical element, which generates the diffraction pattern with central lobe in the form of semi-infinite straight line segment. The distance between the lens and the nearest point of this segment is equal to

$$z_{\min} = \frac{R^2}{2\,r_g}, \tag{6.31}$$

where R is the radius of the lensing star. For $z < z_{\min}$ we cannot see the Einstein ring. For our Sun $z_{\min} = 8.3 \cdot 10^{10}$ km. As the distance to the nearest star is $4.1 \cdot 10^{12}$ km (4.3 l.y.), the gravitational lens effect can be observed for all stars. For a source star of the Sun dimension, $R_\odot = 6.95 \cdot 10^5$ km, the angular Einstein ring radius is equal to $\theta_\circ \approx 2.8 \cdot 10^{-7}$ rad. The visible angular dimension of the source star of the width of the Einstein ring is equal to $\phi_\circ = R_\odot / l_\circ \approx 10^{-8}$ rad, and the ratio $\theta_\circ / \phi_\circ \approx 30$, where the distance $l_\circ = 8$ l.y. $\approx 7.6 \cdot 10^{12}$ km. As the modern optical telescope has the angular resolution $\sim 5 \cdot 10^{-6}$ rad, this ring cannot be resolved.

To detect the gravitational lens effect from stars of our Galaxy, an indirect method of observation has been used [236, 238]. In the configuration of perfect alignment of the source star and the lens star, the brightness of the source star will be more than in other configurations. The magnification factor $A_{\max}$ is equal to the ratio of the Einstein ring area to the area of the visible disc of the source star:

$$A_{\max} \approx \frac{2\pi \cdot \theta_\circ \cdot \phi_\circ}{\pi \phi_\circ^2} = 2\left(\frac{\theta_\circ}{\phi_\circ}\right). \tag{6.32}$$

For the configuration with eccentric angles α,

$$\phi_\circ \leq \alpha \leq \theta_\circ, \tag{6.33}$$

the magnification factor A is equal to

$$A \approx \frac{\theta_\circ}{\alpha} \qquad (\alpha < \theta_\circ), \tag{6.34}$$

and for the configuration when the source star is well outside the Einstein ring, the magnification factor A can be approximated as

$$A \approx 1 + 2\left(\frac{\theta_\circ}{\alpha}\right)^2, \qquad (\alpha > \theta_\circ). \tag{6.35}$$

The magnification factor A for each of two meso–optical images, $I_1/I_\circ$ and $I_2/I_\circ$ can be expressed through the angles α and β, where α is the source–image angle and β is the source–lens angle (Fig. 6.5). We have [409, 428]

$$\left.\begin{aligned} \frac{I_{1,2}}{I_\circ} &= \frac{1}{2}\left(\pm 2 + \frac{\alpha}{\beta} + \frac{\beta}{\alpha}\right), \\ \frac{I_{tot}}{I_\circ} &= \frac{1}{2}\left(\frac{\alpha}{\beta} + \frac{\beta}{\alpha}\right). \end{aligned}\right\} \tag{6.36}$$

In the approximation of thin gravitational lens the deflection angle ϵ depends only upon the surface mass density of the lens.

The magnification of the star brightness, induced by the gravitational lens, can be detected as time symmetric variation of the visible brightness of the source star. These radial "spikes" of the magnified light are oriented over the whole sphere. The average distance between the points, where radial "spikes" intersect the projecting plane, is equal in our Galaxy to $d \approx 10^8$ km. The core of maximum intensity in this "spike" have an area of the average radius comparable to that of our Sun. So the average distance L between two adjacent cores is of the order of

$$L = \frac{\sigma}{2R_\odot n} \approx 10^{11}\,\text{km}, \tag{6.37}$$

where σ is area of the Galactic disk , $\sigma \approx 10^8(\text{l.y.})^2$, and n is the number of the radial "spikes" which penetrate the central area,

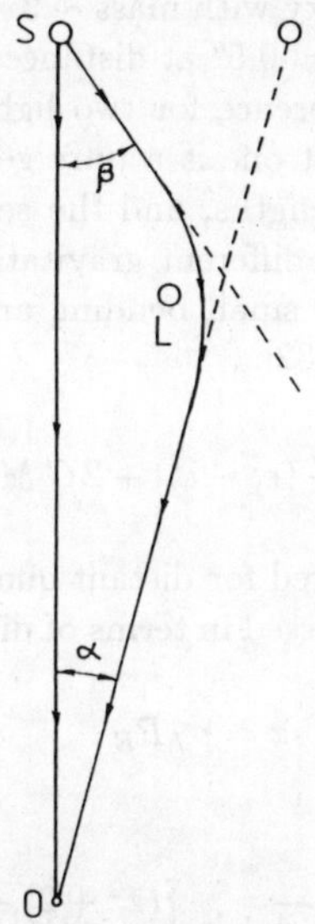

Fig. 6.5. Geometry of the lensing event by using only the angles α, β, α_1 and α_2.

$$n = \frac{N^2}{10}. \tag{6.38}$$

From Eg. (6.37) we may conclude that the mean time τ between two successive observations of the radial "spikes" is equal to $\tau = L/v \approx 100$ years. The average duration of the light intensity burst would be 6 mouths [236].

Of the particular interest is the close binary star system, composed, for example, of two neutron stars [237]. The magnification effect in such a system is repeated with constant period. As $D_{OL} \gg D_{SL}$, the Einstein ring radius is equal to

$$R_E^b = \left(r_{gL} \frac{D_{SL}}{D_{OL}} \right)^{1/2}, \tag{6.39}$$

where r_{gL} is the gravitational radius of the lens,

$$r_{gL} = \frac{4\,GM_L}{c^2 D_{OL}} \tag{6.40}$$

and M_L is the lens mass.

As was explained in [239], the angular Einstein radius of the gravitating galaxy is several order larger than that of individual gravitating star. For example, the angular Einstein radius of the galaxy with mass $\sim 3 \cdot 10^{10} \cdot M_{\odot}$ and of galaxy radius $R_L \approx 0.5 \cdot 10^4$ l.y., will be equal $\theta_* = 0.5''$ at distances $D_{OS} = 2.7 \cdot 10^9$ l.y.

Now we consider the time difference for two light paths [257]. There are two causes for this difference. The first one is a pure geometrical one, defined by the difference ΔL between the path lengths, and the second cause is due to the fact that the light rays is subjected to different gravitational potential during moving with different light velocities. For small bending angles the total time difference (time delay) is equal to [257, 413, 432]:

$$\Delta t = \frac{D_{OL}\, D_{LS}}{2\, D_{OS}} \left(\epsilon_1^2 - \epsilon_2^2\right) + 2\, G\, M \ln \left(\frac{a_1^2}{a_2^2}\right). \tag{6.41}$$

The second term cannot be neglected for distant quasars [414].

The time delay Δt can be expressed in terms of dimensionless impact parameter

$$x = p\,/\,R_E \tag{6.42}$$

in the form

$$\Delta t = 2GM \left\{ x\sqrt{x^2+4} + \ln \left[\frac{(x^2+2) + x\sqrt{x^2+4}}{(x^2+2) - x\sqrt{x^2+4}} \right] \right\}. \tag{6.43}$$

As magnification factor, expressed in x, is equal to

$$A(x) = \frac{x^2+2}{x\sqrt{x^2+4}}, \tag{6.44}$$

then we have also

$$\Delta t = 2\, GM \left[2\,(A^2+1)^{-1/2} + \ln \left(\frac{A+1}{A-1} \right) \right]. \tag{6.45}$$

The time delay Δt can be expressed in units of $4\, GM/c^3$ [456] as:

$$d\tau = \frac{1}{2}(\vec{u}_i - \vec{u}_s) - \ln |\vec{u}_i|, \tag{6.46}$$

where $\vec{u}_i$ and $\vec{u}_s$ vectors define the angular positions of the source and its image:

$$\left.\begin{array}{c} \vec{\theta}_s = \theta_* \vec{u}_s, \quad \vec{\theta}_i = \theta_* \vec{u}_i, \\ \theta_*^2 = \frac{4GM}{c^2 D}, \quad D = \frac{D_{OL} D_{OS}}{D_{LS}}. \end{array}\right\} \tag{6.47}$$

From Eq. (6.43) we see that the geometrical term dominates at large x. From Eq. (6.45) we see that the lens mass M can be estimated directly from the time delay Δt and from the total magnification factor A. The time delay effect explained here for light rays, or photons, exists also for other elementary particles such as relativistic neutrinos.

6.3. Moving Gravitational Lens

The gravitational lens effect provides important astronomical information about mass and spatial density of the source objects such as distant quasars, galaxies and stars. The observation of the moving gravitational lens offers additional ability to determine separately masses and distances to the celestial objects [240]. To explain the properties of the moving gravitational lens we introduce the coordinate system, in which the projected coordinates of the source object are

$$x = v_x t, \quad y = y_\circ, \tag{6.48}$$

where v_x is the velocity of the source object in this coordinate system. The projected position vector $\vec{r}(x,y)$ of the source object and the projected position vector $\vec{r}_1(x_1,y_1)$ of its image are given by the equation

$$\vec{r}_1 = \vec{r} + \frac{\zeta}{r_1^2}\vec{r}_1, \tag{6.49}$$

where ζ is the strength parameter of the gravitational lens,

$$\zeta = \theta_\circ^2 \cdot D_{OL}^2. \tag{6.50}$$

The velocity vector $\vec{v}_1(v_{x1}, v_{y1})$ of the image $\vec{r}_1(x_1,y_1)$ is defined by their components [240]:

$$\left.\begin{aligned} \frac{v_{x1}}{v} &= A_{v,x} = \tfrac{1}{2}\left[1 \pm (1+4\eta s)^{1/2} \mp 4\eta s(1-s)(1+4\eta s)^{-1/2}\right], \\ \frac{v_{y1}}{v} &= A_{v,y} = -2\eta s^{3/2}(1-s)^{1/2}(1+4\eta s)^{-1/2}, \end{aligned}\right\} \tag{6.51}$$

where $A_{v,x}$ and $A_{v,y}$ are two velocity magnification factors, and

$$\eta = \zeta/y_\circ^2, \quad s = (1+\xi^2)^{-1}, \quad \xi = v_x t/y_\circ. \tag{6.52}$$

For $\eta \gg 10$ and for $0 \le \xi \le 1$ the velocity magnification factor $A_{v,x}$ can be as large as $A_{v,x} \approx 100$. Thus the velocity of the gravitating object is magnified by the gravitational lens by the different scales in different directions. The velocity components (v_{x1}, v_{y1}) of the source object and the image coordinates (x_1,y_1) are coupled by equation

$$v_{x1} = \frac{x_1}{t} + v_{y1}\frac{x_1}{y_1}. \tag{6.53}$$

The absolute value of the projected velocity $v_1 = \sqrt{v_{x1}^2 + v_{y1}^2}$ depends on the image coordinates (x_1, y_1) according to the equation

$$v_1 = \frac{x_1}{t}\left[1 - \frac{2x_1}{\zeta}\frac{(x_1 - vt)^2}{(x_1 - vt/2)} + \frac{x_1^2 + y_1^2}{\zeta^2}\frac{(x_1 - vt)^4}{(x_1 - vt/2)^4}\right]^{1/2}. \quad (6.54)$$

The second feature of the moving gravitational lens is the ability to convert the linear movement of the object source into the rotary movement of the image. This can be explained by the fact, that velocity magnification factors $A_{v,x}$ and $A_{v,y}$ along two mutually orthogonal directions are different. Due to this some part of the image can move with superlight apparent velocity. The effect of rotary phenomenon can be especially pronounced if we express the image trajectory in the polar coordinate system (r_1, θ_1):

$$r_1 = \frac{y_\circ}{2}\left\{[1 + (\tan\theta_1)^{-2}]^{1/2} + [1 + (\tan\theta_1)^{-2} + 4\zeta/y_\circ^2]^{1/2}\right\}, \quad (6.55)$$

where

$$\tan\theta_1 = y_\circ/v_x t. \quad (6.56)$$

We see that undistorted movement occurs only in the case of $\zeta = 0$.

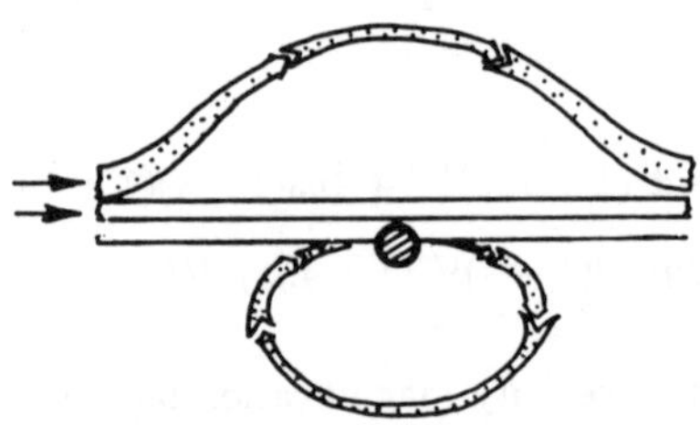

Fig. 6.6. Trajectory of the projected meso–optical images of the moving source.

The trajectory of the meso–optical images of the extended source star is shown in Fig. 6.6 for the case when the source object is moving from the left to the right along x-axis with velocity v_x and with eccentricity parameter $y_\circ$ relative to the lens-deflector D. The impression of an apparent rotation induced by the non-uniform acceleration is seen evidently.

The third feature of moving gravitational lens is the ability to convert the radial component of the source movement with respect to the deflector into a transversal component of the image movement. This novel astrophysical effect can be observed when the source-lens distance is changing along the line of sight with time derivation defined by the equation:

$$\frac{d\zeta}{dt} = 4\, r_g \left(\frac{D_{OL}}{D_{OS}}\right) \frac{dl_{OS}}{dt}. \tag{6.57}$$

The velocity components of the image v_{x1} and v_{y1}, given by Eq. (6.51), acquire additional terms [240]:

$$\left.\begin{aligned} \frac{\Delta v^r_{x1}}{v_r} &= \pm \,\text{const} \cdot \frac{\xi s}{y_\circ (1 + 4\,\zeta s)^{1/2}}, \\ \frac{\Delta v^r_{y1}}{v_r} &= \pm \,\text{const} \cdot \frac{s}{y_\circ (1 + 4\,\zeta s)^{1/2}}, \end{aligned}\right\} \tag{6.58}$$

with

$$v_r = \frac{dD_{OS}}{dt} \tag{6.59}$$

as the radial component of the source velocity. For small ξ and for large η the additional velocity terms, defined by Eq. (6.52), can be as large as v_{x1} or v_{y1}.

The observation of the image movement offers the means for measurement of distance and mass of the gravitating object in the case of very distant sources, when $D_{OS} \gg D_{OL}$, and also for prediction of the source position from the positions of the meso–optical images. Therefore, if we know the coordinates of two images $\vec{x}_1^{\,+}[x_1^+(t), y_1^+(t)]$ and $\vec{x}_1^{\,-}[x_1^-(t), y_1^-(t)]$, we may estimate

$$\left.\begin{aligned} x = v_x t &= x_1^+(t) + x_1^-(t), \\ y = y_\circ &= y_1^+(t) + y_1^-(t), \end{aligned}\right\} \tag{6.60}$$

Then we may estimate the strength parameter ζ, the distance D_{OL} and the mass M.

The quantity η can be deduced from the measurements of the relative intensities at different time moments. Independent estimations of the quantities v, ζ and η can be made from velocity measurements at four different time moments. All these new astrophysical effects can be used to solve the problem of massive black holes with masses of the order of $10^6 M_\odot$ in the nucleus of our Galaxy by observation of the deformation in the background sources pattern. The effective time scale τ of the corresponding velocity variations can be estimated from the relation [240]:

$$\tau = \frac{\theta_* D_{OL}}{v} = 500 \left(\frac{M}{10^6 M_\odot}\right)^{1/2} \left(\frac{D_{LS}}{D_{OS}}\right)^{1/2} \left(\frac{v}{100\,\text{km}\,\text{s}^{-1}}\right)^{-1} yr\,. \tag{6.61}$$

For example, the time scale $\tau \approx 15 y_r$ is expected for observation of the distant background sources behind the black hole of the mass $M = 10^8 M_\odot$ and of the velocities $v \approx 10^4$ km s^{-1}.

Now we list the features of the moving gravitational lens [240, 241]:

1) the disturbance of the central projection in the gravitational imaging and, as a result, the absence of the geometrical similarity between the source trajectory and the image trajectory,

2) the acceleration effect in the trajectory of the source image observed at constant source velocity,

3) the image velocity exceeds substantially the source velocity,

4) the conversion of the linear source movement into the rotating image movement,

5) the conversion of the radial source movement into the rotating image movement,

6) the redshift difference between multiple source images.

6.4. Examples of the Gravitational Lens

We describe some typical examples of the gravitational lenses discovered recently. The first one is observed as split images of a distant quasar. The twin quasar 0957+ 561 A, B [242 – 244] provide identical optical spectra [243, 244], and have angular separation 6.2″. Their absorption line velocities differ no more than 15 km s^{-1}. The redshift of these objects is equal to $z = 1.41$ which corresponds to the distance $2 \cdot 10^9$ l.y. The time delay between two images is equal to $\Delta t = 536 \pm 12$ days [248, 264]. The deflected galaxy is absent on the optical photograph and also on the radio map, recorded by Very Large Array (VLA) of a diameter 27 km with angular resolution 0.8″ [245]. Only recently a faint lensing galaxy has been detected by 5-m telescope equipped with charge-coupled device (CCD) as photodetector [246]. This matrix multielement photodetector gives no saturation effect, typical for traditional photographic plate, and offers high dynamic range. The image of this faint elliptic galaxy was 1″ north from the southern twin or 3″ south from a line connecting the undeflected quasar and observer.

The elliptic galaxy, acting as a nonsingular deflector of extended mass, produce odd number of secondary images, normally three. Hence a third component has to be detected. Meanwhile it was shown [247] that the deflector system includes not only a faint, elliptic galaxy, but also a cluster of galaxies with redshift $z = 0.39$. From the calculations based on the model of spheroidal transparent lens, it was predicted that B-image should be a double system with separation of the order 0.2″, and with comparable components brightnesses. Thus, a multiple image, produced by this complex gravitational lens, consists of three components. However this was not the end of a story. The subsequent calculations [247] have changed the image model [246]. It was shown that the brightness of third B-component is of the order of 1/44 of the brightness of the main B-component. The effect of the random field perturbation from stars of the galaxy has to be taken into account [248].

The expected time scale of these non-periodic variations of the B-2 component is equal to 100 yr.

The gravitational lens $PG\,1115+08\ A, B, C$ has three images, separated by comparable distances of the order of $3''$. The redshift of this distant quasar is $z = 1.73$. The optical spectra of all three components A, B and C were identical. Due to weak radio brightness, this gravitational lens could not be detected even by VLA radio system [249].

The next gravitational lens was a triple radio source $MG\,2016+112$, which has been detected by VLA radio telescope [246]. The components A and B are separated by $3.4''$, and the components C and B by $2.0''$. Three radio components A, B, and C had the same radio spectral indexes. The optical photo of this radio source, detected by CCD camera on 5-m telescope, can be explained with some ambiguities [250].

The structure of the radio source $MG\,1131+0456$ can be interpreted as a classical Einstein ring [251]. The radio map of this source was detected at frequency 5 GHz ($\lambda \approx 1.96$ cm). The gravitational image consists of an elliptic ring of low surface brightness with axes $2.2''$ and $1.6''$ and two more compact components: doublet $A1 + A2$, singlet component B and a weak source C. No emission sources can be seen at the center of this ring. The optical image of $MG\,1131+0456$, taken on 4-m telescope with CCD camera, reveals an elliptical object with axes $2.5''$ and $2.1''$. The absence of the emission lines in the spectrum of this object does not enable to estimate the redshift. The radio spectral index is equal to 0.9 ± 0.2, and the morphology of $MG\,1131+0456$ shows that this image is produced by gravitating object with elliptic gravitational potential.

The next example of the gravitational radio source is a system $MG\,1654+1346$ [251, 252] with two components separated by $7''$ in the radio map, detected at frequency 8.4 GHz ($\lambda \approx 3.5$ cm). This system represents a simple gravitational lens with single galaxy of moderate brightness as a lens (deflector). The redshift of a distant quasar is equal to $z = 1.75$ and deflector is much closer to the observer than to the quasar ($z = 0.254$). The first observed Einstein radio ring of the radius $0.98'' \pm 0.02''$ is circularly symmetric. From the polarized flux density distributions we may conclude that the source is composed of three components: a weakly polarized hotspot, a strongly polarized hotspot and a diffuse emission component with separations $1.9''$ and $2.0''$ between the first and the second components, and between the second and the third components, respectively. The enclosed mass of the lensing galaxy $M = (9.5 \pm 1.9)10^{10} M_{\odot}$. By adding the mass outside the images this yields total mass of the galaxy: $M_g = (2.97 \pm 0.32)10^{11} M_{\odot}$.

6.5. Gravitational Lens as a Beamsplitter

As was explained in Section 6.1, the image of the off-axis celestial object, producing by the gravitating object, consists of two virtual parts. This splitting of the object by the gravitational lens into two virtual images we can compare with analogous splitting of the wavefront observed in the coherent optics and holography. The most simple example is Fresnel biprism (Fig. 6.7). A spherical wavefront from the point source S of the monochromatic radiation with wavelength λ is split by Fresnel biprism into two spherical wavefronts issued from virtual images S_1 and S_2 of the primary source S. In the region of overlapping of these two spherical waves a system of Young interference fringes with period

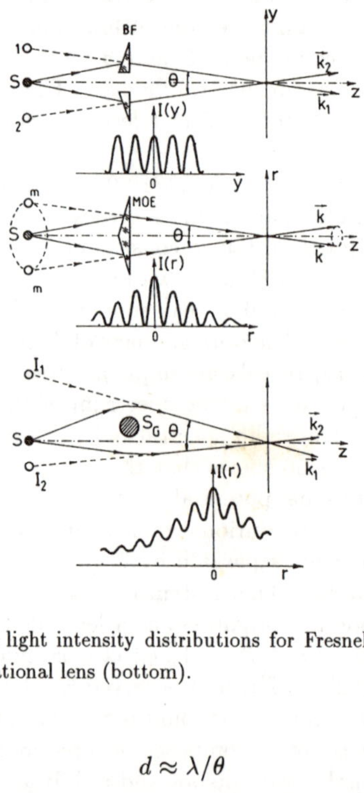

Fig. 6.7. Comparison of the light intensity distributions for Fresnel biprism (top), meso-optical element (middle), and gravitational lens (bottom).

$$d \approx \lambda/\theta \tag{6.62}$$

is generated. Here θ is an angular distance between two virtual images S_1 and S_2, viewed by the observer O. Two extreme cases are conceivable: 1) angle θ is greater than the angular resolution $\Delta\theta$ of the optical telescope, $\theta \gg \Delta\theta$, 2) angle θ is smaller than the angular resolution $\Delta\theta$ of this instrument, $\theta \ll \Delta\theta$. In the second case the observer cannot see separately two virtual images and they will be seen as one unresolved spot. Meanwhile the information about the angle θ can be withdrawn from the structure of the large scale interference pattern, detected by the observer as a function of time. For extremely small angle, $\theta \sim 10^{-15}$ rad, the fringe period d amounts to the macroscopic or even the cosmic scales. Just this

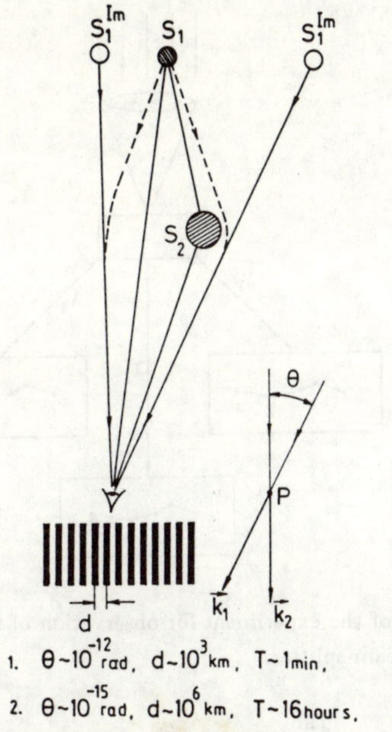

Fig. 6.8. The expected structure of the interference pattern.

interference pattern in space produces the gravitational lens (Fig. 6.8). To detect such an interference pattern it is sufficient to use a narrow spectral filter.

The idea of this experiment is shown in Fig. 6.9. The light from two unresolved virtual images, produced by the gravitational lens, is split into m narrow spectral components each of the width $\Delta\lambda_i$. Each color image is projected onto one of m $2D$ photodetectors such as CCD. Due to relative movement of the observer with respect to the axis of the gravitational lens the unresolved object spot is converted into quasi-periodic variable star. The effective period T_λ of this brightness variations is equal to

$$T_\lambda = \frac{d_\lambda}{v_t}, \tag{6.63}$$

where d_λ is the period of the interference fringes for m_λ-th filter and v_t is the relative transverse velocity.

Let us compare the time derivative of the single color light intensity variation with that of the total light intensity magnification $A(t)$ (Fig. 6.10). We have

$$\frac{dI_1}{dt} = I_\circ \frac{2\,\theta_\circ}{\phi_\circ}\frac{v}{p} = I_\circ \frac{v\,\theta_\circ\theta}{\phi_\circ\lambda}, \tag{6.64}$$

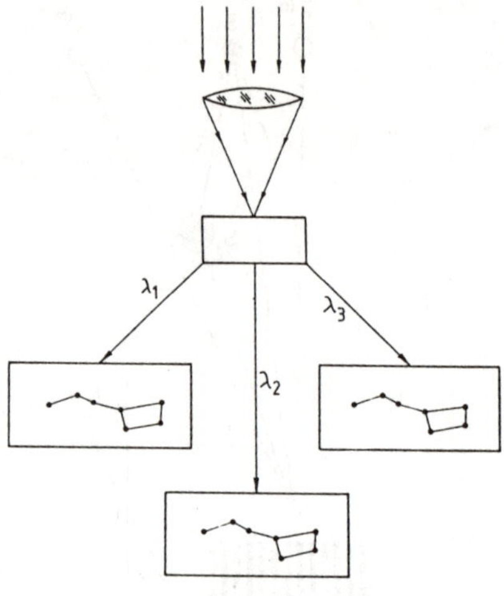

Fig. 6.9. The principal scheme of the experiment for observation of the interference effect induced by the gravitational lens as a beam-splitter.

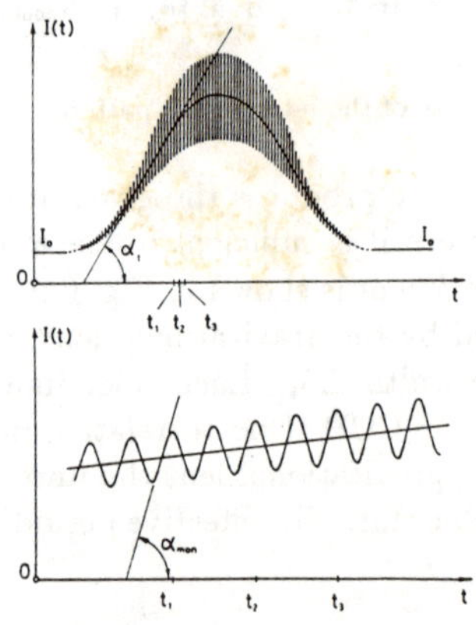

Fig. 6.10. The expected time variation of the light intensity in the event observed through a color window, and the total light curve.

and

$$\frac{dI_2}{dt} = I_\circ \frac{2\,\theta_\circ}{\phi_\circ} \frac{v}{\theta_\circ D_{OL}}, \tag{6.65}$$

where $I_\circ$ is the brightness of the source in the absence of the lens. The ratio of these two time derivatives is the factor of merit of the interferometric survey of the unresolved gravitational lenses. As $\theta \approx \theta_\circ$, we have

$$\kappa = \left(\frac{dI_1}{dt}\right) \Big/ \left(\frac{dI_2}{dt}\right) = \frac{\theta_\circ^2 D_{OL}}{\lambda} . \tag{6.66}$$

For $\theta_\circ = 10^{-10}$ rad, $D_{OL} = 10^{20}$ cm, $\lambda = 10^{-4}$ cm we get

$$\kappa = \frac{10^{-20} \cdot 10^{20}}{10^{-4}} = 10^4 : 1 . \tag{6.67}$$

Thus for a given threshold sensitivity the probability of detecting of the unresolved gravitational lens by means of the interferometric technique will be much higher than in traditional approach. The one color brightness time variations of the source can be detected as well in the form of moving modulation picture over the total radiation spectrum of the source [256]. Both these methods suit well for effective detection of the gravitational lens with angular splitting of the source object up to 10^{-6} arc.s.

It is worthwhile to recall that it was Fizeau [253] who made for the first time the indirect multiplex measurements in the optical spectroscopy. This was 18 years before the interferometer of A.Michelson and 49 years before the interferometer of H.Rubens and R.W.Wood [254]. By measuring the modulation period of the interference fringe contrast, Fizeau estimated the splitting value of the spectral doublet without any wavelength spectrometer. The spectral resolution of the Fizeau interferometer was better than that of the conventional prism spectrometer. The same multiplex approach is now used in the Fourier transform optical spectroscopy [254] and in some other physical devices [255].

If we take into account the magnification effect, then the interference pattern will be described by the equation [256]:

$$\frac{I}{I_\circ} = \frac{I_+ + I_-}{2} + 2\frac{\sqrt{I_+ I_-}}{I_\circ} \cos(2\pi dl/\lambda) , \tag{6.68}$$

where the length difference

$$dl = \frac{1}{2} \, \frac{D_{OS}}{D_{OL}(D_{OS} - D_{OL})} \, b_c^2 x \, \sqrt{x^2 + 4} \tag{6.69}$$

and

$$\left.\begin{aligned} I_\pm &= I_\circ |1 - b_\pm^{-4}|^{-1} , \\ b_c &= \beta_{\text{crit}} \, D_{OL} , \\ x &= p \, / b_c , \\ \beta_{\text{crit}} &= \left[\frac{4 G M (1 + z_1)(D_{OS} - D_{OL})}{c^2 D_{OL} D_{OS}}\right]^{1/2} , \end{aligned}\right\} \tag{6.70}$$

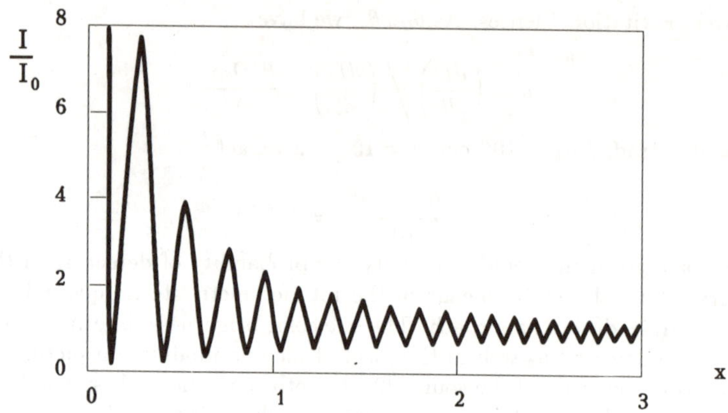

Fig. 6.11. The calculated interference pattern for a point-like source.

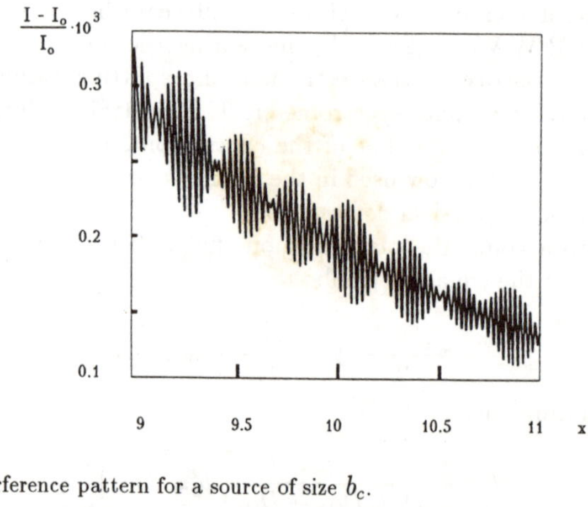

Fig. 6.12. The interference pattern for a source of size b_c.

with z_1 — redshift of the lensing object.

Fig. 6.11 demonstrates the calculated interference pattern for a point source: $M = 10^{30}\,\text{g}$, $\lambda = 100\,\text{cm}$, $z_1 = 0.26$, $z_2 = 1.0$.

When the source has finite size, such as quasar of $10^{15}\,\text{cm}$, then the contrast of the interferometric pattern will be < 1. The effect of finite source size with disk surface brightness variation was treated in [256]. An example of the interference pattern for a source of size b_c is shown in Fig. 6.12.

We may imagine one fantastic situation, when the birth of the Galactic supernova will be microlensed and the time delay between two neutrino bursts from supernova will be detected [257]. This would be the first experiment in which the gravitational lensing of the elementary particles will be detected. The nonzero neutrino mass, $m_\nu \neq 0$, will increase the time delay effect by a factor m_ν^2/E^2, where E is the total energy. Besides we may detect the magnification effect of the neutrino luminosity from distant supernova as the apparent discrepancy with optical data about this supernova. However the probability of such an event in our Galaxy is extremely small. Only extragalactic supernova may get the chance to be detected along the mentioned neutrino program.

6.6. Gravitational Microlensing

Now we return to the observation of the variable stellar objects and consider the effect of gravitational microlensing induced by the individual stars or by the isolated compact dark matter. As this effect gives unresolved secondary images, it superimposes on the main lensing effect of an entire galaxy [259, 263]. It manifests itself as unexpected burst of the apparent luminosity of the source object. The angular splitting of the microlensing effect of one star is of the order of 10^{-5} arc.s. The reason why the microlensing effect can be detected as noticeable intensity variation is that these variations are defined by the spatial derivative of the deflection angle in the deflector plane, but not by the deflection angle itself [259]. Therefore even one isolated star can induce the unresolved but detectable splitting of the source image.

The light rays in the gravitational microlensing are given in Fig. 6.13 for the case of large deflector G (galaxy) and small deflecting mass S (star). The light ray QDO from the light source Q to the observer O experiences the simultaneous influence of the galaxy G and of the star S. The light ray QSP_D is defined only by the galactic deflection. The light ray QSP_{DS} is defined by both the galactic and star deflection. The deflection by the galaxy G in the observation plane $\vec{x}(x, y)$ is a function of $\vec{\xi}(\xi, \eta)$. In the first order of approximation the mapping of the point $\vec{\xi}$ into the point $\vec{x}$ can be written as

$$\left.\begin{aligned} x &= \xi + \gamma\,\xi\,, \\ y &= \eta - \gamma\,\eta\,, \end{aligned}\right\} \tag{6.71}$$

where γ is a parameter which defines the galactic deflection. Such an approximation is valid for light rays which pass star S at distances ≤ 1 l.y. As was shown in [259], there are two or four meso–optical images of the source Q, the number of images depending on the observation point $\vec{x}(x, y)$ and on the location of the point locus, where the Jacobian of the mapping $\vec{x} \to \vec{\xi}$ is equal to zero:

$$J = \frac{\partial(x, y)}{\partial(\xi, \eta)} = \begin{vmatrix} \dfrac{\partial x}{\partial \xi} & \dfrac{\partial x}{\partial \eta} \\[2mm] \dfrac{\partial y}{\partial \xi} & \dfrac{\partial y}{\partial \eta} \end{vmatrix} = 0\,. \tag{6.72}$$

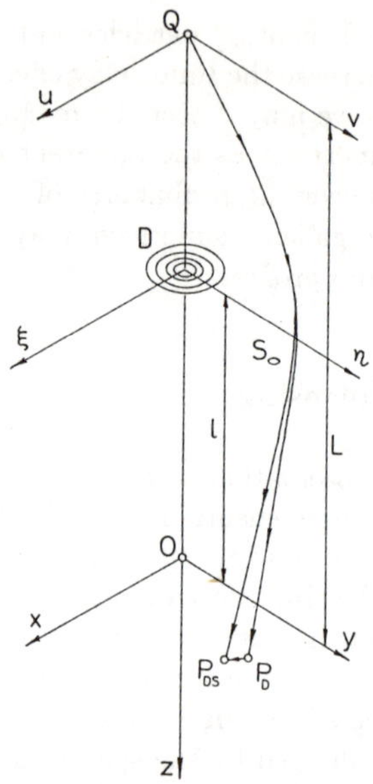

Fig. 6.13. Explanation of the gravitational microlensing event.

This equation defines the "critical curves" on the $\vec{\xi}$-plane. The rapid change of the apparent luminosity of the source occurs when the observer crosses one of the critical curves [260]. The brightness burst may be of the order of $\Delta m \approx 3\,\mathrm{mag}$.

The typical sequence of the meso–optical images is shown in Fig. 6.14. At the start time $t_\circ$ the observer sees two split meso–optical images of the source Q, which are moving along the velocity vectors marked by arrows. At the time moment $t = t^*$ a new bright image appears for $\xi > 0$ and immediately splits into two images, which are moving in the opposite directions. At the time moment $t = t^\dagger$ an annihilation of one of these two images, which is moving up, with one original image occurs.

A program to detect the dark objects in the mass range $10^{-7}M_\odot - 10^2 M_\odot$ of the deflector object has been announced [263]. The characteristic length scale of the microlensing and the deflection angle are

$$a_\circ \approx 5 \cdot 10^{16}\sqrt{M/M_\odot}\ \mathrm{cm}\,, \qquad \theta_\circ \approx 10^{-6}\sqrt{M/M_\odot}\ \mathrm{arc\ sec}\,. \tag{6.73}$$

For effective microlensing the angular size of the source object θ_S must be smaller than $\theta_\circ$. The expected time scale $t_m \approx 10\,\mathrm{yrs}$.

The high magnification event, which occurs when compact source Q crosses the critical curve, can be detected even at small displacements, $> 0.1\,\mathrm{AU}$, of the observer relative to the critical curve, in particular, as a result of the annular oscillations

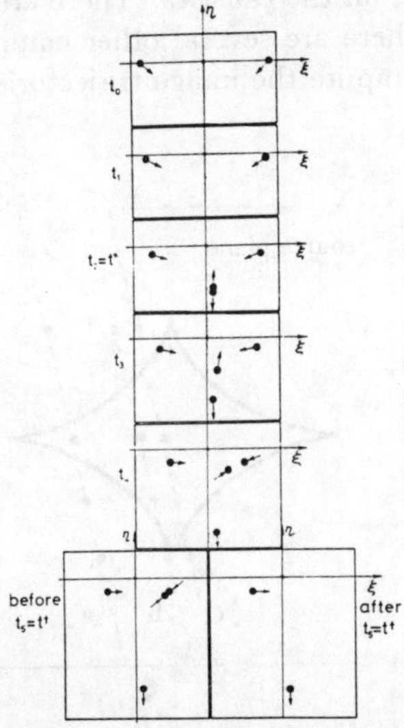

Fig. 6.14. A typical time sequence of the appearance and disappearence of the meso–optical images at the time moment, when the source star crosses the critical curve.

induced by the Earth orbital moving. In the case of a double star the displacement of the observer by only 6 km induces the brightness change up to 0.1 mag. However orbital velocity of the Earth ($< 50\,\mathrm{km\ s^{-1}}$) is smaller than the typical transverse velocity in the observer plane ($\sim 500\,\mathrm{km\ s^{-1}}$) and due to this we may neglect the effect of the Earth movement. The astrophysical applications of the gravitational microlensing are discussed in [258, 261, 262].

6.7. Caustics

Critical curve (caustics) separates the source plane into areas with different number of gravitational images. The typical caustics of a point mass lens has the shape of diamond. When the source trajectory crosses the caustics, a pair of lensed images appears or disappears. Each image generates its definite light curve. In the case of the source trajectory "*a*" we have four parts of image trajectories (Fig. 6.15). When we cross a critical curve in the lens plane for the first time, we are moving backwards in the source plane. There are four partial light curves which must be added to get a total light curve. When we cross the critical curve once more, we turn around the motion direction in the source plane again. In the case "*b*" two image trajectories are merged together into one image. A merging or separation of two image trajectories is taking place only when the source trajectory in the source plane is tangential to a point on the caustics. There are source trajectories which cross caustics four times. There are several other empirical features described in [442] which enable one to compute the image trajectories and the light curves for an array of point masses.

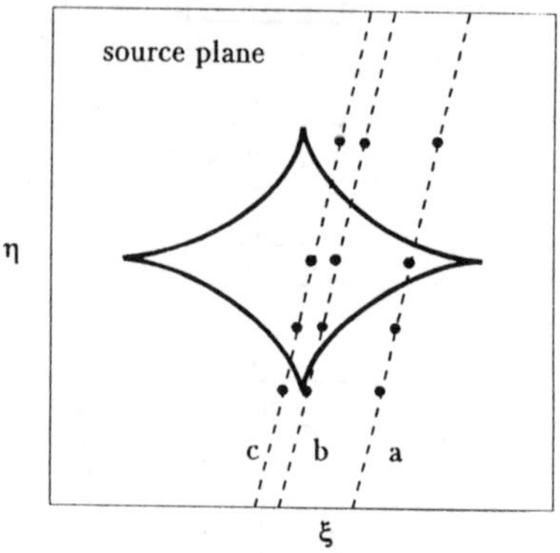

Fig. 6.15. The "diamond" caustics crossed by the three source trajectories, "a", "b", and "c".

When the critical curve is crossed by the compact source object, we observe very high spatial magnitude gradient, $\nabla m \approx 0.3\, m/AU$, and very fast brightness changing, $\dot{m} \approx 0.1\,\mathrm{day}^{-1}$. Thus, any small displacement of the observer induces detectable brightness changes, $\delta m \sim 0.^{m}025$ for the lensed quasar [260].

If the source has finite dimensions, the magnification factor of the event has more complex structure than for the point mass. In this case a phenomenon like eclipse is taking place [430, 444]. The angular source radius is equal to

$$\rho = \frac{R}{\theta_* D_{SO}}, \tag{6.74}$$

where R is the physical radius of the star, θ_* is the angular Einstein ring radius, and D_{OS} is the source–observer distance. The magnification of such an event is

$$A_{\rm fin}(x|\rho) = \frac{1}{2\pi\rho^2}\int\limits_0^{2\pi} d\theta \int\limits_0^{\rho} d\rho' \rho' A\left[(x+\rho'\cos\theta)^2 + (\rho'\sin\theta)^2\right] = A_\circ(x)\cdot B(z), \tag{6.75}$$

where $A_\circ(x)$ is the magnification factor for the point source as a function of $x = \theta_s/\theta_*$, and θ_s is the angular source–lens separation, and $B(z)$ — deviation factor, defined by equation:

$$B(z) = \frac{z^2}{\pi}\int\limits_0^{2\pi} d\theta \int\limits_0^{1/z} dq \cdot q\,(1+q^2+2\,q\cos\theta)^{-1/2}. \tag{6.76}$$

When the center of the Einstein ring is outside the star disk, we have $z \gg 1$, and $B(z) \to 1$. For $z \ll 1$ the deviation factor $B(z)$ manifests itself as a linear-like function. In this region we may estimate the angular Einstein ring radius θ_* of the lensed object and its transverse velocity v_T of the proper motion:

$$v_T = \omega\, \theta_*\, D_{OL} \cdot \tag{6.77}$$

For the source of finite dimensions the required sampling time interval,

$$\Delta t_{\text{fin}} \approx R\, D_{OL}/5\, D \cdot v_T\,, \tag{6.78}$$

corresponds to the transit time of the star source face, which equals to 700 s for Galactic objects and 2.5 hr for LMC objects.

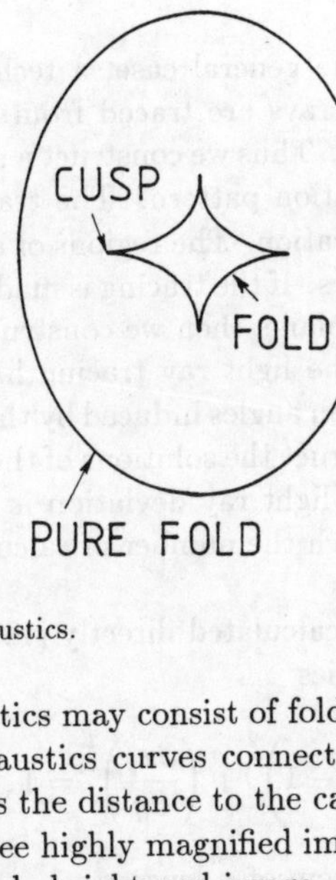

Fig. 6.16. General case of the caustics.

In the general case caustics may consist of folds (the proper caustic curves) and cusps (points where two caustics curves connect) (Fig. 6.16). The magnification factor $A \sim d^{-1/2}$, where d is the distance to the caustics. If the source is just inside of the cusp, we observe three highly magnified images. If the source is just outside the cusp, we observe a single brightened image. The structure of the caustics for elliptic lens is defined by the core radius of the lens. For small angular core radius we have four-cusps caustics as an inner part and fold caustics as an outer part. As the core radius increases, the inner caustics breaks through the outer caustics, and the outer caustics shrinks (Fig. 6.17). Two caustics penetrate each other, and separate again, forming two nested lips caustics each with two cusps. The inner lips caustics continues to shrink and then it disappears completely, leaving only a single lips caustics with regions, which define one or three images. And finally the remaining caustics disappears.

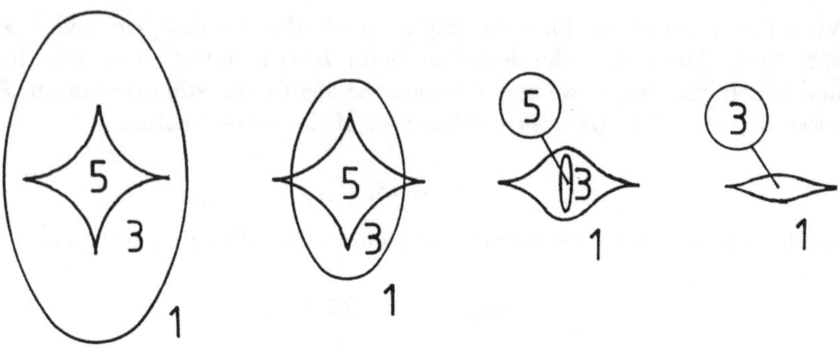

Fig. 6.17. Evolution of the caustics structure with increasing of the core radius, from the left to the right.

To construct a caustics in the general case, a technique of the backward ray tracing is used [430, 471]. Light rays are traced from a point source through the lens forward to the observer plane. Thus we construct a ray plot diagram which gives the distribution of the magnification pattern. The tracing ray density is directly proportional to the local magnification. The regions of extremely high ray densities indicate the position of the caustics. If the tracing is made from the observer through the lens backward to the source plane, then we construct anti-caustics [430].

The more power variant of the light ray tracing has been developed [475], in which the summing of the deviation angles induced by the individual stars is omitted and instead of this stage we construct the solutions of the Poisson equation and then $2D$ gravitational potential. The light ray deviation is calculated as a gradient of this potential. In such an approach the number of calculations does not depend on the number of gravitating objects.

The caustics network can be calculated directly [421]. For this aim we use the analytical expression of the caustics

$$\left(\frac{\partial x_s}{\partial \xi} - 1\right)^2 + \left(\frac{\partial x_s}{\partial \eta}\right)^2 = 1, \tag{6.79}$$

and caustics equation for the system of n lenses:

$$\left[\sum_{i=1}^{n} \frac{f_i^2}{d_{in}^2} \cos(2\,\theta_{in})\right]^2 + \left[\sum_{i=1}^{n} \frac{f_i^2}{d_{in}^2} \sin(2\,\theta_{in})^2\right]^2 = 1, \tag{6.80}$$

where $d_{ik} = P_i\, I_k$, $\theta = (P_i\, I_k\, , u)$, u — arbitrary axis. If the number of the independent caustics equations is equal to the number of the deflecting objects, the problem is soluble.

The decreasing of the magnification factor with the increasing of the angular dimension of the source can be used to estimate the angular dimension of the stars with errors of the order of small part of the Einstein ring radius. It was shown that in these conditions a weak chromatic effect may be detected by means of special spectrometric facility [469].

There are three different causes for chromatic effects in microlensing. The first cause is taking place when we have the lens object with comparable or even greater brightness than the background object [462]. These color-shifted microlensing events provide information about masses, distances to the background and lensing objects, and the transverse velocity of the lens. For this aim it is sufficient to measure the light curves of the color-shifted event in two or more spectral bands. The observed brightness in the λ-spectral band is equal to

$$I_\lambda = I_\lambda^f + A \cdot I_\lambda^b \,, \tag{6.81}$$

where "f" — foreground and "b" — background suffixes. The observed λ-magnification

$$\mathrm{A}_l(t) = \frac{I_\lambda^f + A \cdot I_\lambda^b}{I_\lambda^f + I_\lambda^b} = (1 - r_\lambda) + A(t)\, r_\lambda \,, \tag{6.82}$$

where $r_\lambda = I_\lambda^b/(I_\lambda^b + I_\lambda^f)$. If $A \gg 1$, the observed color will be the color of the background object. The time scale of the color-shifted event is smaller than the time scale of the achromatic event. Color-shifted events, induced by this cause, are rare compared with the achromatic events, because the mass function is a decreasing function of mass.

The second achromatic effect in microlensing is induced by the finite dimension of source object [469]. We observe differential magnification of the lens star of nonzero size [447], and a shifting of the spectral line centroid during the lensing event. The observed centroid as a function of time enables one to estimate the Einstein ring radius and proper motion of the lensed object.

In contrast to the rotation of the star, which induces the broadening of spectral line profile the different parts of a finite stellar disk are subjected to the magnification which are varying during the event [447]. The averaged spectral line profile is equal to

$$f(t, v) = C \int_{-\sqrt{\rho^2 - z_1^2}}^{\sqrt{\rho^2 - z_1^2}} dz \cdot S(z)\, A[\,|\vec{x}(t) + \vec{z}|\,] \,, \tag{6.83}$$

where $A(\bullet)$ is the magnification factor, with $x(t) = \sqrt{\beta^2 + (\omega t)^2}$ and

$$v = \frac{z_1}{\rho}\, v_{\mathrm{rot}} + v_\circ \,, \tag{6.84}$$

where ρ is the angular stellar radius, v_{rot} is the projected rotation velocity of the star, and $v_\circ$ is the velocity of the stellar center of mass. The axis z is aligned along the projection of the stellar rotation axis. The shift of the spectral line is expressed in the spectral line centroid, which is invariant to the intrinsic rotation broadening and which is defined by the apparent radial velocity of star $\langle v(t)\rangle$. For $\rho \ll x \ll 1$ and $S(z) = const$, we have

$$f(t,v) = C\frac{\rho}{v_{\rm rot}}\sqrt{v_{\rm rot}^2 - (v-v_\circ)^2}\left[1 + \frac{v-v_\circ}{v_{\rm rot}}\,\rho\,\frac{\beta\cos\alpha + \omega t\sin\alpha}{\beta^2 + (\omega t)^2}\right], \qquad (6.85)$$

where

$$\langle v(t)\rangle \;=\; v_\circ + \frac{\xi}{4}\,v_{\rm rot}\,\rho\,\frac{\beta\cos\alpha + \omega t\sin\alpha}{\beta^2 + (\omega t)^2}, \qquad (6.86)$$

α is the angle between the lens velocity vector $d\vec{x}/dt$ and the projected rotation axis of the star. The shift of the spectral line centroid

$$\delta\langle v(t)\rangle \;=\; \langle v(t)\rangle - v_\circ, \qquad (6.87)$$

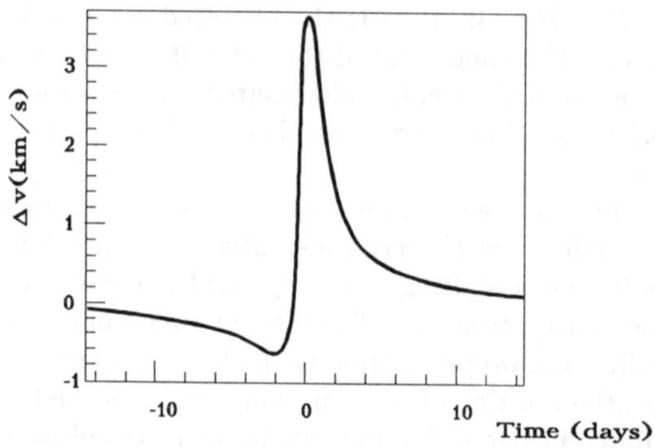

Fig. 6.18. The shift of the spectroscopic line centroid $\Delta v({\rm km\,s^{-1}})$ as a function of time t for $\rho = 0.01$; $\alpha = 45°, \beta = 0.05$.

is expressed in km s^{-1}. We may conclude that this spectroscopic signature (Fig.6.18) of the lens event can be considered as an additional feature of the lens event against to the intrinsic star variability. From the time dependence of the chromatic line shift we may estimate the Einstein ring radius.

The third chromatic effect in the gravitational lensing could be observed for a quasar as a source. The cause of this chromatic effect is that a quasar may have different sizes in different wavelengths. When a quasar crosses a caustics, the color change in the light curve may occur [431]. The time scale of such color event is very short, about $\sim 1/5$ of the magnitude event time scale. For small sources this chromatic effect is negligible.

The structure features of the gravitational images produced in the vicinity of the caustics cusp point have been treated in [473]. The analytical solution of the lens equation is given for this region. The mutual interference coherence of the lensed images has been calculated [474].

To get the parametric representation of the caustics it is useful to use complex quantities [449].

6.8. Parallax Effect

Two observers at two different positions see different light curves. Thus they may detect a microlensing parallax effect [260], and get additional astronomical information [426, 444, 445, 461]. Two observers have projected separation vector $\vec{r}$. The time difference Δt between maximum magnifications in two light curves is to be measured. To describe the lensing parallax effect a difference between two impact parameters Δp must be introduced.

Each light curve is describes by the magnification factor

$$A(x) = \frac{x^2+2}{[x(x^2+4)]^{1/2}}, \tag{6.88}$$

where

$$x(t) = [\omega^2(t-t_\circ)^2 + \tilde{p}^2]^{1/2}, \tag{6.89}$$

and $\tilde{p}$ is the impact parameter in units of the Einstein ring radius R_E. The $2D$ displacement vector is

$$\vec{\Delta x} = (\omega \Delta t, \Delta \tilde{p}) = \frac{\vec{r}}{\tilde{R}_E}. \tag{6.90}$$

The time scale ω^{-1} of the microlensing event is described by three physical parameters M, D_{OS} and v:

$$\omega^{-1} = \frac{[4\,G\,M\,D_{OS}\,d\,(1-d)]^{1/2}}{v\,c}, \tag{6.91}$$

with $d = D_{OL}/D_{OS}$, v is a transitive velocity of the lensing object relative to the observer–source line of sight, $\tilde{R}_E$ is the reduced Einstein ring radius

$$\tilde{R}_E = \sqrt{\frac{4\,G\,D_{OS}}{c^2}\,\frac{M\,d}{1-d}}. \tag{6.92}$$

The difference in the impact parameter Δp

$$(\Delta p_\circ)_\pm \; = \; |p' \pm p| \,, \tag{6.93}$$

leads to the ambiguity in the coordinate x:

$$\Delta x_\pm \; = \; \{[\omega \Delta t]^2 + [(\Delta p_\circ)_\pm]^2\}^{1/2} \,, \tag{6.94}$$

and in the reduced velocity $\tilde{v} = v/(1-d)$

$$\tilde{v}_\pm \; = \; \frac{\omega \, r}{\Delta x_\pm} \,. \tag{6.95}$$

The angle Φ, which defines the transverse orientation of the separation vector $\vec{r}$, assumes also two possible values:

$$\Phi = \pm \tan^{-1}[\Delta p/\omega \, \Delta t] \,. \tag{6.96}$$

By measuring the difference in time scales for two observers, $\Delta\omega = \omega' - \omega$, we can break the degeneracy in Δp.

The projected separation vector $\vec{r}$ and the projected velocity vector $\vec{u}$ are defined by $3D$ separation vector $\vec{R}$ between two observers, by $3D$ relative velocity vector $\vec{U}$ between two observers, and direction source vector $\hat{\vec{s}}$:

$$\left. \begin{aligned} \vec{r} &= \vec{R} - (\vec{R} \cdot \hat{\vec{s}}) \, \hat{\vec{s}} \,, \\ \vec{u} &= \vec{U} - (\vec{U} \cdot \hat{\vec{s}}) \, \hat{\vec{s}} \,. \end{aligned} \right\} \tag{6.97}$$

For small velocity u relative to the lens velocity $\hat{v}$, $u \ll \hat{v}$, the relative difference between inverse time scales $\Delta\omega$ is equal to

$$\frac{\Delta\omega}{\omega} = \frac{\hat{v}_{||} \, u_{||} + \hat{v}_\perp \, u_\perp}{\hat{v}^2} \,, \tag{6.98}$$

where

$$\left. \begin{aligned} & u_{||} = \frac{1}{r} \, (\vec{r}, \vec{u}) \,, \quad u_\perp = \frac{1}{r} \, (\vec{r} \times \vec{u}) \cdot \hat{\vec{s}} \,, \\ & \tilde{v}_{||} = \frac{r \, \omega^2 \, \Delta t}{(\Delta x)^2} \,, \quad \tilde{v}_\perp = \frac{r \, \omega \, \Delta p}{(\Delta x)^2} \,, \\ & \tilde{v} = \frac{r \, \omega}{\Delta x} \,. \end{aligned} \right\} \tag{6.99}$$

With very useful relation

$$u_\perp \Delta p = r \, \Delta\omega - u_{||} \, \omega \, \Delta t \,, \tag{6.100}$$

we may estimate unambiguous value of Δp.

By means of the parallax effect [260, 463] we may estimate the transverse velocity of galaxy in the event quasar plus star in the galaxy disk with error $\sim 60\,\mathrm{km\ s^{-1}}$.

For this aim we must lunch a satellite into the solar Neptune-like orbit ($\sim 30\,\mathrm{AU}$). The velocity error in the parallax effect measurements is defined by the velocity dispersion of the order of $40\,\mathrm{km\ s^{-1}}$ and does not depend on distance. So we have no systematic errors in the distance determination. By means of this technique we may estimate the Hubble's parameter with error ~ 5 per cent.

The parallax effect in the gravitational microlensing can be used for estimation of the relative velocity v_T. For this aim we need three observers A, B and C [260] (Fig. 6.19). In this experiment we have two time lags δt_{AB} and δt_{AC}, two position differences r_{AB} and r_{AC}, and angle $\alpha = \widehat{CAB}$. The transverse velocity v_T is equal to:

$$v_T = |\sin\gamma| \left\{ \left(\frac{\delta t_{AB}}{r_{AB}}\right)^2 + \left(\frac{\delta t_{AC}}{r_{AC}}\right)^2 - 2\cos\alpha\,\frac{\delta t_{AB}\cdot\delta t_{AC}}{r_{AB}\cdot r_{AC}} \right\}^{-1/2} . \tag{6.101}$$

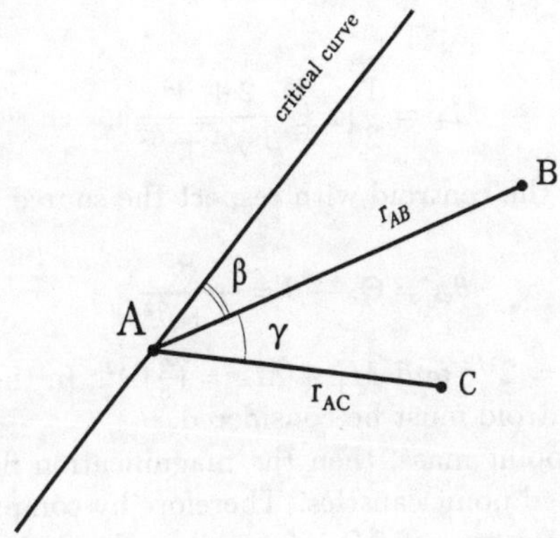

Fig. 6.19. Position of three observers, projected onto the observer plane, in the complete parallax experiment.

In the case of two observers, A and B, we may estimate only an upper limit of v_T:

$$v_T = r_{AB}\sin\beta \,/\, \delta t_{AB} \le r_{AB} \,/\, \delta t_{AB} . \tag{6.102}$$

The annual motion of the Earth induces annual oscillation in the light curve of the high magnification event. The orbital Earth motion induces an apparent motion of the Einstein ring with velocity [426, 445]:

$$\vec{v}_{\rm induced} = \left(\frac{D_{LS}}{D_{OS}}\right)\vec{v}_\oplus , \quad v_\oplus = 30\,\mathrm{km\ s^{-1}} , \tag{6.103}$$

and of radius

$$r_{\rm induced} = \left(\frac{D_{LS}}{D_{OS}}\right)\vec{a}_\oplus , \quad a_\oplus = 1\,\mathrm{AU} , \tag{6.104}$$

Thus we may estimate the reduced transverse velocity $\tilde{\vec{v}}_T$.

The Einstein ring radius may be estimated from apparent image motions even in conditions, when the ring structure of the gravitational image cannot be resolved [418]. For this aim we must measure the centroid of the whole image, the angular position of this centroid is defined by the equation

$$\theta_c = \frac{M_+\,\theta_+ + M_-\,\theta_-}{M_+ + M_-}\,, \tag{6.105}$$

where the angular locations of the gravitational images are equal to

$$\Theta_\pm = \frac{1}{2}\,[\beta \pm \sqrt{4+\beta^2}] \tag{6.106}$$

where β is the dimensionless angular parameter, and magnification factors M_+ and M_- are equal to

$$M_\pm = \frac{1}{2}\,[1 \pm \frac{2+\beta^2}{\beta\,\sqrt{4+\beta^2}}]\,. \tag{6.107}$$

The angular location of the centroid with respect the source, θ_{cs}, is equal to

$$\theta_{cs} = \Theta_c - \beta = \frac{\beta}{2+\beta^2}\,, \tag{6.108}$$

with $\theta_{cs}^{\max} = 8^{-1/2}$ for $\beta = 2^{1/2}$ and $M_+ + M_- = (\frac{2}{3})^{1/2}$. In the case of an extended source, the weighted centroid must be considered.

If we have isolated point mass, then the magnification field is circularly symmetric around the isolated point caustics. Therefore by comparing two light curves in the parallax experiment we get $2D$ information about the lens motion relative to the separation vector of two observers [426, 445, 460]. Meanwhile in the case of highly magnified quasar field we have a linear caustics, and information about relative motion will be one-dimensional [260]. This difference in topologies of the caustics structure is of principal value.

6.9. Optical Depth

The probability of microlensing or the **optical depth** of the gravitational microlensing is defined as [425]

$$\tau = \int_0^{D_{OS}} \frac{4\,\pi\,G\,D}{c^2}\,\rho_d\,dD_d\,, \tag{6.109}$$

where ρ_d is the mass density. Fig. 6.20 shows the contributions to the optical depth of microlensing by the Galactic disk stars, located at various distances from the

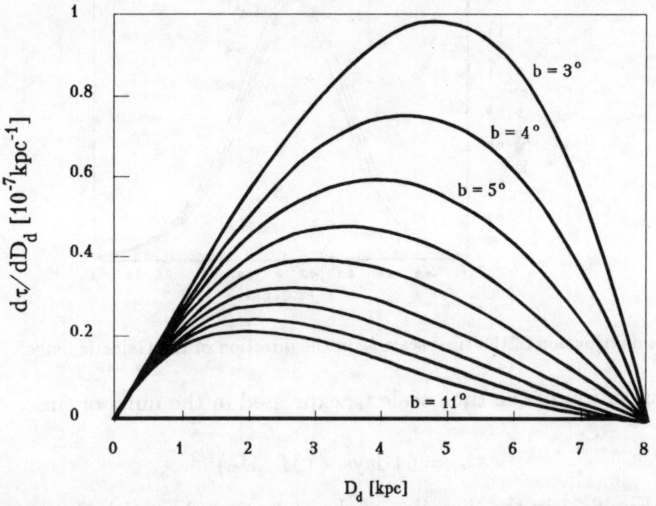

Fig. 6.20. Contribution to the optical depth of the microlensing events by the Galactic disk stars at various distances from the observer D_{OL} toward the Galactic longitude $l = 0°$ at various Galactic latitudes "b".

observer, D_d, along the lines directed toward the Galactic longitudes b, from $b = 3°$ to $b = 11°$. If stars of mass M are distributed along the line of sight in proportion to the Galactic **disk** density, and have random velocity with *rms* values

$$\langle V_{dl}^2 \rangle^{1/2} = 21\,\text{km s}^{-1}, \qquad \langle V_{db}^2 \rangle^{1/2} = 17\,\text{km s}^{-1}, \tag{6.110}$$

and the Galactic **bulge** stars have random velocity with *rms* values

$$\langle V_{bl}^2 \rangle^{1/2} = 100\,\text{km s}^{-1}, \qquad \langle V_{bb}^2 \rangle^{1/2} = 120\,\text{km s}^{-1}, \tag{6.111}$$

then the relative proper motion of these stars may be written as

$$\frac{\dot{\varphi}}{\dot{\varphi}_m} = [(1 + v_{dl}\frac{D_s}{D_d} + v_{bl})^2 + (v_{db}\frac{D_s}{D_d} + v_{bb})^2]^{1/2}, \tag{6.112}$$

with

$$\left.\begin{aligned} v_{dl} &= V_{dl}/v_{rot}, \quad v_{db} = V_{db}/v_{rot}, \\ v_{bl} &= V_{bl}/v_{rot}, \quad v_{bb} = V_{bb}/v_{rot}, \end{aligned}\right\} \tag{6.113}$$

where

$$\dot{\varphi}_m = V_{rot}/D_s, \quad V_{rot} = 220\,\text{km s}^{-1}. \tag{6.114}$$

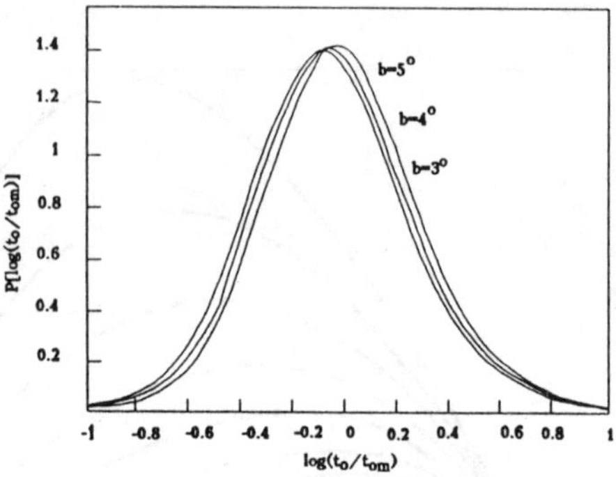

Fig. 6.21. The distribution of the time scale $t_\circ$ in the direction of the Galactic bulge.

The distribution of the time scale $t_\circ$, expressed in the units of time,

$$t_{om} = 64\,\text{days} \times (M\,/M_\odot)^{1/2} \tag{6.115}$$

is shown in Fig. 6.21 in the direction of the Galactic bulge, with Galactic longitude $l = 0°$ for three values of Galactic latitude $b = 3°, 4°, 5°$. The average time interval between microlensing events with time scale $t_\circ$ and with impact parameter p, which is smaller than Einstein ring radius, $p < R_E$, is equal to

$$\langle \Delta t \rangle = \frac{t_\circ}{\tau}\frac{\pi}{2}\,. \tag{6.116}$$

The average number of events per 10^6 Galactic bulge stars per year is equal to

$$N_{10^6,\,yr.} = 10^6\,\tau\,\frac{1\,yr}{t_{om}}\,\frac{2}{\pi}\,\langle\frac{t_{om}}{t_\circ}\rangle\,, \tag{6.117}$$

with $\langle t_{om}/t_\circ \rangle = 1.47$.

In the real situation we must take into account the mass distribution function $n(M)dM \sim M^{-2}dM$ and $d\tau \sim M^{-1}dM$. The number of the microlensing events per year per 10^6 Galactic bulge stars as a function of the event time scale $t_\circ$ is shown in Fig. 6.22 for the mass range $0.1 - 1.0\,M_\odot$ and in Fig. 6.23 for the mass range $0.01 - 1.0\,M_\odot$.

The frequency of the gravitational microlensing event [472] is defined as

$$\Gamma = \langle \Delta t \rangle^{-1} = \frac{2\,\tau}{\pi\,t_\circ} = \frac{2\,v\,\tau}{\pi}\,\left[\frac{4\,G\,M}{c^2}\,D\right]^{-1/2}\,. \tag{6.118}$$

The transverse lens velocity is equal to

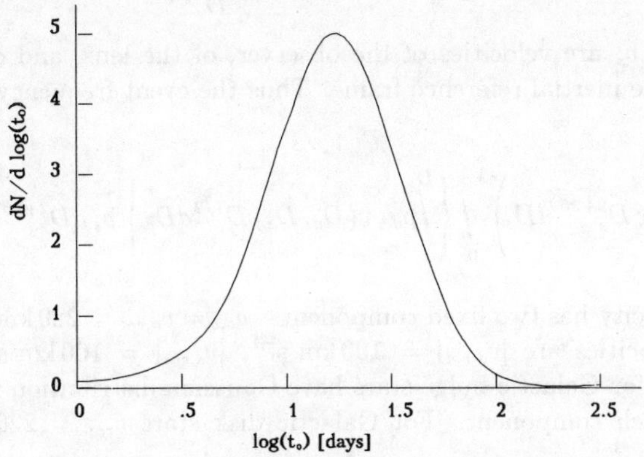

Fig. 6.22. The number of microlensing events per year per 10^6 Galactic bulge stars as a function of the time scale t_o (days) for the lens mass range $0.1 - 1.0M_{\odot}$.

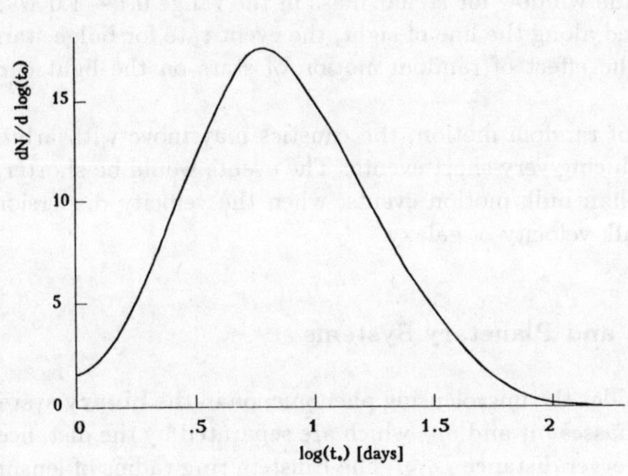

Fig. 6.23. The number of microlensing events per year per 10^6 Galactic bulge stars as a function of the time scale t_o (days) for the lens mass range $0.01 - 1.0M_{\odot}$.

$$v = (v_b^2 + v_l^2)^{1/2} \tag{6.119}$$

with velocity components v_b and v_l along the Galactic coordinates b and l. Each component can be calculated according to the equation

$$v_{b,l} = [v_d - v_\circ + (v_\circ - v_s)\,\frac{D_d}{D_s}]_{b,l}\,, \tag{6.120}$$

where $v_\circ$, v_d and v_s are velocities of the observer, of the lens, and of the source, respectively, in the inertial reference frame. Thus the event frequency

$$\Gamma = \frac{4\,G^{1/2}}{c\,M^{1/2}}\left(\int\limits_0^\infty \rho_{s,b}\,D_s^{2+2\beta}\,dD_s\right)^{-1}\int\limits_0^\infty\left\{\int\limits_0^{D_s}\rho_{d,b}\,v(D_d,D_s)\,D^{1/2}dD_d\right\}\rho_{s,b}\,D^{2+2\beta}dD_s. \tag{6.121}$$

The observer velocity has two fixed components: $v_{\circ,l} = v_{\circ,rot} = 220\,\mathrm{km\ s^{-1}}, v_{\circ,b} = 0$. The rotation velocities are $|v_{d,rot}| = 100\,\mathrm{km\ s^{-1}}$, $|v_{s,rot}| = 100\,\mathrm{km\ s^{-1}}$. All random components for Galactic bulge stars have Gaussian distribution with variance $100\,\mathrm{km\ s^{-1}}$ for each component. For Galactic disk stars $v_{rot} = 220\,\mathrm{km\ s^{-1}}$, and dispersion $20\,\mathrm{km\ s^{-1}}$.

The event rate for Galactic bulge stars is higher than for Galactic disk stars. The disk lenses may dominate only for long events, while the bulge lenses dominate at short time scales. The distribution of the microlensing time scales has maximum at $t_\circ \approx 10$ days, and the total event rate is equal to $\Gamma \approx 12$ per year per 10^6 bulge stars in the Baade window for stellar mass in the range $0.1 - 1.0\,M_\odot$. As Galactic bulge is elongated along the line of sight, the event rate for bulge stars may be still higher [472]. The effect of random motion of stars on the light curves has been studied in [471].

In the case of random motion, the caustics may move with arbitrary high velocity, thus producing very short events. The events would be shorter, steeper, and more frequent than bulk motion events, when the velocity dispersion of the stars equals to the bulk velocity of galaxy.

6.10. Binaries and Planetary Systems

Now we consider the microlensing phenomenon in the **binary** system, consisting of two stars of masses m_1 and m_2, which are separated by the distance $r(t)$, smaller than source–observer distance D_{OS}. The Einstein ring radius of lensing m_2 is equal to [465]

$$R_{E,1}(t) = \frac{[4\,G\,m_1\,r(t)]^{1/2}}{c}\,. \tag{6.122}$$

The corresponding optical depth τ_1, averaged over the orbit, is equal to

$$\tau_1 = \frac{m_1}{M}\,\frac{\langle v^2\rangle}{c^2}\,, \tag{6.123}$$

where the total mass of the binary system $M = m_1 + m_2$, and $\langle v^2 \rangle$ is the mean square relative velocity. The optical depth τ_2 for lensing m_1 is equal to

$$\tau_2 = \frac{m_2}{M} \frac{\langle v^2 \rangle}{c^2}, \tag{6.124}$$

and the total optical depth

$$\tau = \tau_1 + \tau_2 = \frac{\langle v^2 \rangle}{c^2}. \tag{6.125}$$

The maximum magnification of the event for circular orbit and for finite radii ρ_1 and ρ_2 of two stars is equal to

$$A_{\max} \approx 1 + 2\, R_{E,1}^2 / \rho_2^2. \tag{6.126}$$

The majority of the Galactic disk stars are the binary systems with periods ranging from 1 day to 10^9 days. These periods are distributed uniformly along the axis $\log P_{bin}$. The microlensing of the Galactic bulge stars by the Galactic disk binaries has been treated with the aim to search for **planetary** systems [422, 452, 456]. As the planet velocity is of the order 10 km s^{-1} and the source velocity relative to the planetary system is of the order of 250 km s^{-1}, the position of the planet may be accepted as fixed during the event. The projected position of the planet,

$$x_p = r / D_{OL}\, \theta_*(z), \tag{6.127}$$

has the cumulative distribution

$$F(x_p|z) = \left[1 - 4\, z\, (1 - z)\, \frac{x_p^2}{x_*^2}\right]^{1/2}, \tag{6.128}$$

where x_* is the orbital radius parameter

$$x_* = a\, c / (G\, M\, D_{OS})^{1/2}. \tag{6.129}$$

The probability distribution $p(x_p)$ is strongly peaked near $x_p = x_*$ [452].

The geometry of the star plus planet lens system is shown in Fig. 6.24. The typical light curve for planetary lens system for $m = 0.001M$ with $M = M_\odot$ is given in Fig. 6.25. The magnification factor is very high, and this can be explained by the coincidence of the Einstein radius of the Sun with Jupiter orbital radius. This coincidence as a typical one for any planetary system. The planet signal on the general light curve is lasting about 1 day or less. The triggering program is now in progress to search for planetary system. The author hopes that the reader of the book MESO–OPTICS would probably be acquainted with the "second Jupiter" in our Galaxy detected by means of the microlensing effect.

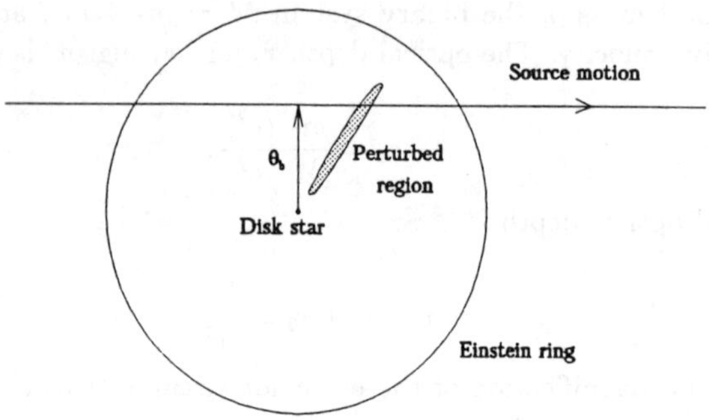

Fig. 6.24. Geometry of the star plus planet lens system.

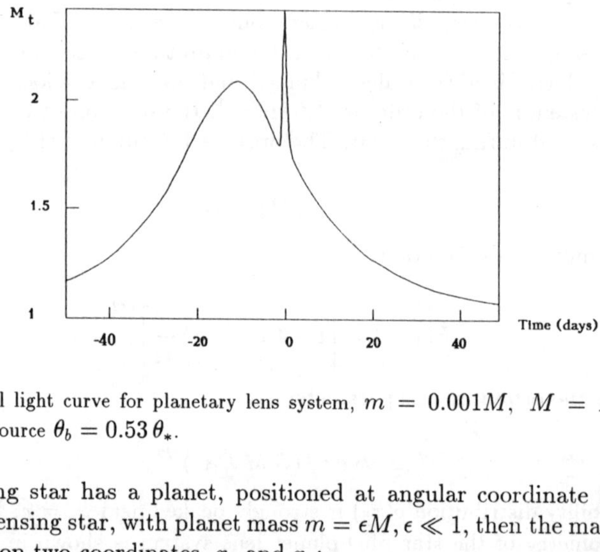

Fig. 6.25. Typical light curve for planetary lens system, $m = 0.001M$, $M = M_\odot$, impact parameter of the source $\theta_b = 0.53\,\theta_*$.

If the lensing star has a planet, positioned at angular coordinate $\vec{\theta}_p = \vec{x}_p\,\theta_*$ relative to the lensing star, with planet mass $m = \epsilon M, \epsilon \ll 1$, then the magnification factor depends on two coordinates, x_s and x_p:

$$A(x_s, x_p) = (1+\delta)a(x_z)\,, \tag{6.130}$$

where δ is the excess magnification induced by the planet. If the planet moves near one of the lensed images, it perturbs the light curve strongly. This planetary perturbation must be added to the initial time delay, given by Eq. (6.43). The time delay between two light rays for the planetary system will be equal to

$$d\tau_p = \frac{1}{2}(\vec{x}_i - \vec{x}_s)^2 - \ln|\vec{x}_i| - \epsilon \ln|\vec{x}_i - \vec{x}_p| . \qquad (6.131)$$

If we introduce the coordinate of the unperturbed image $\vec{x}_{io}$ $(\epsilon \to 0)$, and relative coordinates

$$\left.\begin{aligned} \vec{\xi}_i &= \vec{x}_i - \vec{x}_{io} , \\ \vec{\xi}_p &= \vec{x}_p - \vec{x}_{io} , \end{aligned}\right\} \qquad (6.132)$$

then the time delay for $\epsilon \ll 1$ assumes the form:

$$\tau = \frac{1}{2}[(1+\gamma)\xi_i^2 + (1-\gamma)\eta_i^2] - \frac{\epsilon}{2}\ln[(\xi_i - \xi_p)^2 + (\eta_i - \eta_p)^2] , \qquad (6.133)$$

with $\gamma = (x_{io})^{-2}$. The solution of the algebraic equation of four degrees for ξ_i values is given in [452].

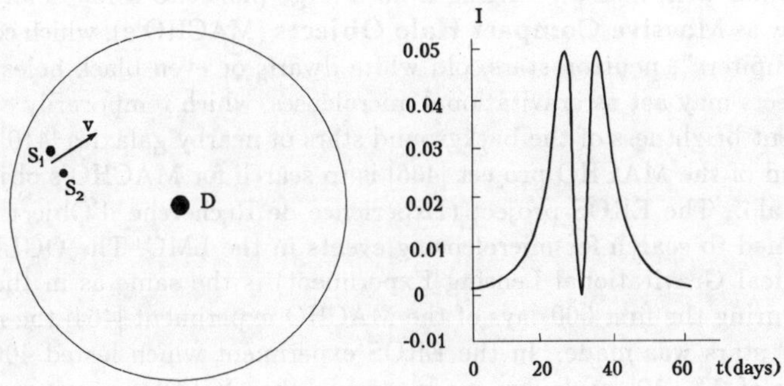

Fig. 6.26. Geometry of the binary source gravitational lens system.

In contrast to [422], where the binary system was the gravitational lens, the binary system treated in [468] was the **source** object (Fig. 6.26). As both components of the planetary source object are moving along non-straight trajectories, the light curve in this case would be strongly non-symmetrical, and the color effect would be present.

The partial magnification of the event for binary source system with circle orbit is equal to

$$A_i = \frac{1}{2}(u_i + 1/u_i) , \qquad (6.134)$$

where

$$u_i = \sqrt{1 + 4\,\theta_o^2\,\theta_i^{-2}} , \qquad (6.135)$$

and θ_i ($i = 1,2$) being the angular vector of the i-component. The magnification of the whole binary source is defined by equation

$$A(t) = I_1\, A_1(t) + I_2\, A_2(t)\,, \tag{6.136}$$

and the color $(R - B)$ shift is equal to

$$R(t) - R_\circ = \frac{(I_1^r\, I_2^b - I_2^r\, I_1^b)\,[A_1(t)A_2(0) - A_2(t)A_1(0)]}{[I_1^b\, A_1(t) + I_2^b\, A_2(t)]\,[I_1^b\, A_1(0) + I_2^b\, A_2(0)]}\,. \tag{6.137}$$

6.11. Applications

6.11.1. Problem of the dark halo objects

The bodies with masses, ranging from a large planet to a few solar masses, is termed now as **Massive Compact Halo Objects** (MACHO's), which cover brown dwarfs, "Jupiters", neutron stars, old white dwarfs or even black holes [448, 432]. These objects may act as gravitational microlenses, which temporarily can amplify the apparent brightness of the background stars of nearby galaxies [416].

The aim of the MACHO project [436] is to search for MACHO's objects in the Galactic halo. The EROS project (Expérience de Recherche d'Objects Sombres) [437] is aimed to search for microlensing events in the LMC. The OGLE program [459] (Optical Gravitational Lensing Experiment) is the same as in the MACHO project. During the first 300 days of the MACHO experiment [464] the monitoring of $2.6 \cdot 10^6$ stars was made. In the EROS experiment which lasted 400 days, the monitoring of $2.8 \cdot 10^6$ stars was made, and in the OGLE experiment there were monitored $1.1 \cdot 10^6$ stars during 200 days. All these experiments were performed with electronic photodetector CCD, which gave total 10^8 photon measurements. A complex computer processing system of the primary data was used [425].

There were detected **three** events [436] with maximum magnification $A_{\max} = 6.8, 3.3$, and 2.5, and with time scales $\omega^{-1} = 17, 30$, and 26 days. The events were achromatic and the light curves fit well the expected three parameter expression (Eq. 6.88). The observed optical depth to microlensing estimated on **nine** microlensing events and equal to $\tau = (3.3 \pm 1.2) \cdot 10^{-6}$ [453], surpasses the theoretical predicted value $\tau_\circ \sim (0.5 - 1.0) \cdot 10^{-6}$ [425]. This discrepancy cannot be explained by any model of the Galactic disk and also by unique Galactic bulge model. Thus we may conclude that the observed events are produced by the ordinary stars of the Galactic bar, the axis of which is elongated toward us. There are also no evidence that the OGLE events are induced by the dark matter.

There are several distinguishing features between events caused by stars in the LMC and that caused by MACHO's in the Galactic disk or halo. The first group of events will be clustered toward the central region of the LMC, whereas the second group of events will be uniformly distributed over the whole LMC [443]. For perfect alignment of the source and the lens the impact parameter has to be $\lesssim 0.001$ for the Galactic halo or disk stars whereas it has to be $\lesssim 0.1$ for LMC lens. The typical Einstein ring radius for Galactic disk or halo stars is at least two orders of magnitude greater than that for the LMC stars [443].

Meanwhile some OGLE events are very interesting, and two of them, OGLE No 6 and OGLE No 7, will be described here [455, 458]. These events cannot be explained by a single lens, and are induced indeed by a binary system. The parameters of the event OGLE No 6 (Fig. 6.27), extracted from the light curve [458] are: $q = 0.26$, $a = 2.38\,R_E$, $b = 0.16\,R_E$, $\theta = 6.2°$, $t_E = 9.0\,\text{days}$, $t_g = 822.7\,\text{days}$, $m_\circ = 18.07$.

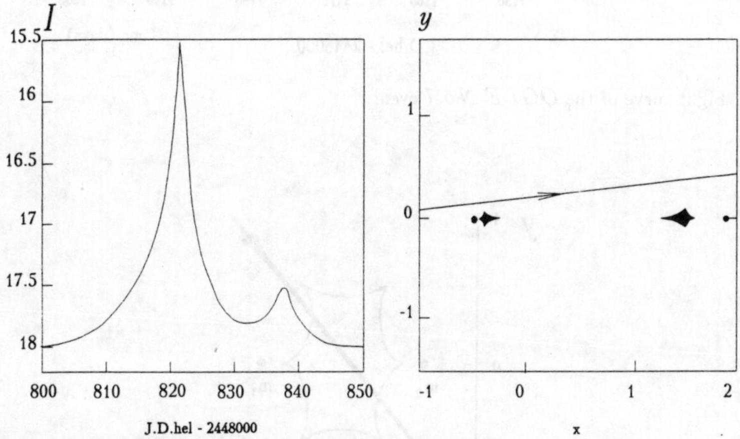

Fig. 6.27. Light curve and the lensing geometry of the *OGLE No* 6 event.

The strange light curve of the event OGLE No 7 with two sharp maxima (Fig. 6.28) [455] corresponds to the following parameters: $q = 1.02$, $a = 1.14\,R_E$, $b = 0.05\,R_E$, $\theta = 48.3°$, $t_E = 80\,\text{days}$, $t_g = 1172.5\,\text{days}$. The relative part of the light, contributed by the lensed star at the minimum brightness $I_\circ$, is equal to $f = 56\%$. The lensing geometry of the event OGLE No 7 and caustics are sketched in Fig. 6.29.

The problem, whether the compact lensed objects, observed in the experiments MACHO, EROS and OGLE, lie in the thin Galactic disk, in the thick Galactic disk or in the Galactic halo, is discussed in [446].

As was proved in [453], the **nine** microlensing events of the Galactic bulge stars are likely to be ordinary stars in the Galactic bar. The long axis of the stellar orbit is elongated toward the observer. The event time scales correspond to the ordinary stellar masses rather than very low or very high masses of the expected compact

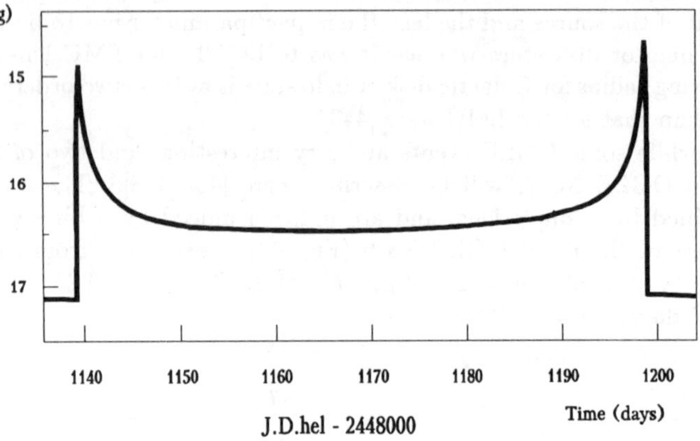

Fig. 6.28. Light curve of the $OGLE\ No\ 7$ event.

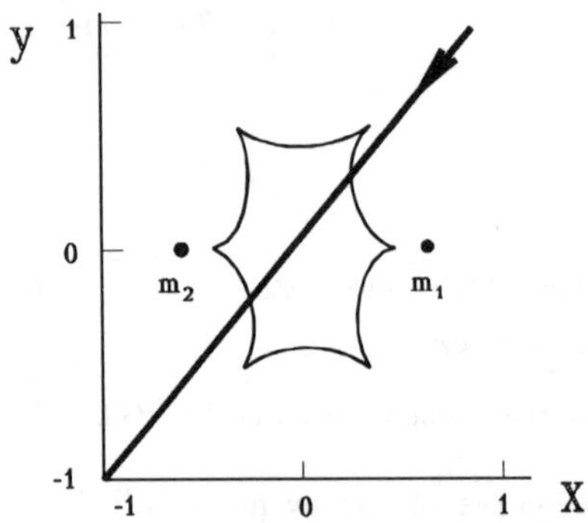

Fig. 6.29. The lensing geometry and the sketch of the caustics of the binary microlens model of the $OGLE\ No\ 7$ event, projected onto the source plane.

dark objects. There is no evidence that the detected microlensing events could be related to the dark matter in our Galaxy, and that they are likely induced by the ordinary stars in the Galactic bar. A very high optical depth of the events toward the Galactic bulge, together with a low rate toward LMC, indicate that these events are generated by the Galactic disk stars. The observed high optical depth toward LMC shows that the compact objects are in the thick disk of LMC [446].

Meanwhile the dark matter in the Galactic halo may be in the form of dark clusters of small mass objects, such as brown dwarfs, "Jupiters", black holes with total mass $M_c \approx 10^6 M_\odot$ [438]. These events should be concentrated in several isolated small regions of the angular size $\approx 20''$. To detect this dark clusters we must know the spatial and the temporal distributions of the corresponding events. If the events are generated by several MACHO's within the same cluster, these lenses would be at identical distances and would demonstrate strong spatial correlation, over the angular scale $\approx 20''$, roughly the same event durations and velocity vectors [450]. In the case of high density of the dark clusters, the projected separation between the observed MACHO's would be comparable with the Einstein ring radius, and the light curves would be time non-symmetrical.

In the case, when the source star and the lens star belong to the Galactic disk, we may observe self-lensing [461]. When these two stars have the same distribution function, the optical depth for such pure self-lensing would be equal to

$$\tau = 2\,\frac{\langle v^2\rangle}{c^2}\,\sec^2 i\,, \tag{6.138}$$

where i is the inclination angle and $\langle v^2\rangle$ is the mass-weighted vertical velocity dispersion,

$$\langle v^2\rangle = 2\int\limits_0^\infty dz\,\rho_m(z)\,\overline{v^2(z)}\Big/\int\limits_0^\infty dz\rho_m(z)\,, \tag{6.139}$$

and $\rho_m(z)$ is the mass density of MACHO's as a function of height z above the plane.

The searching program of the dark objects in our Galaxy with background stars from M31 Galaxy (Andromeda) has been proposed in [470]. As the distance to M31 Galaxy is more (690 kps), than that to LMC or SML (57 kps), this searching program would cover the whole halo of our Galaxy, whereas LMC and SMC themselves lie in this halo. The number of stars projected onto M31 field with brightness more than 25^m, is estimated as 10^6 stars. The number of the lensing dark objects in our Galaxy, projected over M31 Galaxy has been estimated as $7.3\cdot 10^9$. The expected number of events is about 5 at exposure time 14 months for mass $\sim 10^{-3}M_\odot$. In the proposed program a traditional photomethod would be used with exposure time $\sim$ 30 min. The angular resolution is equal to $\approx 0.6''$ and photometric threshold sensitivity $\Delta m \approx 0.1^m$.

6.11.2. Giant luminous arcs

An example of giant luminous arc is shown in Fig. 6.30. These nontraditional gravitational images can be explained by caustics of the multipoint mass system with arc thickness defined by the size of the source [419, 421, 434]. If a luminous arc has uniform thickness, it implies that the source is circular. However for arc in the A370 system the source is likely to be elliptic. The angular size of this and many other giant luminous arcs, more than $\approx 10''$, indicates that we observe a

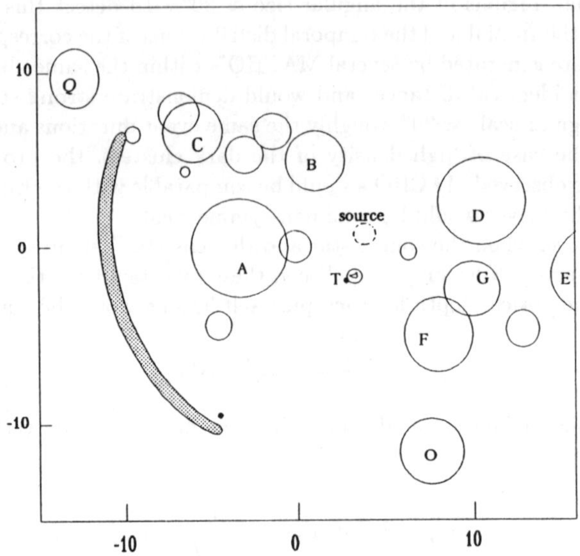

Fig. 6.30. Giant luminous arc (gray area) in the $C12244-02$ system, accepted position of the source S (dotted) and two groups of galaxies.

gravitational image of very distant source, with lens as distant clusters of galaxies. Most galaxies in cluster radiate as a point source with effective angular radius $\sim 1''$ [415].

It was proved [435] that luminous arc in the A370 system consists of three merging parts, but the gravitational lens C1 2244-02 is composed of two twin gravitational images of the spiral galaxy at redshift $z > 1$. These luminous arcs are produced by the background source, which lies at the focus of a "gravitational telescope". As a rule, giant luminous arcs copy a part of the caustics, which depends on the mass distribution inside the cores [421].

In the center of the rich cluster of galaxies [427] A2218 there were identified 32 luminous arcs in the field $4' \times 3'$. The most of the arcs are compatible with high redshift ($z > 0.6$) galaxies.

The investigations of highly magnified gravitational images, produced by cluster of galaxies [439, 440] have shown that the curvature and orientation of the long luminous arcs are defined by the radial profile of the mass distribution of the clusters. The probability distribution of luminous arcs lengths was estimated for lenses with spherical and quadrupole components [440]. The cluster of galaxies induced the gravitational lens deflection of the order of $30''$ in contrast to $1.75''$ which gives our Sun [411, 412].

Meanwhile it was shown [454] that observed timescales of the quasar variability at a fixed wavelength "should not be expected to correlate linearly with $(1+z)$, contrary to common perception". The authors [454] conclude that "the evidence at this time does not require the gravitational microlensing hypothesis to explain high-redshift quasar variability". This conclusion in [454] is based on the statistical analysis of about 300 quasar light curves, interpreted earlier as microlensing effect on substellar mass objects.

6.11.3. Large scale inhomogeneities of the Universe

When light rays from a distant galaxy pass through the region with mass overdensity, the observed gravitational images would be elongated slightly tangentially with respect to the center of perturbation [424]. Perturbations along the line of sight are added noncoherently. For nearby structures such as Coma Cluster and Local Supercluster, we may observe the image polarization and thus estimate the density of the Universe as a function of critical density. This effect is a coherent one over the place of tens square degrees. Thus a weak lensing effect, detected as a distortion of the images by a gravitational field without the creation of multiple images [467], can be considered as potentially powerful technique for studying large scale inhomogeneities of the Universe [451, 434].

We know that some part of the Universe consists of the dark matter, having great influence on the gravitational lens observations. Any dark inhomogeneities should manifest themselves in overall lensing statistics. Beside the angular minute scale correlation of lensed images around cluster of galaxies, we must observe large scale correlation effects. According to the model adopted in [433], the probability distribution of magnification factor, in the noncompact lenses, will be broader with increasing of the source redshift, but the correlation, induced by the dark matter, would be weaker for $z_s \leq 1$.

The possible effect of magnification by very large scale density fluctuations, including dark matter, in the range of redshifts $z < 5$, when the thin lens approximation and the Einstein linearized methods do not hold, was treated in [420]. The focusing effect was calculated in the absence of any definite gravitational lens along the light of sight at moderate density excess. There were found new solutions at intrinsic gravitational redshifts in the center of the galaxy clusters. Due to magnification effect, the rich clusters of galaxies may be used as a real "gravitational telescopes" for observation the faintest end most distant quasars, galaxies, radio sources and other extremely distant objects [420]. If we take into account the higher order terms in redshifts, which do not depend on the cosmological model chosen, the gravitational lens effect can be used for testing several cosmological theories [411]. To estimate redshifts of the faint blue galaxies and the mass distribution of foreground clusters of galaxies, a geometrical estimation technique [466] was proposed. It was found that the lensing magnification produces distortions of the background galaxies, increasing with redshift, and that the surface brightness of galaxies decreases steeply with redshift. Thus we get a functional relation between relative distances to galaxies and surface brightness. We may label galaxies by means of their surface brightness which is conserved under the gravitational lensing. The surface brightness is likely to be more tightly correlated with the galaxy redshift, than with its magnitude. This technique does not depend on the model chosen.

The method for general nonlinear cluster lens reconstruction by using the observed distortion of the background galaxies is given in [457].

It was proved [417] that the ratio $\Delta t/M_\circ$ with $M_\circ$ being the gravitational lensing mass, does not depend on the cosmological model. Thus we have estimated, for example, the mass $M_\circ = 1.17 \cdot 10^{12} M_\odot (\pm 20\%)$ from QSO 0957+561 event [242, 244, 246, 247].

The gravitational lensing effect depends on the mass of the deflecting galaxy [239] through the light travel time difference Δt of two lensed images. Namely, the ratio $\Delta t/M_\circ$ is a linear function of the normalized deflection vector $\hat{\vec{\alpha}}_1$ and does not depend on the cosmological models.

The relationship between structures of real space and redshift space was treated [423] with the aim to understand the dynamics and properties of caustics in the redshift space. The main point of these problems is that galaxies exist in $6D$ phase space, and the redshift would be the exact measure of the distance only if galaxies move along the direction of general expansion of the Universe. This metrological relation is distorted by the galaxies motion with peculiar velocities relative to the general expansion movement. The analysis of the relationship between structures in the real space and in the redshift space [423] had shown that in the condition of large-scale streaming movement of galaxies both spaces should be considered as $3D$-projections of $6D$-phase space.

The redshift surveys of galaxies locate them in the redshift space, not in the real space. Due to large-scale motion of galaxies, the redshift space is not isotropic. Peculiar velocities alter galaxy positions only along the line of sight. The anisotropy of the galaxy distribution is defined by the anisotropy of the galaxy movement.

6.11.4. The Hubble's parameter

The Hubble's parameter can be estimated independently from the time travel difference in the gravitational lensing effect [410, 429, 434]. The source object in this experiment must be supernova S behind the distant galaxy G of the small redshift. Simultaneously we may estimate the mass of the lensed galaxy. In the case of the linear relation between the distance a and the redshift z, we have two equations, one for galaxy and one for supernova:

$$\left.\begin{aligned} a_G &= z_G\, c\, H^{-1}\,, \\ a_s &= z_s\, c\, H^{-1}\,, \end{aligned}\right\} \tag{6.140}$$

The Hubble's parameter H is defined by:

$$H = \frac{z_s\, z_G\, \alpha\, (\alpha_1 - \alpha_2)}{\Delta t\, (z_s - z_g)}\,, \tag{6.141}$$

and the galaxy mass is equal to

$$M_G = \frac{\Delta t\, \alpha\, c^3}{16\, G\, (\alpha_1 - \alpha_2)}\,. \tag{6.142}$$

Both Hubble's parameter H and the galaxy mass M_G are defined by the directly observable quantities. The Hubble's parameter can be estimated by means of the wavefront technique [429] (Fig.6.31). The time delay in this technique is equal to distance between the wavefronts at the observation points:

$$\Delta t = \frac{x\, \theta_{AB}}{c}\,, \tag{6.143}$$

where x is the distance from the symmetry axis to the observer. As

$$\theta_s = x\, \frac{D_{LS}}{D_{OL}\, D_{OS}}\,, \tag{6.144}$$

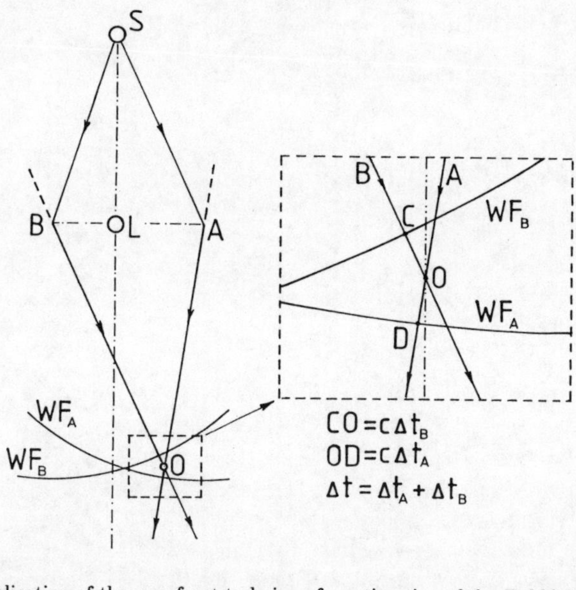

Fig. 6.31. Application of the wavefront technique for estimation of the Hubble's parameter.

then from the Hubble's relations for source and for lens

$$\left.\begin{array}{l} D_{OS} = c\, z_d\, H_\circ^{-1}\,, \\ D_{OS} = c\, z_s\, H_\circ^{-1}\,, \end{array}\right\} \tag{6.145}$$

we get the Hubble's parameter $H_\circ$, expressed in terms of the observable quantities

$$H_\circ = \frac{z_d\, z_s\, \theta_{AB}\, (\theta_A - \theta_B)\, (2 - \epsilon)}{2\, (z_s - z_d)\, \Delta t}\,, \tag{6.146}$$

where the parameter ϵ defines a deflection law

$$\hat{\alpha} \sim \xi^{\epsilon - 1}\,, \tag{6.147}$$

with ξ-impact parameter, and $\epsilon = 1.36$ or $\epsilon = 1.28$. The possibility of observing this phenomenon is discussed in [410].

Chapter 7

INVERSE MESO–OPTICS

7.1. Introduction

The term "meso–optics" has been introduced in [3, 4] to cover a wide class of nontraditional optical devices, which accomplish the homomorphic (non-reversible) transformations of the object space, under which the dimension of the image space is getting greater than the dimension of the object space. For example, an axicon or a conical lens transforms a point on the optical axis into the straight line segment. In the meso-optical Fourier transform microscope [129] a point is transformed into a short shirt, giving an example of the homomorphic transformation of the type $(0 - D) \to (2 - D)$, for which the dimension of the image space is greater by two than the dimension of the object space. In this chapter we consider a new class of transformations accomplished by means of the inverse meso–optics, namely, when an object in the form of a long straight line segment is converted into a single point in the image space. We give the definition of the inverse meso–optics and we present the necessary conditions under which this transformation can be made in the frame of the geometro-optical approximations.

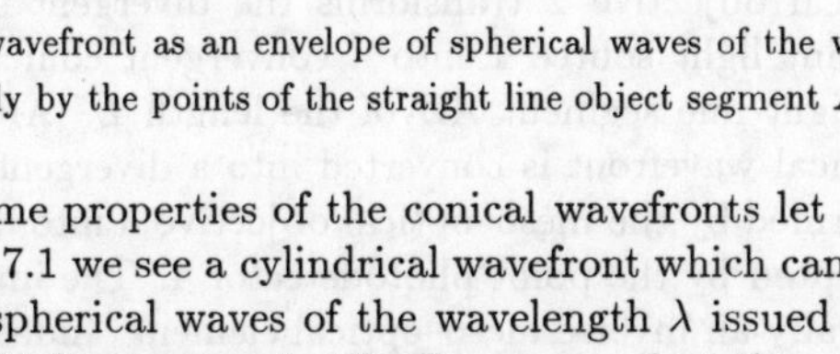

Fig. 7.1. Cylindrical wavefront as an envelope of spherical waves of the wavelength λ, which are generated synchronously by the points of the straight line object segment AB.

To illustrate some properties of the conical wavefronts let us consider two pictures [265]. In Fig. 7.1 we see a cylindrical wavefront which can be considered as an envelope of many spherical waves of the wavelength λ issued coherently from the points of the straight line segment AB. Here we neglect the radiation which propagates along z-axis. An example of the convergent cylindrical wavefront produced by the right-angle reflaxicon was shown in Fig. 2.31. If spherical waves are radiated coherently with phases that are changing linearly along the straight line segment CD (Fig. 7.2), a conical wavefront with the same symmetry axis is generated. This

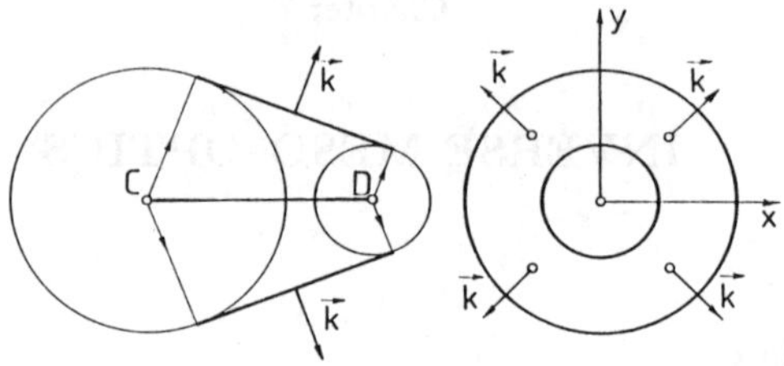

Fig. 7.2. Conical wavefront as an envelope of spherical waves of the wavelength λ, which are generated by the point of the straight line segment AB with phase delay changing linearly along z-axis.

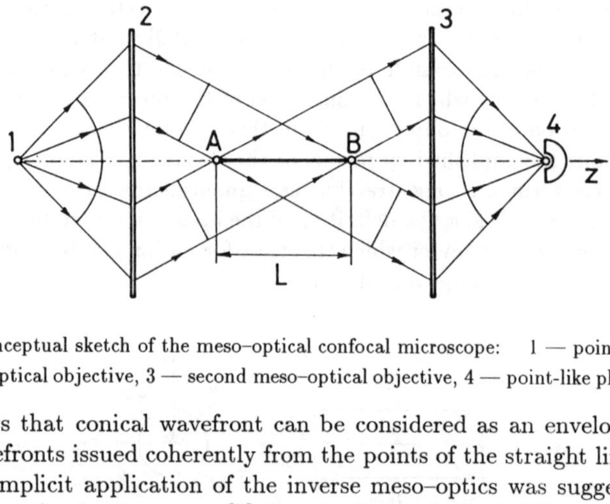

Fig. 7.3. A conceptual sketch of the meso–optical confocal microscope: 1 — point light source, 2 — first meso–optical objective, 3 — second meso–optical objective, 4 — point-like photodetector.

example shows that conical wavefront can be considered as an envelope of many spherical wavefronts issued coherently from the points of the straight line segment.

The first implicit application of the inverse meso–optics was suggested in the meso–optical confocal microscope [3] shown schematically in Fig. 7.3. The illuminating meso–optical objective 2 transforms the divergent spherical wavefront produced by the point light source 1 into a convergent conical wavefront which collapses into a straight line segment AB of the length L. After reversing in the optical axis this conical wavefront is converted into a divergent conical wavefront. The latter is transformed by the meso–optical objective 3 into a convergent spherical wave, which accepted by the point photodetector 4. The imaging meso–optical objective 3 is essentially an inverse meso–optical element. Indeed it transforms the divergent conical wavefront generated by the segment AB into a convergent spherical wave and then into a point. The described sequence of transformations can be accomplished because the radiation from the straight line segment AB is coherent.

Some other inverse meso–optical devices are shown in Fig. 7.4 [265]: a) a coherently radiating segment AB is transformed by the element $IMOO_1$ (inverse–meso–

optical–objective–1) into a point C; b) a coherently radiating ring R is transformed by the element $IMOO_2$ into a point C; c) a coherently radiating segment AB is transformed by the element $IMOO_3$ into a ring R at the first stage and then by the element $IMOO_4$ into a point C at the second stage.

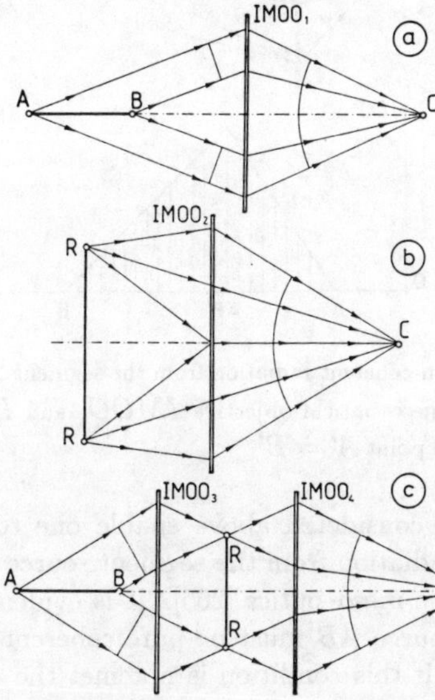

Fig. 7.4. Inverse meso–optical devices which accomplish: a) transformation of the conical wavefront, coherently radiated by the segment AB, into a point; b) transformation of the coherent radiation from a narrow ring-shaped source R into a point C by means of the inverse meso–optical objective $IMOO_2$; c) two-stage transformation of the coherent radiation from the segment AB into the ring-shaped focus R, by means of the inverse meso–optical objective $IMOO_3$, on the first stage, and thereupon into the point C, by means of inverse meso–optical objective $IMOO_4$, on the second stage.

Now we consider the properties of the non-coherent radiating segment AB (Fig. 7.5), when phases are randomly distributed over this segment. Each point of the segment AB generates a divergent spherical wave which is transformed by the imaging meso–optical element 3 into a segment on the optical axis. One extreme point A goes into the segment $A'A''$, and another extreme point B goes into the segment $B'B''$ on the same z-axis. The resulted light intensity distribution on the optical axis can be approximated by the non-coherent sum of many brilliant segments. We get a triangle intensity distribution between points A'' and B'' with intensity maximum at point $A' = B'$.

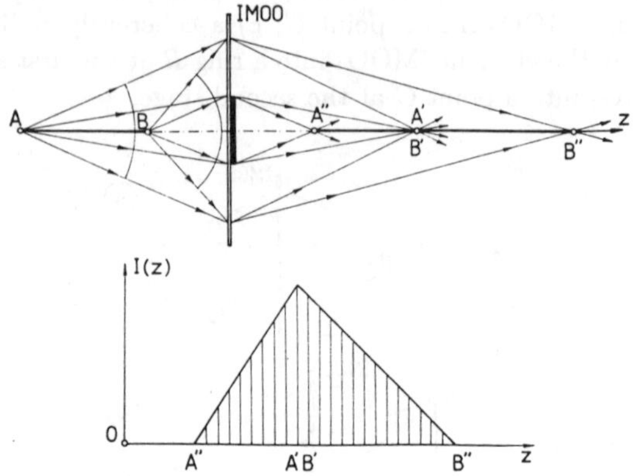

Fig. 7.5. Transformation of the non-coherent radiation from the segment AB into another segment $A''B''$ by means of the inverse meso–optical objective $IMOO$, and $I(z)$ is the light intensity distribution with maximum at the point $A' = B'$.

These two limited cases considered above enable one to state the necessary condition under which the radiation from the segment source can be focused into a point by means of the inverse meso–optics [265]. It is evident, that the wavefront generated by the segment source AB must be pure coherent with linear variation of the phases along z-axis. If this condition is not met the meso–optical element does not work well. The only operation that can be made in this case consists in isolating one conical wavefront from many others by placing a pinhole before the photodetector in the point $A' = B'$. All other non-coherent conical components will be suppressed. A similar filtration occurs in the traditional optics. The pinhole placed in the focus F isolates one spherical wavefront and rejects all other.

Below we consider in detail two examples in which coherent condition is fulfilled completely: a) an undulator radiation and b) Cherenkov radiation used in the differential gas Cherenkov counter. In both cases the radiation source has the form of a straight line segment which generates pure coherent conical wavefronts of various wavelengths. Any device designed for transporting and focusing of this radiation must involve meso–optical or inverse–meso–optical elements.

7.2. Undulator Radiation

Relativistic electrons moving in the magnetic field produce synchrotron radiation directed along velocity vector of electrons $\vec{v}$ in the angular aperture of the order of $\theta_{SR} \sim 1/\gamma$, where γ — relativistic factor of the electrons, $E = \gamma m_o c^2$ — its energy, typically 1–10 GeV and m_o — its rest mass [266, 267]. Frequency spectrum of synchrotron radiation is continuous with maximum at frequency $\omega = \omega_o/(1 - v^2/c^2)^{3/2}$, where ω_o is the revolution frequency of electrons in the ring accelerator. The intensity of synchrotron radiation increases with energy of electrons as E^2.

To increase the intensity of the synchrotron radiation sources, an undulator with transversal periodic magnetic field is used nowadays. The traditional undulator consists of the periodically mounted magnetic blocks which involve two permanent magnets, such as $SmCo_5$, and two elements of soft magnetic material located alternatively (Fig. 7.6) [268]. The magnetization vectors of the permanent magnets are directed along undulator axis with regular alternation. The transversal periodic (sinusoidal) component of the magnetic field is generated along the trajectory of relativistic electrons. If d is the period of the transversal magnetic field of the undulator and N is the number of these periods over the length L, then the wavelength of radiation $\lambda(\theta)$ at the observation angle θ is defined by the equation [267]

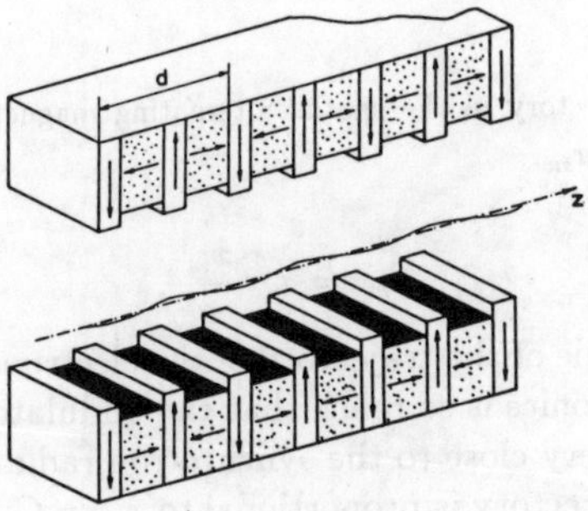

Fig. 7.6. Design of the conventional undulator with periodic transversal magnetic field.

$$\lambda(\theta) = \frac{d_o}{2\gamma^2}(1 + \gamma^2\theta^2 + K)\,, \tag{7.1}$$

where magnetic deflection parameter K is equal to

$$K = 0.934\, B_o\, d_o\,, \tag{7.2}$$

where peak magnetic field B_o is expressed in Tesla, and d_o in cm. The radiation cone half angle is

$$\theta_{1/2} = \frac{1}{\gamma\sqrt{N}}\,. \tag{7.3}$$

For undulators $K < 4$, and for wigglers $K > 10$.

Spectrum and polarization characteristic of the undulator radiation was treated in [269]. In the frame of the dipolar approximation the undulator of the length

$L = N\, d_\circ$ produces line spectrum for each harmonic k with relative spectrum line width $\sim (N\,k)^{-1}$. In more general case the structure of the undulator radiation spectrum depends on the parameter

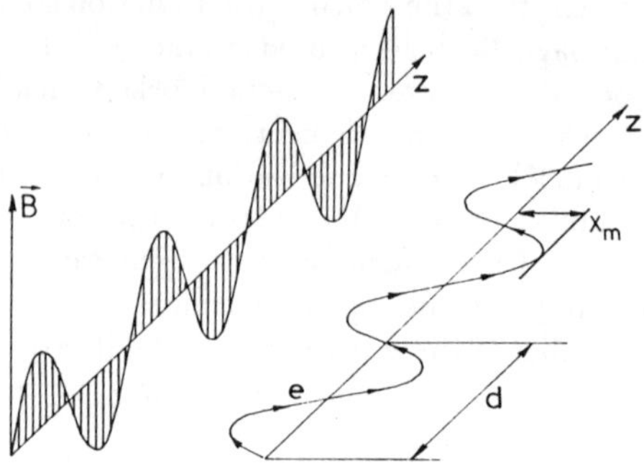

Fig. 7.7. The electron trajectory in the spatial alternating magnetic field of the period d with amplitude of this trajectory x_m.

$$\beta_m = \omega_\circ \frac{x_m}{c}\,, \tag{7.4}$$

where x_m is the amplitude of the trajectory of the electrons (Fig. 7.7). For $\beta_m \gamma \simeq 2$ the overlapping of harmonics is so much that the undulator radiation spectrum becomes continuous and very close to the synchrotron radiation spectrum. As amplitude of the electrons trajectory is proportional to $x_m = C_1 B_\circ$, where C_1 is constant, we may conclude that line spectrum is observed only at small magnetic field, when magnetic deflection parameter $K \approx 0$. By varying the magnetic deflection parameter K we can change the undulator radiation spectrum and have the spectrum with desired overlapping of spectral lines. In more detail the broadening of the undulator radiation spectrum lines has been analyzed in [270].

Due to bunch structure of the electron beams and due to small angular spreading of the relativistic electrons in the modern electron storage rings, the phase space volume of electrons can be made very small to have undulator radiation of high lateral and longitudinal coherence for use in off-axis X-ray holography [268, 271].

To control polarization state of the undulator radiation there have been proposed several exotic undulators [272] such as helical undulator [273, 274], undulator with variable period [275], including electrostatic designs and irregular structure undulator [276] for suppressing of various harmonics. It has been proved that fixed field undulator provides more light power then fixed-period devices and that tunability of this device is also superior to that of the fixed field system with variable period undulator [277]. The helical undulator [273, 274] consists of two collinear planar undulators with common period and with orthogonal orientations: one with a vertical magnetic field and another with a horizontal magnetic field (Fig. 7.8). By

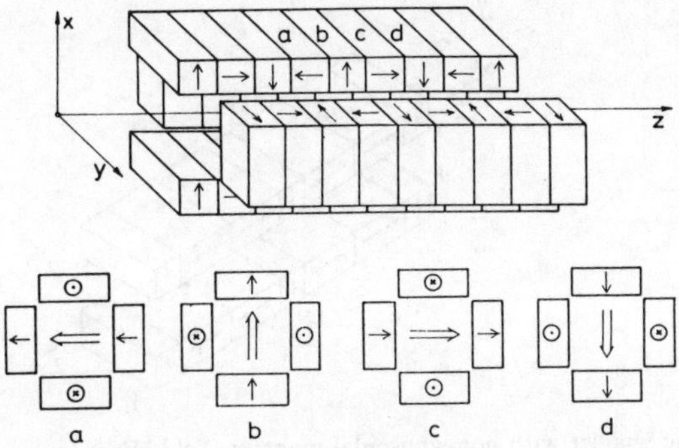

Fig. 7.8. Helical undulator, composed of two collinear orthogonal planar undulators of the common spatial period, which have translating degree of freedom.

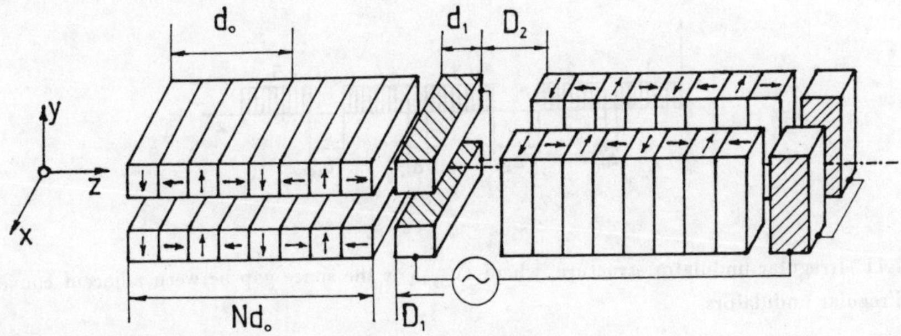

Fig. 7.9. System of two collinear crossed oriented undulators, which permits to control the polarization state of radiation.

translation one undulator with respect to the other over a distance smaller than a period-length d_o, we change the polarization state of the undulator radiation from linear polarization to circular polarization.

In the system with two crossed undulators [278] (Fig. 7.9) an electron path length modulator producing variable phase delay is mounted between two undulators. To increase photon energy at the price of lower brightness and brilliance an asymmetric wiggler has been proposed [279] (Fig. 7.10). Apart from these exotic undulators is the irregular structure undulator [276] which consists of a chain of identical conventional plane regular undulators, separated by free spaces of different lengths (Fig. 7.11). At appropriate separated lengths we may enhance or suppress the intensity and the spectral shape structure of desired radiation harmonics.

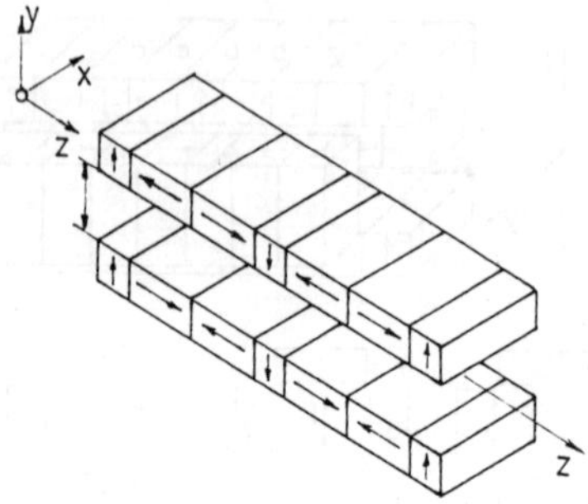

Fig. 7.10. Asymmetric wiggler with non-sinusoidal magnetic field. Each magnetic block consists of six permanent magnetic elements.

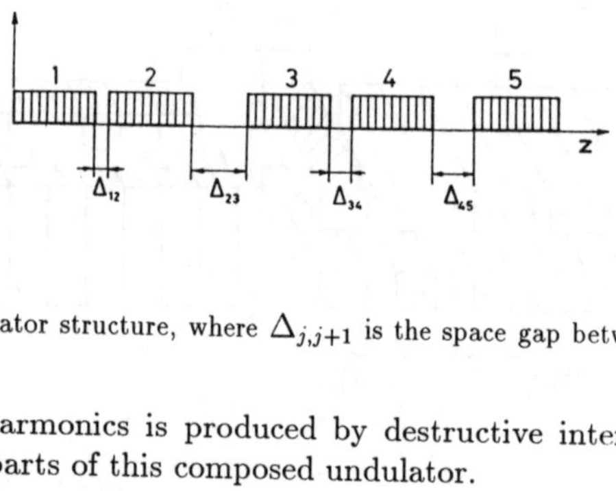

Fig. 7.11. Irregular undulator structure, where $\Delta_{j,j+1}$ is the space gap between adjacent conventional regular undulators.

The suppression of harmonics is produced by destructive interference of partial waves from different parts of this composed undulator.

7.3. Meso–optical Undulator

To describe wave-optical properties of undulator radiation it is appropriate to use a virtual grating model [280] (Fig. 7.12). As was shown the spacing between grooves of a diffraction grating "a" corresponds to the period $d_\circ$ of an undulator. As there is no real incident light falling in the undulator, there incident grazing angle α in this phenomenological model is a pure imaginary and is defined by the relation

$$\frac{c}{v_{eff}} = \cos\alpha \simeq 1 - |\alpha|^2/2 \,, \tag{7.5}$$

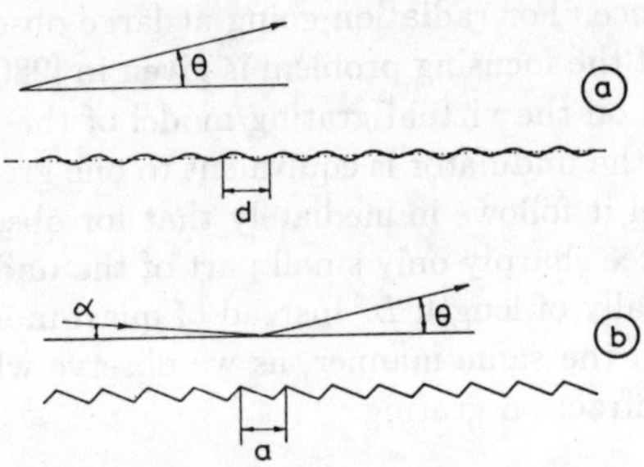

Fig. 7.12. Virtual grating model for undulator with $d_\circ = a$: a) undulator, b) grating.

where v_{eff} is the averaged velocity of electrons moving along z-axis in the magnetic field of the undulator

$$v_{eff} \;=\; c\,[1 - \frac{1}{2\gamma^2}\,(1 + K^2/2)]\,, \tag{7.6}$$

the retardation effect being induced by the sinusoidal structure of the electron trajectory and defined by the magnetic deflection parameter K.

By taking into account the equation of the real diffraction grating

$$m\,\lambda \;=\; a\,(\cos\alpha \;-\; \cos\theta)\,, \tag{7.7}$$

where m — harmonic order of the diffraction, we arrive at the equation

$$m\,\lambda \;=\; \frac{d_\circ}{2\gamma^2}\,[1 \;+\; K^2/2 \;+\; \gamma^2\theta^2]\,, \tag{7.8}$$

which is identical to Eq. (7.1). The sinusoidal trajectory of electrons and the finite number of periods N in the fixed-period undulator induce additional angular divergence of the undulator radiation which for $\theta = 0$ is given by

$$\Delta\,\theta_{add} \;=\; \frac{1}{\gamma}\left[\frac{1 + K^2/2}{2N}\right]^{1/2}\,. \tag{7.9}$$

For $\theta \neq 0$ and $\gamma\theta \gg 1$ we have

$$\frac{\Delta\,\theta_{add}}{\theta} \;=\; \frac{1}{4\,N\,m}\,. \tag{7.10}$$

The finite length L of the undulator provides some problems in focusing of the undulator radiation and in matching grating monochromators with undulator radiation. On considering these problems we must adopt that according to the uncertainty principle the minimum optical emittance W_m of radiation with wavelength λ is equal to $\lambda/2$, for example, $W_m = 10^{-9}$ m rad for $\lambda = 20$ Å. The coherent part of the undulator radiation for small observation angle θ, $\theta\gamma \ll 1$, corresponds to the minimum optical emittance. For radiation going at large observation angle θ with $\theta\gamma \gg 1$ the description of the focusing problem is given in [280]. The main point of this explanation is based on the virtual grating model of the undulator, according to which every period of the undulator is equivalent to one groove of the diffraction grating. From this model it follows immediately that for observation system with small depth of focus we see sharply only small part of the undulator and the effective source region is actually of length L. Instead of quasi-monochromatic light we detect white radiation, in the same manner, as we observe white light by viewing only one groove of the diffraction grating.

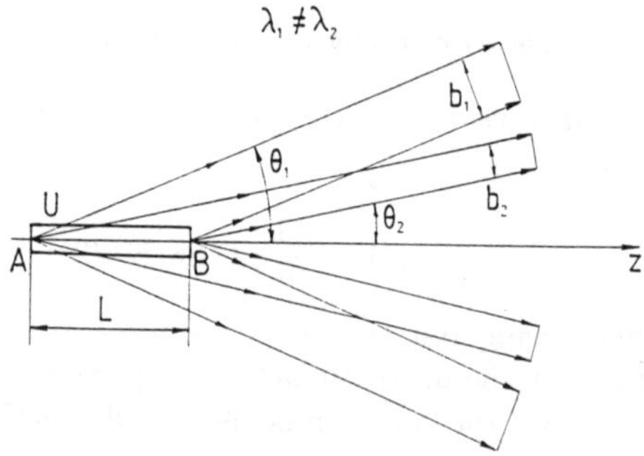

Fig. 7.13. Two side ways conical wavefronts of the quasi-monochromatic radiation, generated by the undulator of the length L.

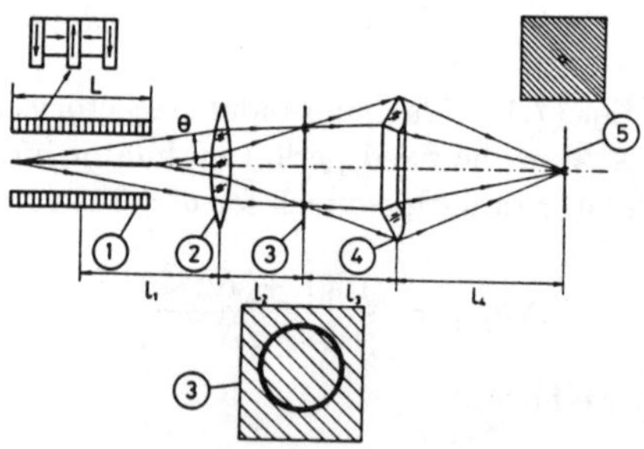

Fig. 7.14. Inverse meso–optical device, which focus into a point the side ways going radiation from undulator ($\theta \neq 0$): 1 — undulator, 2 — first inverse meso–optical objective, 3 — ring-shaped slit diaphragm, 4 — second inverse meso–optical objective, 5 — pinhole diaphragm.

Fig. 7.13 shows two conical wavefronts with different wavelengths λ_1 and λ_2 emitted at various observation angles θ_1 and θ_2 by the straight light segment AB of the length L [280]. Each conical wavefront represents coherent quasi-monochromatic radiation. The wavelength dispersion of the undulator is defined by Eq. (7.1).

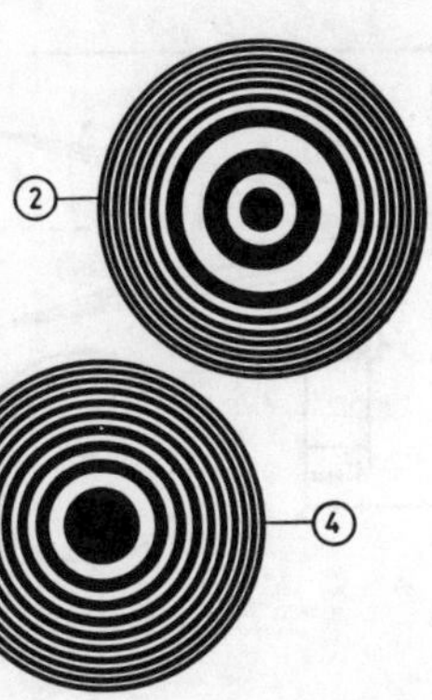

Fig. 7.15. Binary optical relief of the inverse meso-optical elements (2) and (4) (Fig. 7.14).

To focus the side ways going undulator radiation we use inverse meso-optical elements shown in Fig. 7.14 [281]. The first meso-optical element 2 transforms the conical wavefront of radiation with wavelength λ_1 into a ring-shaped focus. Thus focused radiation is transmitted through the screen with narrow ring-shaped slit 3, which accomplishes premonochromatization of the radiation, by rejecting any other wavelengths with $\lambda \neq \lambda_1$. The quasi-monochromatic coherent radiation, generated by **full length undulator**, is focused into the point on the optical axis by means of the second inverse meso-optical element 4. The angular aperture of the first inverse meso-optical objective 2 is much greater than the effective angular aperture for the conventional lens and the diameter of the ring-shaped slit is much grater than the diameter of the pinhole used in all conventional undulators. Due to this the diameter of the central focusing spot near the pinhole diaphragm 5 in Fig. 7.14 will be much smaller than the diameter of the focusing spot in the traditional system. Another advantage of the meso-optical undulator shown in Fig. 7.14 is that radiation before the diaphragm 5 is partially monochromatized. Radiation issued from the sources, evenly distributed over the undulator length L, is focused by this inverse meso-optical system without any focusing problems. It should be emphasized that some technical problem of cooling of the ring-shaped slit 3 is facilitated due to large length of the heat exchanger. The inverse meso-optical objectives 2 and 4 for radiation with wavelength $\lambda \sim 10$ nm can be fabricated in the form of the diffraction elements [282] with special optical profiles shown in Fig. 7.15 as binary equivalents.

7.4. Self-focusing Undulator-wiggler

The most remarkable feature of the undulator radiation is the dependence of the wavelength λ from the observation angle θ, expressed by Eq. (7.1). By changing

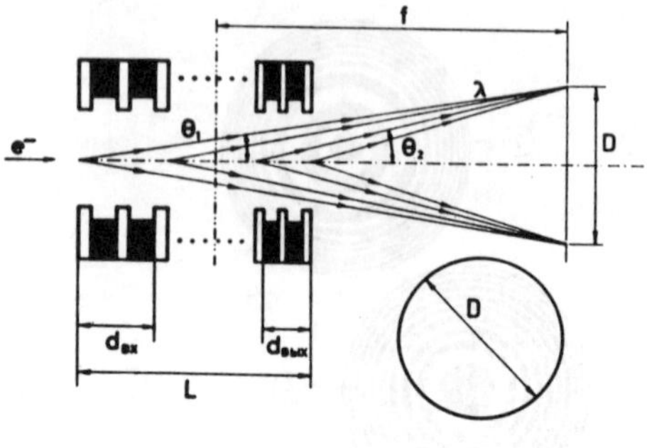

Fig. 7.16. Self-focusing undulator.

the period of the undulator d_o we control the wavelength observed at angle θ. To keep constant the wavelength λ we must change simultaneously both d_o and θ.

Fig. 7.16 shows a self-focusing undulator [265, 283, 284] in which the period length d_o of the transversal magnetic field is a slowly decreasing function of z-coordinate. If f is the distance from the center of the self-focusing undulator to the plane where a ring-shaped focus is to be observed for chosen wavelength λ, the periods of the undulator blocks at the input, d_{in}, and at the output, d_{out}, of the undulator must be chosen according to the equations

$$\left.\begin{aligned} d_{in} &= 2\gamma^2\lambda/(1+\gamma^2\theta_1^2+K)\,, \\ d_{out} &= 2\gamma^2\lambda/(1+\gamma^2\theta_2^2+K)\,, \\ \text{where}\qquad \theta_1\gamma &= 1-L/2f\,, \\ \theta_2\gamma &= 1+L/2f\,, \end{aligned}\right\} \tag{7.11}$$

the period decreasing as linear function of z between maximum d_{in} and minimum d_{out} along the velocity vector of relativistic electrons.

7.5. Undulator with Apodized Spectral Lines

As was shown in Fig. 3 of Ref. [267], the base of the first harmonic peak for $K = 0.515$ exhibits small interference ripples near the main peak which are induced by the sharp truncation of the periodic structure of the undulator magnet. Sometimes this spectral response of the undulator can hinder the running of spectral precision experiments, because of the small dynamic range of the whole device. To overcome this disadvantage of the conventional undulator we must suppress these side lobes in the spectral response of the undulator. For this purpose the sharp truncation of the periodic structure of the undulator magnet is to be withdrawn. With this aim in mind we vary the thickness t of element made of soft magnetic material over the undulator length L, for example, according to the equation

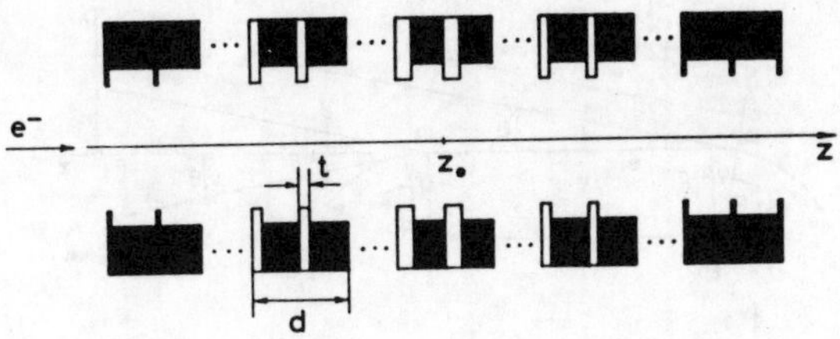

Fig. 7.17. Nontraditional undulator which provides the spectral lines of the radiation without side lobes.

$$t(z) = t_{\circ}\, exp[-A(z - z_{\circ})^2 / L^2], \tag{7.12}$$

where $z_{\circ}$ — the coordinate of the undulator center, and A — a constant, equal to $\approx 2.2 \div 2.4$. This undulator is shown in Fig. 7.17 in the form of a model with five periods. The amplitude of the electrons trajectory and the intensity of the undulator radiation are changing non-monotonously along undulator axis, being minimal at the extreme points and being maximal in the center of the undulator. As the amplitude of the electrons trajectory is varying according to the truncated Gaussian law, Eq. (7.12), the shape of the spectral line will be close to Gaussian without intense side lobes. This feature of the suggested undulator offers high dynamic spectral range and permits to accomplish the measurements in the spectral region very close to the central absorption line.

7.6. Interference of Synchrotron Radiation

The first investigation of the interference pattern produced by synchrotron radiation was made in [285]. The coherent sources were two extreme points of the straight line gap in the synchrotron "Syrius" with energy of electrons 1.5 GeV. The geometry of the experiment is shown in Fig. 7.18. Two radiation fans produced by electrons on the circular trajectories in the vicinity of the points A and B are overlapping in the observation plane as shown in Fig. 7.18. The transverse interference picture was detected versus orientation angle θ and azimuthal angle φ. The phenomenon is defined by the retardation time Δt_{AB} between electrons and radiation wavefront of the wavelength λ

$$\Delta t_{AB} = t_e - t_{\lambda}. \tag{7.13}$$

The moving time t_e of electrons between points A and B is equal to

$$t_e = \frac{1}{\beta c}\, l_e, \tag{7.14}$$

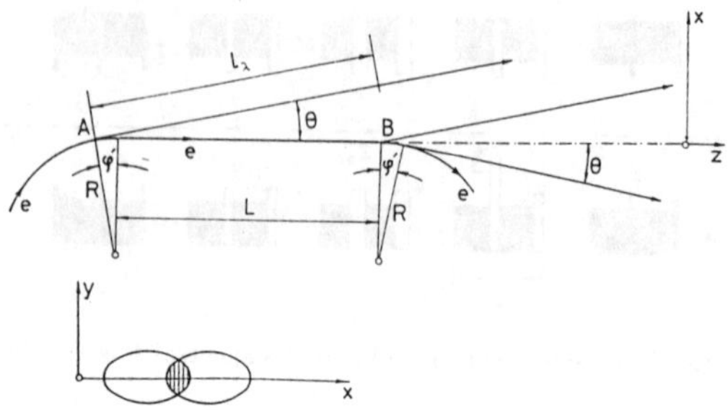

Fig. 7.18. Geometry of the first experiment for observation of the interference of the synchrotron radiation.

where the trajectory length

$$l_e = L + 2R(\varphi' - \sin\varphi') \tag{7.15}$$

and R is the curvature radius of the electron trajectory. The radiation propagating time

$$t_\lambda = \frac{l_\lambda}{c} = \frac{L}{c}\cos\theta\,\cos\varphi\,. \tag{7.16}$$

In the overlapping zone $2\varphi' \approx \gamma^{-2}$.

The angular distribution and the spectrum of the interference pattern were detected with a pinhole diaphragm at energy spread of the accelerated electrons 0.049 MeV and at effective wavelength $\lambda_{eff} = (500 \pm 10)$ nm.

Another type of the interference pattern is generated in the system of several coaxial undulators with free straight line gaps of different lengths (see Fig. 7.11). By choosing the lengths of free spaces it is possible to change the intensity and the shape of the individual radiation harmonics [276].

To provide effective overlapping of two conical waves from two short coaxial undulators, the period of the second undulator in Fig. 7.19 must be chosen smaller than that of the first one. The analogous demands were accepted in the self-focusing meso-optical undulator, shown in Fig. 7.16, Eq. (7.11), which was turned at fixed wavelength λ. The period of the circular interference fringes in this system is equal to

$$A = \left(\frac{\theta_1 - \theta_2}{2}\right)^{-1}\lambda\,. \tag{7.17}$$

As

$$(\theta_1 - \theta_2)\gamma = -L/f\,, \tag{7.18}$$

we have

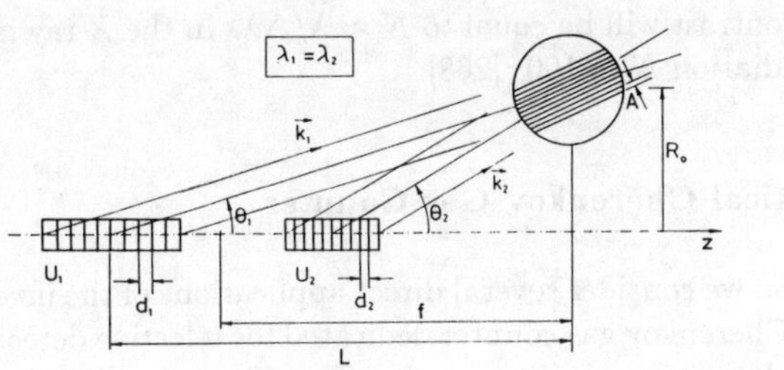

Fig. 7.19. Geometry of the interference experiment with two coaxial undulators of different periods; R_o is the average radius of the ring-shaped interference fringes.

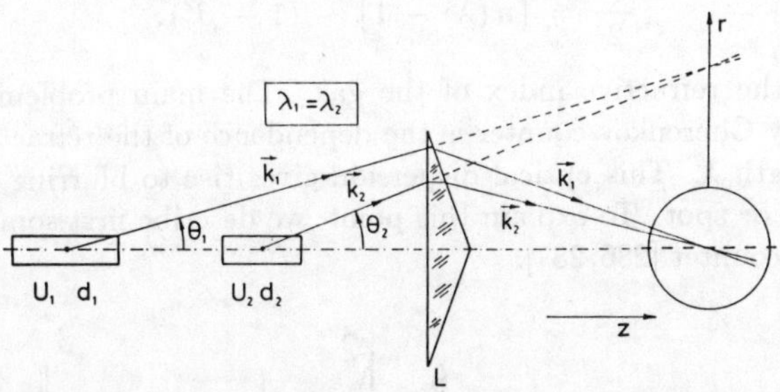

Fig. 7.20. Geometry of the experiment for observation of the longitudinal interference, produced by two coaxial conical wavefronts from two undulators of different periods: L — axiconic lens.

$$A = \lambda \cdot \gamma \cdot \frac{f}{L}, \tag{7.19}$$

where L is the length between the centers of two short undulators and f is the distance from this center to the observation plane. To produce the longitudinal interference of these two coaxial conical wavefronts, a system with axiconic lens must be used (Fig. 7.20).

So far we anticipated that the whole system generates a coherent radiation. In the real experiment we use a partially coherent radiation. The spectral purity of radiation, $\lambda/\Delta\lambda$, defines the longitudinal (temporal) coherence length

$$\Delta\, l_{\parallel} \cong \frac{\lambda^2}{\Delta\lambda}. \tag{7.20}$$

The transverse coherence length $\Delta\, l_{\perp}$ is defined by the diameter of the electron beam D and the angular divergence of the photon beam $\Delta\theta$ [268]. The phase-space volume of the electronic beam must be small with respect to the phase-space volume of the light beam $\frac{1}{2}\, D \cdot \Delta\theta$. In the general case the number of interference fringes of high contrast will be equal to $N \approx \lambda/\Delta\lambda$. In the X-ray microscope driven by undulator radiation $N \approx 100$ [268].

7.7. Meso–optical Cherenkov Gas Counter

In this Section we consider several direct applications of the inverse meso–optical elements in the Cherenkov gas counter dedicated for selective detection of the charge particles with relativistic velocity $v = \beta c$. The Cherenkov light emitted by charge particle in gas medium represents a conical wavefront with Cherenkov angle θ which is given for small θ by the equation:

$$\frac{\theta^2}{2} = [n(\lambda) - 1] - (1 - \beta^2), \tag{7.21}$$

where $n(\lambda)$ is the refractive index of the gas. The main problem arisen in the designing of any Cherenkov counter is the dependence of the refractive index $n(\lambda)$ on the wavelength λ. This optical dispersion gives rise to blurring of the focused Cherenkov ring or spot. To explain this point, we describe first some conventional Cherenkov gas counter [286, 287].

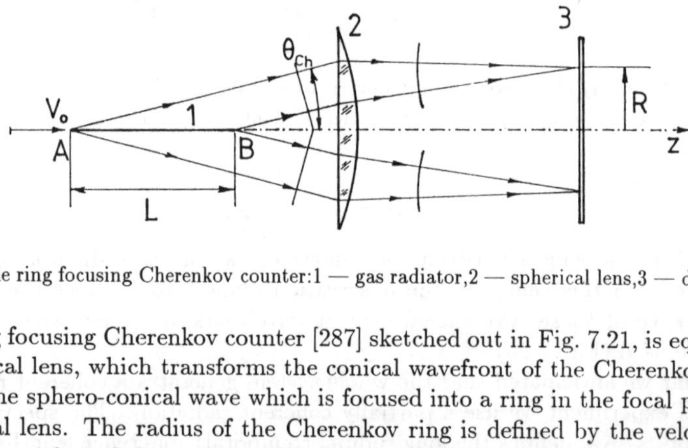

Fig.7.21.Simple ring focusing Cherenkov counter: 1 — gas radiator, 2 — spherical lens, 3 — detector.

The ring focusing Cherenkov counter [287] sketched out in Fig. 7.21, is equipped with spherical lens, which transforms the conical wavefront of the Cherenkov radiation into the sphero-conical wave which is focused into a ring in the focal plane of the spherical lens. The radius of the Cherenkov ring is defined by the velocity of the charge particle and the position of the center of this ring is given by orientation angle τ of the charge particle velocity vector with respect to the optical axis.

The spot focusing Cherenkov counter, sketched out in Fig. 7.22, involves an axicon lens and a spherical lens. By means of the axicon lens the conical wavefront of Cherenkov radiation is converted into a plane wave and this plane wave is transformed by a spherical lens into a convergent spherical wave. The latter is collapsed into a point-like focus in the focal plane, if the velocity of the charge particle is equal to β_o. For velocities $\beta \neq \beta_o$, a ring image is produced.

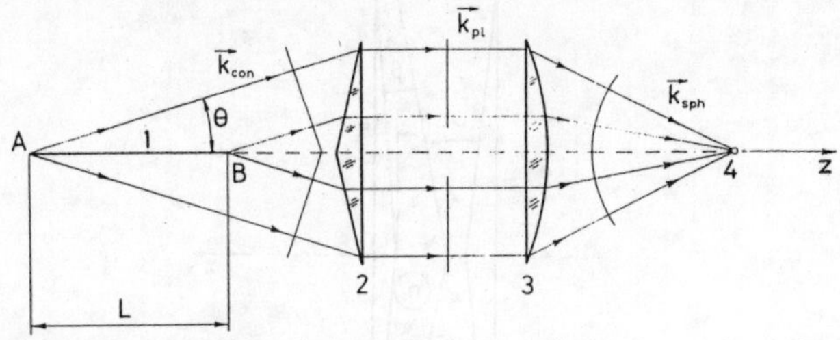

Fig. 7.22. Simple spot focusing Cherenkov counter: 1 — radiator, 2 — axiconic lens, 3 — spherical lens.

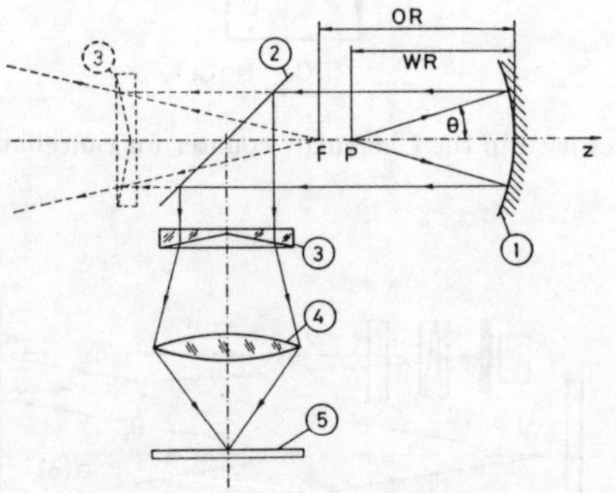

Fig. 7.23. Spot focusing Cherenkov counter for particle beam of small angular divergence: 1 — spherical mirror, 2 — plane mirror, 3 — divergent axicon, 4 — positive lens, 5 — screen, F — focal plane of the spherical mirror 1.

A better solution is to use a spherical mirror and optics free from coma and chromatic aberrations [287], Fig. 7.23. The radiated target of the particles is located at the center of the spherical mirror, and the axicon lens is placed such that its virtual image produced by means of the inclined plane mirror is at the target position. The reflected Cherenkov light rays are nearly parallel to the optical axis of the counter and the spot-like focus as the virtual image is produced in F, on the focal plane of the spherical mirror.

Wavelength variation of the refractive index $n(\lambda)$ over the chosen spectral range induces chromatic dispersion of the Cherenkov angle $\Delta\theta_{chr}$. The chromatic dispersion of the axicon lens can be compensated by chromatic dispersion of the gas and thus to cancel the resulting chromatic dispersion of the whole system. This occurs if the optical dispersion of the axicon is twice the optical dispersion of the gas radiator,

$$\nu_{axi} = 2\,\nu_{gas}\,, \tag{7.22}$$

for $\gamma_1 \to \infty$, where optical dispersion ν (Abbe number) is defined by the equation

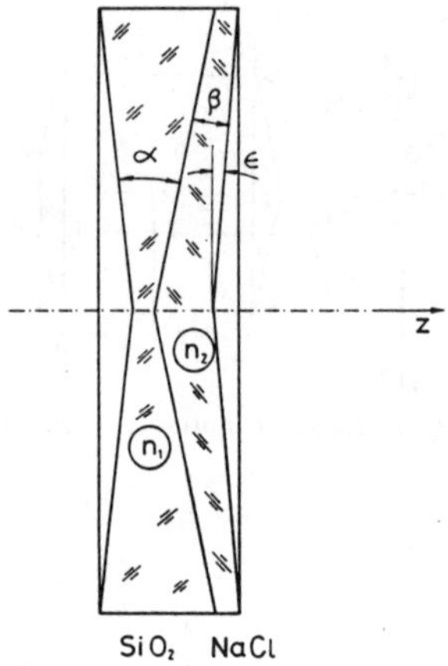

Fig. 7.24. Axiconic doublet used in the Cherenkov counter for cancellation of the chromatic aberrations.

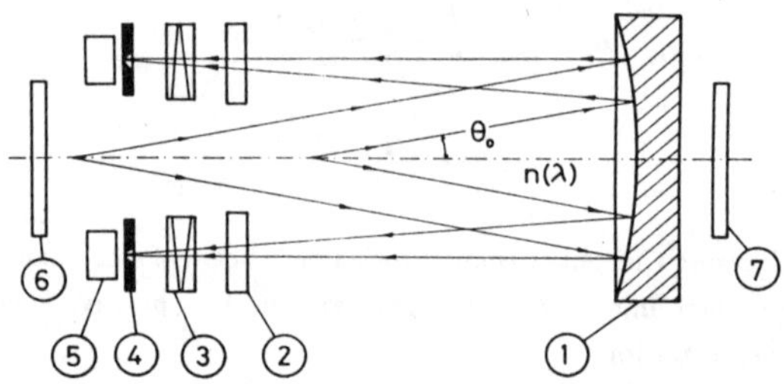

Fig. 7.25. Differential isochronous self-collimating Cherenkov counter: 1 — spherical mirror, 2 — coma corrector, 3 — chromatic corrector, 4 — ring slit, 5 — photomultipliers, 6 and 7 — electronic system with scintillation counters.

$$\nu = \frac{n(\lambda_2) - 1}{n(\lambda_1) - n(\lambda_3)}, \tag{7.23}$$

with λ_1 and λ_3 being the extreme wavelengths of the chosen spectral range and λ_2 being the average wavelength.

However, it is practically impossible to select an optical material and gas which meet this condition, Eq. (7.22). Therefore for cancellation of the chromatic aberrations it is worthwhile to use an axiconic doublet shown in Fig. 7.24 [287]. The wedge angles α and β of each axicon lens must be properly chosen. Such an axicon doublet has been made from fused silica SiO_2 and sodium chloride NaCl which have different optical dispersion parameters. But even with this design it is impossible to cancel completely the chromatic dispersion of the Cherenkov radiation. With triplet chromatic lens corrector the chromatic error $\Delta\theta_{chr}$ can be reduced by a factor of 15.

There are two kinds of geometro-optical aberrations in the Cherenkov counter: spherical aberration and coma aberration. The origin of the spherical aberration of the main mirror is that the source of the Cherenkov radiation in gas represents a straight line segment of the length up to 2 m or more. The origin of the coma aberration is that the axes of the conical waves are distributed over the area in the cross-section of the Cherenkov counter and some charge particles are moving eccentricically with respect to the optical axis of the main mirror of the counter.

The typical design of the differential isochronous self-collimating Cherenkov counter is shown in Fig. 7.25 [286]. This counter, dedicated for particle identification at high energies, consists of spherical mirror 1, coma corrector 2, chromatic corrector 3, opaque screen with narrow ring-shaped slit 4, photomultipliers 5 and electronic master system 6 and 7. The diameter D of the spherical mirror 1 with long focal length is chosen in accordance with inequality

$$D \geq b + 2L\theta, \tag{7.24}$$

where "b" is the diameter of the incident particle beam and L is the length of the gas radiator. The opaque screen with narrow ring-shaped slit 4 transmits the Cherenkov light from charge particles moving with velocity $v_\circ \pm \Delta v$, and the transmitted radiation is detected by a number of photomultipliers put equidistantly over the annulus.

In the design [288] an axicon objective is built up from the axicon doublet and the a spherical mirror is used as a corrector. The light ray traverses the axicon objective twice and the accepted deviation angle must be chosen equal to $\theta_\circ/2$. The material used in this axicon objective was fused silica plus water (SiO_2 + H_2O).

This focusing Cherenkov counter can be considered as one-channel velocity selector tuned at $v_\circ \pm \Delta v$ velocity range and also as one-channel direction selector with maximum angular beam divergence τ_{max}. To extend the angular and velocity acceptance of the focusing Cherenkov counter, the opaque screen with circular slit must be replaced by the meso–optical system, such as toroidal lens (Fig. 7.26). The transfer equation for the negative toroidal lens has the form in the thin lens approximation:

$$\alpha_2 - \alpha_1 = \frac{h - R_\nu}{-f} = -\frac{h}{f} + \frac{R_\nu}{f}, \tag{7.25}$$

where R_ν is the eccentricity at which the center of the generating circle arc of radius r is placed, and f is the equivalent focal length of a spherical lens with curvature $1/r$. This toroidal lens can be considered as an equivalent of the assembly of an axicon and of a spherical lens, both of very strong power. It has been used in the

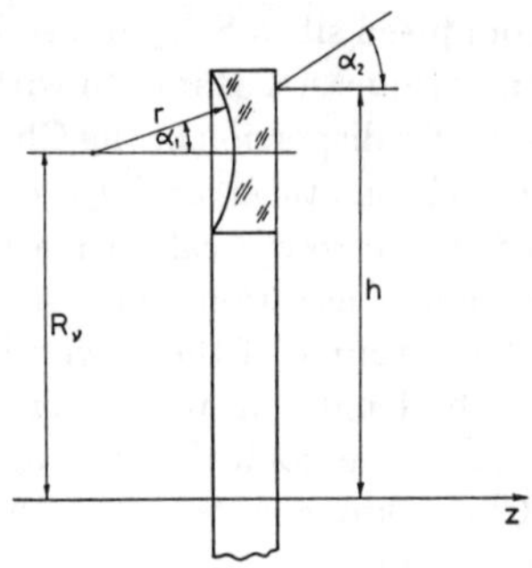

Fig. 7.26. Toroidal lens used in the multiplex Cherenkov counter instead of ring slit.

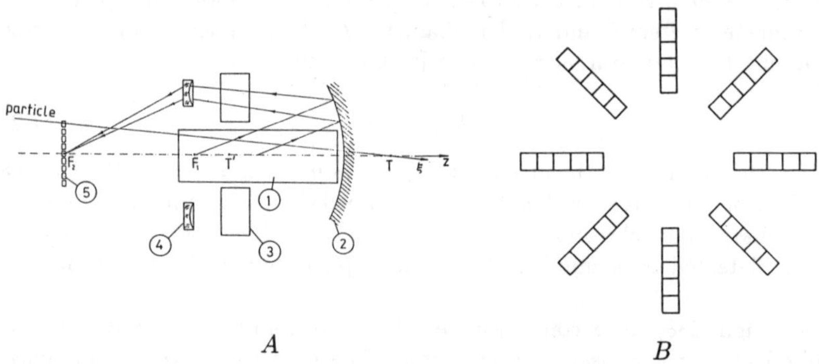

Fig. 7.27. Multiplex focusing Cherenkov counter: A: 1 — gas radiator holder, 2 — spherical mirror, 3 — optical corrector, 4 — toroidal lens, 5 — array of the spherical lenses, 6 — matrix of photodetector, T — particle beam focus, B: sparse $2D$ array of the photodetectors used in this Cherenkov counter.

"multiplexed" Cherenkov counter with optical system shown in Fig. 7.27. A set of spherical lenses are added to produce Cherenkov rings onto the photodetectors.

This Cherenkov counter is designed to detect the converging particles with beam focus at the point T. The parameter Q is a measure of distance from the principal plane H of the focusing mirror to the target position T, expressed in terms of its focal length F:

$$Q = \frac{(H\,T)}{2\,F}. \tag{7.26}$$

For single spherical mirror of curvature R the principal plane H is on the mirror surface and $R = 2F$. The point T' is the optical image of the target position T in the focusing optics with its axicon placed at T'. The point F_1 is the focal plane of the first-stage optics. As we have $T'F_1 \ll F$ the transfer lens must have a short focal length.

Though being spatially sparse, the detection system of this Cherenkov counter is actually equivalent to the traditional $2D$ matrix of photodetectors. This occurs because we wish to evaluate only **three** parameters: the diameter of the Cherenkov ring and the coordinates of the ring by observing only sparse picture of light pixels. With this a priori information about the object to be detected, we may reduce the number of pixels in the sparse array of the photodetectors. In the Cherenkov counter [289] eight radial arrays of photodetectors with five pixels in each radial array ($5 \times 8 = 40$) has proved to be sufficient to distinguish 85 different topological combinations composed from Cherenkov angle θ, particle divergence angle τ and azimuthal angle φ.

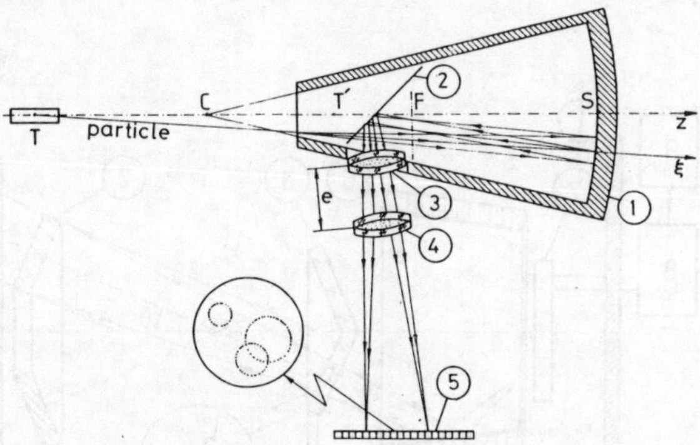

Fig. 7.28. Imaging Cherenkov particle detector with **ring** focusing mode optics: 1 — spherical mirror, 2 — plane mirror, 3 — first transfer lens L_1, 4 — second transfer lens L_2, 5 — matrix of photodetectors, C — center of the spherical surface, F — focal plane, T — particle beam focus, T' — its image produced by the spherical mirror 1.

Imaging Cherenkov Detectors, which have been developed at CERN, Geneva, with **ring** and **spot** focusing modes optics are shown, respectively in Fig. 7.28 and Fig. 7.29. The Cherenkov ring images are produced inside the gas radiator as virtual objects on the focal surface. The spherical mirror S with its center at point C transforms the target point T of the diverging particles into the point T'. The collecting and correcting optics is provided by two lenses L_1 and L_2, which are placed outside the radiation region and separated by the distance e. For Cherenkov angle θ, the diameter of the lens L_1 is chosen equal to $2\theta \cdot ST'$. The lens projects the virtual Cherenkov ring at infinity. The lens L_2 transfers this image on the matrix

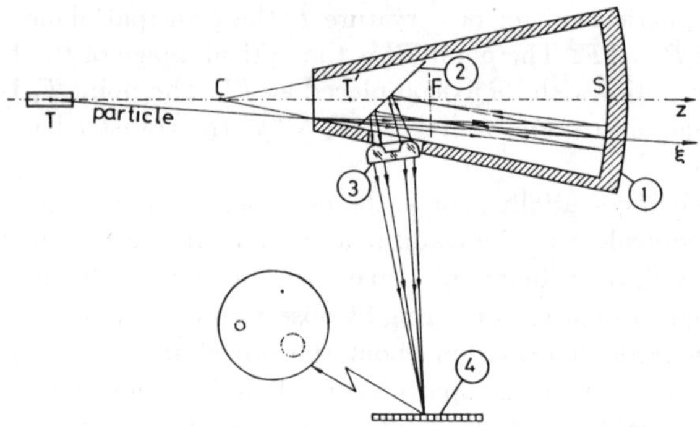

Fig. 7.29. Imaging Cherenkov particle detector with **spot** focusing mode optics: 1 — spherical mirror, 2 — plane mirror, 3 — toroidal lens, 4 — matrix of photodetectors, C — center of the spherical surface, F — focal plane, T — particle beam focus, T' — its image produced by the spherical mirror 1.

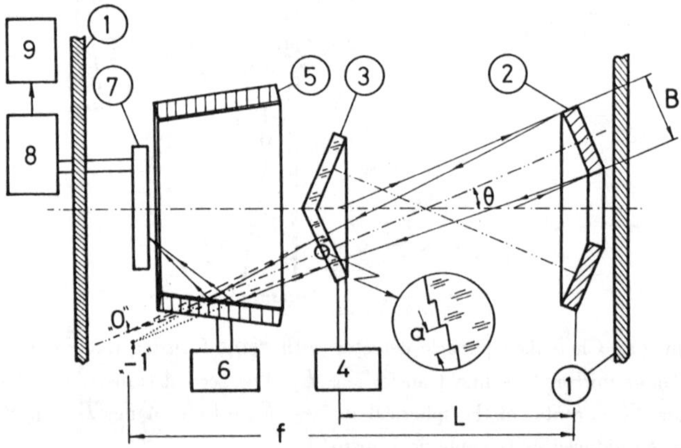

Fig. 7.30. Meso–optical differential Cherenkov gas counter: 1 — gas radiator holder, 2 — main meso–optical mirror, 3 — meso–optical chromatic compensator, 4 — moving stage of the chromatic compensator, 5 — conical mirror, 6 — moving stage of the conical mirror, 7 — matrix photodetector, 8 — moving stage of the matrix photodetector, 9 — computer memory block.

of photodetectors. With this optics it is possible to provide total geometrical and chromatic corrections. The later is a linear function of the separation distance e. By moving the lens L_2, we do not change the focusing or the magnification of the system. The lenses L_1 and L_2 can be made of fused silica and liquid fluorocarbon

FC–72 [290]. In the spot focusing mode optics of the Imaging Cherenkov Detector, shown in Fig. 7.29, a toroidal lens is used.

A circular diffraction grating-axicon has been used for collimating of the Cherenkov radiation, generated as second harmonic in lithium niobate ($LiNbO_3$) waveguide [291]. The circular diffraction grating has been fabricated by diamond turning on acrylic plate as a blazed grating with grating pitch "a" defined by the Cherenkov angle $\theta_\circ$ and by wavelength λ_2 of the second-harmonic wave:

$$a = \lambda_2 / \sin\theta_\circ \tag{7.27}$$

and with blazed angle α, given by equation:

$$\tan\alpha = (n_b - \cos\theta_\circ) / \sin\theta_\circ, \tag{7.28}$$

where n_b is the refractive index of the grating substrate. In the experiment [291] $a = 0.90\ \mu$m, diameter of the grating $D = 5.4$ nm, the total number of grooves $N = 300$ and diffraction efficiency $\approx 50\%$. The aberrations from the grating substrate anisotropy, caused by birefringence, can be compensated by introducing the inclined output facet or by using an elliptic grating instead of a circular one.

Now we describe the meso–optical differential Cherenkov gas counter [265, 292] sketched out in Fig. 7.30. It consists of gas radiator 1, main mirror 2, chromatic compensator 3, moving stage 4 of the chromatic compensator, conical mirror 5, moving stage 6 of the conical mirror, matrix photodetector 7, moving stage 8 of the matrix photodetector and computer memory block 9. The generating line of the meso–optical mirror 1 has the form of a parabola segment with axis of the parabola parallel to the average wavevector of the Cherenkov light. The average orientation angle $\theta_\circ$ with respect to the optical axis of the counter is given by the equation

$$\theta_\circ^2 = 2(n_\circ - 1) - 1/\gamma_\circ^2, \tag{7.29}$$

where $n_\circ$ is the average refractive index of the gas, and $\gamma_\circ$ is the average relativistic factor of the charge particle to be detected. The focal length of the parabola segment is chosen greater than the length of the gas radiator. The chromatic corrector 3 represents a phase circular diffraction grating with first diffraction order directed toward the optical axis. The backing of this phase circular diffraction grating represents the conical surface with opening angle $(90° - \theta_\circ)$. With this chromatic corrector we can suppress **any** linear λ-dependence of the refractive index $n(\lambda)$ of the gas medium. The third meso–optical element of this differential Cherenkov counter is a conical mirror 5 with internal reflecting surface.

The charge particle with relativistic factor $\gamma_\circ$ emits conical Cherenkov wave, which is falling on the main meso–optical mirror 2. After reflection from this mirror the light rays cross the optical axis of the counter and only after that they are focused into a Cherenkov ring. The latter crosses the plane of the figure at the point "0". The main mirror 2 works in the mounting for which the coma aberration is minimal, if $\theta = \theta_o$. These light rays traverse the chromatic corrector 3 perpendicular to the generating line of its backing surface. The feature of the chromatic corrector 3 is chosen in accordance with diffraction grating equation:

$$\sin\theta_d = \frac{\lambda}{a}, \tag{7.30}$$

where θ_d is the diffraction angle of the first order and "a" is the pitch of the phase circular diffraction grating. The conical mirror 5 is dedicated to change at will the diameter of the Cherenkov ring or to focus all Cherenkov radiation into a point.

The relative chromatism of the phase circular diffraction grating is equal to

$$\frac{1}{f}\frac{\partial f}{\partial\lambda} = -\frac{1}{\lambda}, \tag{7.31}$$

where f is the length of the light rays between phase circular diffraction grating 5 and the photodetector 7, the position of which can be changed with moving stage 4. By contrast the relative chromatism of the spherical refracting surface

$$\frac{1}{f}\frac{\partial f}{\partial\lambda} = -\frac{1}{n(n-1)}\frac{\partial n}{\partial\lambda}, \tag{7.32}$$

is about 20 times smaller than the relative chromatism of the diffraction grating. Hence the chromatic correction in this meso–optical Cherenkov counter is attained at smaller distances than in the conventional Cherenkov counter with refractive elements.

All elements of the meso–optical Cherenkov counter can be also fabricated as mirror elements which provide more wide spectral range of Cherenkov counter than in the system with transparent elements.

Now we present an example of the design of the meso–optical Cherenkov counter. If the length of the gas radiator is equal to $L = 2.4$ m, the focal length of the main mirror 2 is equal to $f = 4$ m. For $\theta_o = 5\cdot10^{-2}$ rad the working length of the parabola segment is equal to $B = 120$ mm. For spectral range $\Delta\lambda = \lambda_1 - \lambda_2 = 0.234$ μm the width of the Cherenkov ring is equal to $\Delta x = \frac{\lambda f}{B} = 16$ μm. The opening angle of the chromatic corrector backing is equal to $90° - \theta_o = 87°8'$. The operating width of the phase circular diffraction grating is $A = B\,\frac{f-L}{L} = 50$ mm. The pitch of this circular grating is equal to

$$a = \frac{\Delta\lambda\,(f-B)}{L\cdot\Delta\theta_{chr}} = 52\,\mu\text{m}, \tag{7.33}$$

for $\Delta\theta_{chr} = 3\cdot10^{-3}$ rad. The number of grooves on the working part of the chromatic corrector is equal to $N = A/a = 1000$. The angular width of the diffraction order is equal to $\Delta\theta_o = \theta_o/N = 10^{-5}$ rad for $\theta_o = \lambda_o/a = 10^{-2}$ rad. The width of the Cherenkov ring on the matrix of the photodetectors is equal to $\Delta x_d = 14$ μm.

Chapter 8

MESO–OPTICS AND MICRO-TOMOGRAPHY

8.1. Modern Microscopes for 3D Viewing

The microscopic object to be imaged is referred to as $3D$ one, if its thickness along optical axis is greater than the depth-of-focus of the imaging system. The image of this $3D$ object, produced by any conventional light microscope, represents a $3D$ manifold. To get an impression of the whole $3D$ object, the observer looks at a series of $2D$ images by changing the position of the focus plane inside the object [297]. However $3D$ image of the $3D$ object cannot be detected directly by this manner with a plane detector, because beside the in-focus image the observed signal involves also the contribution from the out-of-focus slices of the object [293, 294, 297]. The detected information represents only an indirect appearance of the object. To reconstruct a real object a stage of reconstruction must be performed. For weak scattering and dilute $3D$ objects we may construct a linear model of imaging, which permits to consider the $3D$ image of the $3D$ object as a stack of intermediate $2D$ images. This input information is used on the reconstruction stage.

By developing general theory of $3D$ viewing, we use some complicated notions such as $3D$ aperture and $3D$ optical transfer function [294, 297, 304, 314]. The starting point of all these modern approaches is Abbe theory of imaging [316] developed for $2D$ microscopic object, which is considered as an array of planar diffraction grating of various periods and orientations. To be well viewed, the first diffraction order of the corresponding diffraction grating is to be transmitted by the objective pupil. The $2D$ manifold of all these elementary diffraction gratings of the $2D$ object is equivalent in the sense of the Fourier transform theorem to the $2D$ input object. As extended version of these ideas of Abbe, we decompose $3D$ object into $3D$ array of the volume diffraction gratings, and accept, that these elementary volume gratings are equivalent in the sense of $3D$ Fourier transform to the $3D$ input object, and there exists a notion $3D$ pupil [297]. To explain $3D$ version of the Abbe theory of microscopic imaging, we consider the principal scheme of the $3D$ imaging system provided with telecentric optics (Fig. 8.1).

Two conditions define the feature of the $3D$ image: the Abbe condition and the Bragg condition. The analogy between simple regular crystal lattice and volume diffraction grating is fairly close. In the same manner as in the case of the crystal lattice, the $3D$ spatial frequency of the volume diffraction grating is indeed a $3D$ vector $\vec{\omega}\,(\omega_x,\, \omega_y,\, \omega_z)$ which can be decomposed into lateral part $\vec{\omega}_\perp\,(\omega_x,\, \omega_y)$ and longitudinal part ω_z. The illuminated light beam is defined also by the $3D$ wave

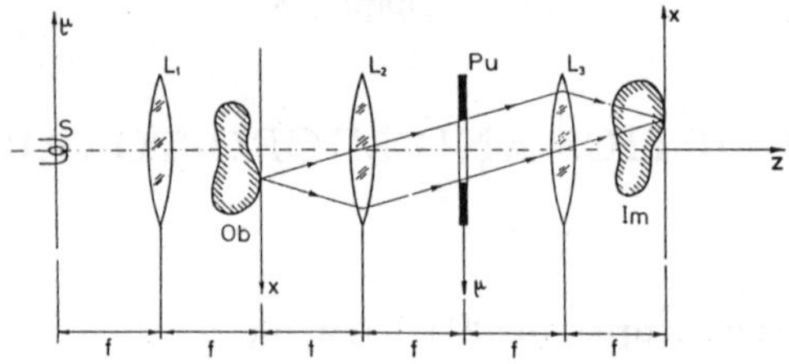

Fig. 8.1. Principal scheme of $3D$-imaging system with telecentric optics: S — light source, L_1 — illuminating lens, Ob — $3D$ object, L_2 — first imaging lens, Pu — pupil diaphragm, L_3 — second imaging lens, Im — $3D$ image of $3D$ object without linear magnification, f — focal length of the L_1, L_2 and L_3 lenses.

vector $\vec{k}_{o}(k_{ox}, k_{oy}, k_{oz})$ with its lateral and longitudinal components $\vec{k}_{\perp o}$ and k_{zo} with $k_{zo} = \sqrt{\lambda^{-2} - \vec{k}^{\,2}_{\perp o}}$. To be effectively interacted with $3D$ volume diffraction grating, the illuminated $\vec{k}_o$ and diffracted $\vec{k}_D$ wave vectors must fulfil the Bragg condition:

$$\vec{k}_{D\perp} = \vec{k}_{o\perp} + \vec{\omega}_{\perp}, \tag{8.1}$$

and the Abbe condition:

$$|\vec{k}_{D\perp}| = |\vec{k}_{o\perp} + \vec{\omega}_{\perp}| \leq R_{\rho}, \tag{8.2}$$

where R_ρ is the radius of the geometrical pupil of our imaging system. If the illuminated light is emerging from a source of finite radius R_s, and if $|\vec{k}_{o\perp}| \leq R_s$, the Abbe's lateral diffraction limit assumes the form

$$|\vec{\omega}_{\perp}| = |\vec{k}_{D\perp} - \vec{k}_{o\perp}| \leq R_s + R_{\rho}. \tag{8.3}$$

The Bragg's longitudinal diffraction limit defines

$$k_{oz} = [\lambda^{-2} - (\vec{k}_{D\perp} + \vec{k}_{o\perp})^2]^{1/2} - (\lambda^{-2} - \vec{k}^{\,2}_{o\perp})^{1/2}. \tag{8.4}$$

Typical $3D$ apertures are shown in Fig. 8.2 for various R_s / R_ρ ratios. Only those $3D$ spatial frequencies, that are located inside the given volumes, are indeed imaged by this imaging system. In the coherent limit with $R_s \to 0$, the object spectrum has the form of a thin sheet (Ewald sphere). The partially coherent illumination with large R_s / R_ρ ratios increases the size of the $3D$ aperture. We see that both the aperture of the imaging system and the spatial coherence of the illumination determine the $3D$ resolving power of the microscope. In the limit of very thin $3D$ object, which can be considered as $2D$ object the z-spatial frequency converges to zero frequency.

As was proved in [294], the wave amplitude near the geometrical focus of the converging wave can be represented as $3D$ Fourier transform of the $3D$ aperture:

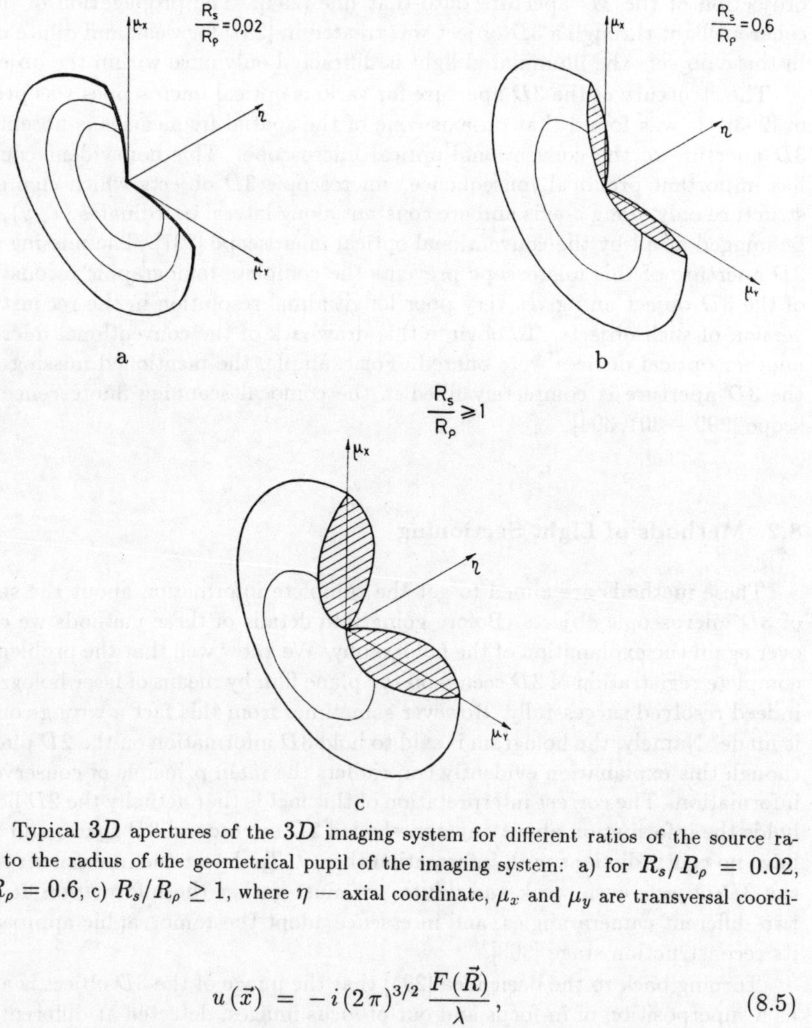

Fig. 8.2. Typical $3D$ apertures of the $3D$ imaging system for different ratios of the source radius R_3 to the radius of the geometrical pupil of the imaging system: a) for $R_s/R_\rho = 0.02$, b) $R_s/R_\rho = 0.6$, c) $R_s/R_\rho \geq 1$, where η — axial coordinate, μ_x and μ_y are transversal coordinates.

$$u(\vec{x}) = -i\,(2\,\pi)^{3/2}\,\frac{F(\vec{R})}{\lambda}\,, \tag{8.5}$$

where $F(\vec{R})$ is $3D$ Fourier transform of the transmitted wave amplitude $A(\vec{q})$,

$$F(\vec{R}) = (2\,\pi)^{-3/2} \int_{\Omega} A(\vec{q}) \cdot exp[-i\,k\,\vec{q} \cdot \vec{R}]\,d\Omega\,, \tag{8.6}$$

$\vec{q}$ is a unit propagating vector, $\vec{R}$ is the position vector and $d\Omega$ is elementary solid angle with center in the point O. The wave amplitude distribution along arbitrarily oriented line, going through the focus, is equal to $1D$ Fourier transform of the projection of the $3D$ aperture onto that line [294]. The propagation of partially coherent light through a $3D$ object was treated in [297] for weak and dilute objects. In these objects the illuminated light is diffracted only once within the object.

The structure of the $3D$ aperture for various optical microscopes was presented in [293]. It was found that on-axis cone of the spatial frequencies is absent in the $3D$ aperture of the conventional optical microscope. This nonevident conclusion has important practical consequence: microscopic $3D$ objects which change their structure only along z-axis and are constant along lateral coordinates (x, y), cannot be imaged at all by the conventional optical microscope [297]. The missing cone in $3D$ aperture of this microscope prevents the complete tomographic reconstruction of the $3D$ object and gives very poor longitudinal resolution in the reconstruction version of such objects. To obviate this drawback of the conventional microscope, another optical devices were offered. For example, the mentioned missing cone in the $3D$ aperture is completely filled in the confocal scanning fluorescence microscope [299 – 301, 304].

8.2. Methods of Light Sectioning

These methods are aimed to get the complete information about the structure of $3D$ microscopic objects. Before going into details of these methods we consider over again the explanation of the holography. We know well that the problem of the complete registration of $3D$ scenes on the plane film by means of laser hologram was indeed resolved successfully. However sometimes from this fact a wrong conclusion is made. Namely, the hologram is said to hold $3D$ information on the $2D$ plane film, though this explanation evidently contradicts the main principle of conservation of information. The correct interpretation of this fact is that actually the $2D$ hologram holds the information about any non-shaded $2D$ **surface** in $3D$ space, and thus we have no contradictions with information theory. To detect the internal structure of the $3D$ object, even weak and dilute, we must gather the information at least at two different camera angles, and in essence adopt the tomographic approach with its reconstruction stage [303].

Turning back to the basic idea [295] that the image of the $3D$ object is accepted as a superposition of in-focus and out-of-focus images, detected at different optical transfer functions of the imaging system, by changing, for example, the position of the focus plane inside the $3D$ object. Each detected image is approximated by linear superposition of contribution from different object layers with weighted coefficients defined by the $3D$ aperture of the microscope. The object reconstruction is formulated as inverse problem with input data as a stack of the intermediate $2D$ images. Due to noise in the image plane, this inverse problem is ill posed and for its solving various regularization methods were introduced. One such method, based on the minimum square estimation, was described in [295]. To get reconstruction of high quality, we need also an exact information about the optical transfer function of the imaging system, about power spectrum of the objects and of the noise. Some new accurate analytical models of the conventional optical microscope are required to formulate adequately the problem of image reconstruction [299].

The relation between the $3D$ object and the $3D$ intensity distribution in the image was derived in [297]. It was shown that in the first-order Born approximation the process of $3D$ image formation for partially coherent illumination is described by **two** optical transfer functions, one for absorptive and one for phase parts of the weak and dilute object. Within $3D$ aperture, the $3D$ optical transfer function is diminished at high spatial frequencies, lateral and longitudinal. This is a reason for out-of-focus noise, which is worse for high degree of coherence of the illuminated light.

The information available in $3D$ image is defined by the sampling theorem [298]. A space of sampled spatial frequencies inside the $3D$ aperture is limited by the lateral spatial frequency $\vec{\omega}_\perp$ and by the longitudinal spatial frequency ω_z with

$$|\vec{\omega}_\perp| < \omega_z / \tan(\frac{\alpha}{2}), \tag{8.7}$$

where α is the angular aperture of the microscope. For lateral part of the spatial frequency $\vec{\omega}_\perp$ the samples must be located at points with lateral coordinates

$$\rho_n = \beta_n / \pi D, \qquad n = 0, 1, 2, ..., \qquad N - 1, \tag{8.8}$$

where β_n is the n-th zero of the Bessel function $J_1(\bullet)$ and D is the diameter of the circle which contains completely the lateral borders of the $3D$ object to be imaged along the lateral coordinate $\vec{x}_\perp$. The array of the spatial frequencies sufficient to reconstruct the object is contained in the space of sampled spatial frequencies if

$$N \equiv D \cdot \Omega_z < \frac{\beta_1}{2\pi} \tan \frac{\alpha}{2}, \tag{8.9}$$

where Ω_z is the border longitudinal spatial frequency of the $3D$ Fourier transform $T(\vec{\omega})$ of the amplitude transparency function $t(\vec{x})$ of the $3D$ object.

The existence of the longitudinal border of the spatial frequency Ω_z implies that $3D$ object may be reconstructed from information in stack of planes perpendicular to the optical axis and that each point in a plane is defocused up to a circle encompassing the entire object in the neighbouring plane. Deconvolution operation on the reconstruction stage reconverts the out-of-focus disks into rings which do not overlap the image reconstructed in the neighbouring planes. An aliasing phenomenon can arise if the object is too large (D) or the distance between the neighbouring planes is too small and due to this the condition of the sampling theorem, given by Eq. (8.9) is violated.

In this Chapter we describe confocal scanning microscopes [300, 301, 304 – 311, 320 – 323], tomographic microscope with oblique illumination [1], meso–optical microscope for straight line objects [313, 324 – 328], stereo-microscope with high depth resolution over large field of view for observation of shallow microrelief on the plane backing. The analysis of the meso-optical Fourier transform microscope as a tomographic device will be given [312].

8.3. Confocal Scanning Microscope

A confocal scanning transmitted microscope [301, 306, 310] shown in Fig. 8.3, is provided with two objectives, the illuminating objective and the imaging (collecting) objective, and with point-like detector. The illuminating system used here is an extreme variant of the critical illumination [296], at which only one small part of the $3D$ object is illuminated in a given moment of time. The axial scanning of a $3D$ object is carried out at audio frequencies, and the lateral scanning along x and y axes is performed with an ordinary mechanical $(X - Y)$ stage. This $3D$ image of the $3D$ object is stored in the computer memory.

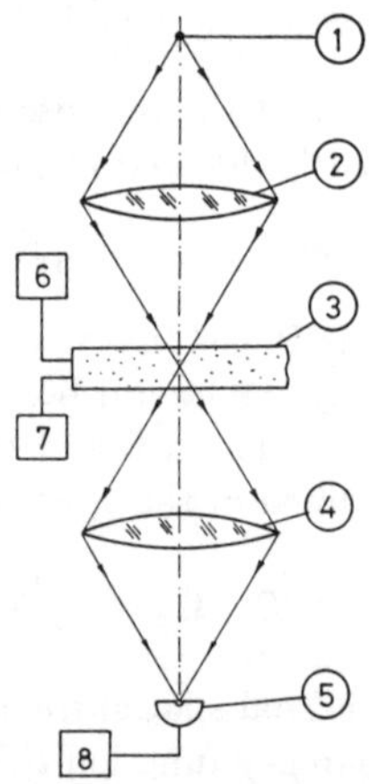

Fig. 8.3. Principal scheme of the conventional confocal scanning transmitted microscope: 1 — point-like light source, 2 — illuminating objective, 3 — object, 4 — imaging objective, 5 — point-like detector, 6 — $(X - Y)$ moving stage, 7 — Z-moving stage.

The performance of the microscope is defined by the point spread function $p(\vec{x}) = p(x, y, x)$. In the conventional microscope with coherent illumination the complex wave amplitude of the image can be represented as $3D$ convolution of the amplitude point spread function $p(\vec{x})$ and the amplitude wave transmittance $t(\vec{x})$ of the $3D$ object [302]:

$$u(\vec{x}) = p(\vec{x}) \underset{3}{\otimes} t(\vec{x}) . \tag{8.10}$$

For spatial invariant point spread function $p(\vec{x})$ we may represent Eq. (8.10) in the spatial frequency space as

$$U(\vec{\omega}) = P(\vec{\omega}) \cdot T(\vec{\omega}), \tag{8.11}$$

where $P(\vec{\omega})$ is the $3D$ optical transfer function of the microscope, equal to $3D$ Fourier transform of the point spread function $p(\vec{x})$:

$$P(\vec{\omega}) = \hat{\mathcal{F}}_3[p(\vec{x})] = \int\int\int p(\vec{x}) \cdot exp(-i\vec{\omega}\vec{x})\, d\vec{x} \tag{8.12}$$

and $T(\vec{\omega})$ is equal to $3D$ Fourier transform of the amplitude wave transmittance $t(\vec{x})$ of the $3D$ object:

$$T(\vec{\omega}) = \hat{\mathcal{F}}_3[t(\vec{x})] = \int\int\int t(\vec{x}) \cdot exp(-i\vec{\omega}\vec{x})\, d\vec{x}. \tag{8.13}$$

In the conventional microscope with incoherent illumination, such as fluorescence microscope, the image intensity is equal to the $3D$ convolution:

$$i(\vec{x}) = |u(\vec{x})|^2 = |p(\vec{x})|^2 \bigotimes_3 |t(\vec{x})|^2, \tag{8.14}$$

where $|p(\vec{x})|^2$ is called the intensity point spread function. In the spatial frequency space

$$I(\vec{\omega}) = [P(\vec{\omega}) \bigotimes_3 P^*(\vec{\omega})] \cdot \hat{T}(\vec{\omega}), \tag{8.15}$$

where

$$\left.\begin{aligned} I(\vec{\omega}) &= \hat{\mathcal{F}}_3[|u(\vec{x})|^2], \\ \hat{T}(\vec{\omega}) &= \hat{\mathcal{F}}_3[|t(\vec{x})|^2]. \end{aligned}\right\} \tag{8.16}$$

The amplitude point spread function of the confocal scanning transmitting microscope, shown in Fig. 8.3, is equal to the product of the amplitude point spread function of the illuminating objective $p_1(\vec{x})$ and that of the imaging objective $p_2(\vec{x})$:

$$p(\vec{x}) = p_1(\vec{x}) \cdot p_2(\vec{x}). \tag{8.17}$$

Due to this property, the lateral and axial spatial resolutions of the confocal scanning transmitting microscope are about twice as large as in the equivalent conventional optical microscope.

In the real system, the detector has finite dimensions, $D(x, y) \neq \delta(x) \cdot \delta(y)$, and the image is described by the lateral intensity distribution

$$I(\vec{x}_\perp) = |p_1(\vec{x}_\perp)|^2 \cdot \{D(\vec{x}_\perp) \bigotimes_2 |p_2(\vec{x}_\perp)|^2\}, \tag{8.18}$$

where $\vec{x}_\perp$ is the $2D$ position vector. As the image of the $3D$ object, generated in the confocal scanning microscope, is stored in computer memory, the depth-of-focus is very high, and the confocal scanning microscope reveals practically unlimited depth-of-focus.

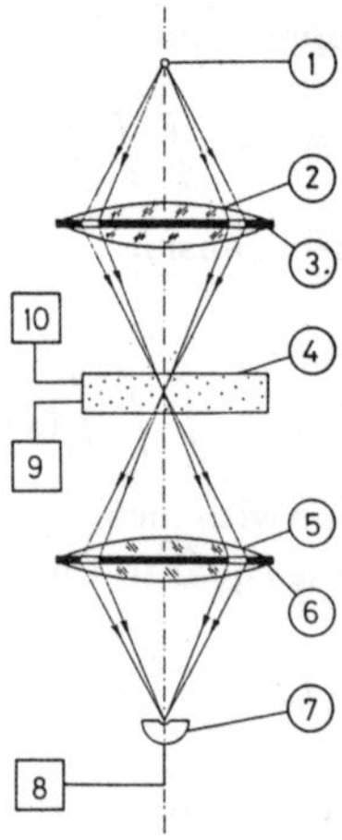

Fig. 8.4. Confocal scanning microscope with annular apertures in both objectives: 1 — point-like light source, 2 — illuminating objective, 3 — annular aperture, 4 — object, 5 — annular aperture, 6 — imaging objective, 7 — point-like detector, 8 — $(X - Y)$ moving stage, 9 — Z-moving stage.

It will be instructive to compare the lateral coherent optical transfer functions of the conventional coherent microscope, of the confocal scanning microscope with full and annular aperture (Fig. 8.4). In the discussion around the report of T. Wilson, D. Rudolph from the University of Göttingen, asked: "did you ever try to apply **two annuli** into your microscope to gain resolution?" [317]. The complete analysis of the confocal fluorescent microscope provided with objective having annular pupils has been done in [320]. The amplitude point spread function of the confocal fluorescent microscope with objectives both having annular apertures is equal to

$$p_\circ(\vec{x}) \equiv p(r, z) \equiv p(\zeta, \eta) = \int A(\rho, \eta)\, J_\circ(\zeta, \rho)\, \rho\, d\rho\,, \tag{8.19}$$

where the pupil function of the annular aperture

$$A(\rho, \eta) = \begin{cases} exp(i\,\frac{\eta}{2}\rho^2)\,, & 0 < \varepsilon \leq \rho \leq 1\,, \\ 0\,, & \text{otherwise}\,. \end{cases} \tag{8.20}$$

written with normalized optical coordinates,

$$\left.\begin{aligned} \zeta &= \tfrac{2\pi}{\lambda} r \sin\alpha\,, \\ \eta &= \tfrac{8\pi}{\lambda} z \sin^2(\alpha/2)\,, \\ \rho &= r/a\,. \end{aligned}\right\} \tag{8.21}$$

Here r is the radial coordinate, a is the outer radius of the objectives and α is the semi-aperture angle of the objectives [320]. For pinhole detector of radius δ the detector intensity sensitivity is equal to

$$D(x_\perp) = D(\zeta) = \begin{cases} 1, & \zeta \leq \zeta_d\,, \\ 0, & \zeta \geq \zeta_d\,. \end{cases} \tag{8.22}$$

with

$$\zeta_d = 2\pi \cdot \delta \cdot a/\lambda l \tag{8.23}$$

and l is the distance between imaging objective and a pinhole detector. The lateral intensity point spread function of this system can be calculated by means of Eq. (8.18). The axial intensity point spread function is equal to $1D$ inverse Fourier transform of the $3D$ optical transfer function $P(\omega_\perp, \omega_z)$ taken at $\omega_\perp = 0$:

$$I_{\|}(\eta) = \int P(\omega_\perp = 0, \omega_z)\, exp(i\omega_z \eta)\, d\omega_z\,. \tag{8.24}$$

The function $I_{\|}(\eta) = I_{\|}(z)$ defines the optical sectioning effect or the ability of the microscope to suppress the out-of-focus images.

The analysis of the $3D$ confocal scanning fluorescent microscope with annular pupils shows that this microscope for large central obstruction reveals high transverse resolution, but for the detector of finite size and for low ε a missing cone, which is typical for conventional microscopes, appears. There is an optimal configuration with central obstruction ε and radius δ of the pinhole detector, at which the axial optical sectioning remains high. The depth of focus of the system with annular pupils is **not** always longer than that with a circular pupil, but the difference is minor. To maintain high signal strength, a system with only **one** objective with annular pupil is recommended.

In the confocal scanning microscope various depth discrimination criteria can be introduced [307]. In particular, an alternating depth discrimination function in the form of the integral of the square of the optical transfer function

$$J(\omega_z) = \int |P(\vec{\omega}_\perp, \omega_z)|^2\, d\vec{\omega}_\perp\,, \tag{8.25}$$

is very useful to consider the pupil of the general form. The function $J(\omega_z)$ gives a measure of the contrast reduction for out-of-focus parts of the $3D$ object. It was shown that the simplest method to perform dark-field observation is to displace the point detector of the confocal microscope laterally [307]. The transfer function in the direction of the displacement is changing as $J_1(2\omega^d)/(2\omega^d)$. For $\omega^d = 1.916$

the transfer function is equal to zero and thus a dark-field observation is achieved. However now the depth discrimination is getting not as sharp as in on-axis position of the point detector. The depth discrimination at 50% is equal to $\pm 0.7\ \lambda$ for objective with numerical aperture 1.0 [309]. This permits the direct transformations of the $3D$ array of samples, such as projections from somewhat different angles constituting a stereo pair, the movie of these projections gives the impression of the object rotation.

To modify the response of the confocal microscope to the straight-edge objects a new transfer function that is constant up to the cut off spatial frequency has been suggested [305]. The spatial edge gradient can be increased up to 2.36 times in comparison to the traditional confocal scanning microscope.

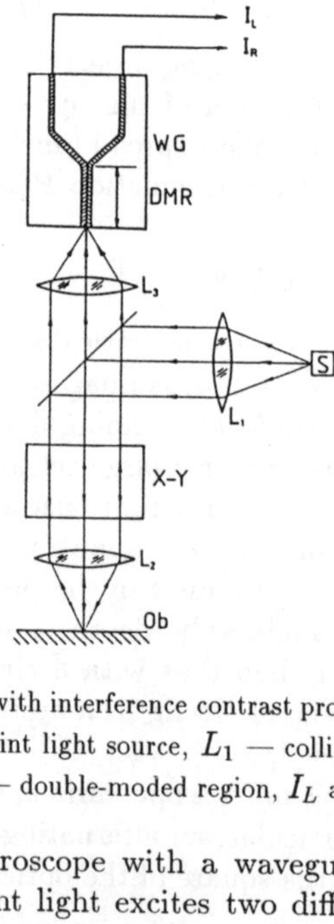

Fig. 8.5. Confocal scanning microscope with interference contrast provided with detector in the form of the waveguide device WG: S — point light source, L_1 — collimator, L_2 — lens, L_3 — lens, $(X - Y)$ — scanning device, DMR — double-moded region, I_L and I_R — two output signals.

In the confocal scanning microscope with a waveguide device instead of the pinhole detector [321] the incident light excites two different modes on the front faces of the double-moded region of this waveguide device (Fig. 8.5). Two output signals from this microscope define the differential interference contrast image, from which we can separately obtain the information about amplitude and phase parts of the object characteristics.

In the fiber-optical confocal scanning microscope [308] two optical single-mode F_1 and F_2 are inserted as illumination source and collection detector (Fig. 8.6). This microscope behaves as a coherent microscope with $3D$ effective point-spread function, which is defined not only by the $3D$ point spread functions of the objectives

L_1 and L_2, but also by the $2D$ amplitude mode profiles of the fibers F_1 and F_2 and by linear magnifications M_1 and M_2 of the objectives L_1 and L_2. The performance of this confocal microscope was estimated for reflection and transmittance modes. The relationship between $3D$ and $2D$ coherent transfer functions has been found [308].

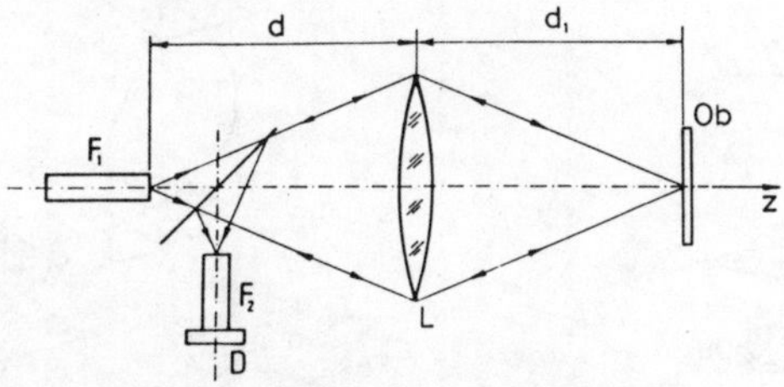

Fig. 8.6. Fiber-optical confocal scanning microscope: F_1 — fiber source, F_2 — fiber detector, L — lens, Ob — object, D — photodetector.

8.4. Microscope with Rotating Oblique Illumination

A transmission optical microscope with rotating oblique illumination [1] is shown in Fig. 8.7. The Köhler illuminator with off-axis point source S provides an eccentrically shifted pupil. The condenser lens L_c generates a beam going at the angle θ with respect to the optical axis. The objective lens L_o produces an image in the image plane. The condenser stop with off-axis pupil is rotating to detect a set of the projection images. The geometry of these projections is equivalent to the tomographic device with viewing axes on the cone. The performance of this microscope is given by the $3D$ optical transfer function. The structure of this function was calculated in the frame of lateral and axial spatial frequencies [1] for coherent and incoherent illuminations. It was shown that the lateral frequency band for incoherent illumination is twice as large as that for coherent illumination. The depth of focus for coherent illumination is much longer than that for incoherent illumination. As well as in all other microscopes, there exists the trade-off between these two parameters. Each $2D$ projected image, $I(\vec{x}_\perp)$, detected in the transmission optical microscope with rotating oblique illumination, represents a linear combination of the $3D$ object transmittance $t(\vec{x}) = t(x, y, z)$:

$$I(\vec{x}_\perp) = \hat{H} \cdot t(\vec{x}) + N(\vec{x}_\perp), \tag{8.26}$$

where $\hat{H}$ is (2×3) operator-matrix of the transformation $t(\vec{x}) \to I(\vec{x}_\perp)$ and $N(\vec{x}_\perp)$ is the image noise. The elements of the operator $\hat{H}$ are defined by the array of the light paths going through the points of the $3D$ object in all (θ, φ)-projections. The $3D$ function $t(\vec{x})$ represents optical density of the object elements.

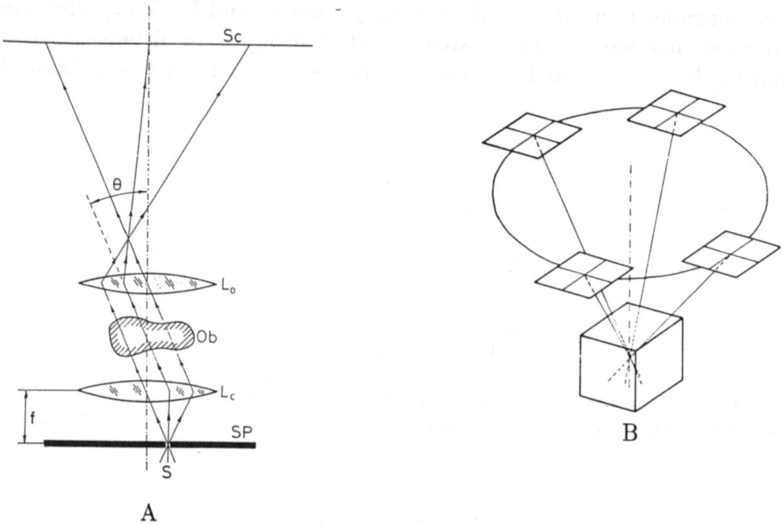

Fig. 8.7. Optical microscope with rotating oblique illumination. "A" — optical scheme: S — light source, SP — shifted pupil, L_C — condenser, Ob — $3D$ object, Lo — objective, S_C — observation screen. "B" — geometry of the tomographic viewing, when the projection axes are on the conical surface.

The inverting of the Eq. (8.26) was made according to the regularization principle of reconstruction by minimization of the least squares. Many other algorithms can be used on the stage of tomographic reconstruction [315, 325, 326]. To improve the spatial resolution of the reconstructed image, **a priori** information about the spatial extent of the $3D$ object was used as the support constraint [1]. The angular range of the projections in this device is limited by the angular aperture of the imaging objective. Here we have a close analogy with problem of image restoration in the tomographic system with limited angle projection data [327]. As operator $\hat{H}$ is quite sparse, this drastically reduces the complexity of the inverting calculations and the demanded computer memory.

8.5. Meso–optical Microscopes as Tomographic Devices

Meso–optical Fourier transform microscope (MFTM) [3] generates two images of the line object viewed simultaneously at two camera-angles. Therefore no wonder that this device can be treated as a tomographic system. Recall that in the reconstructed tomography [328], a point-like object is converted into an intermediate image in the form of sinogram. The analogous sinograms are generated by the straight line objects having common vertex in the Fourier microscope [330], dedicated for estimation of the orientation angles of these objects. A tomographic observation principle with very high productivity can be accomplished with meso-optical illuminating condenser [329] for detecting the microscopic line objects going parallel to the optical axis of the system. Several other designs of these meso–optical microscopes will be described below. There are at least four regions which are common for meso–optical microscopes, on the one hand, and for the reconstructed tomography, on the other hand. It will be shown that the following features of the meso–optical microscopes manifest their tomographic properties: 1) the topological structure of the intermediate data of the straight line objects going at small angles with respect to the normal of the tomographic slice plane, 2) the structure of the Fourier transform of the straight line object which is rotating with respect to the axis oriented at some angle to the optical axis of the system, 3) the structure of the output data which are detected in the course of the $1D$ scanning of the working volume by the fence-like illuminating beam, 4) the theorem that meso–optical images of the straight line objects having common vertex lie on the sinogram [312].

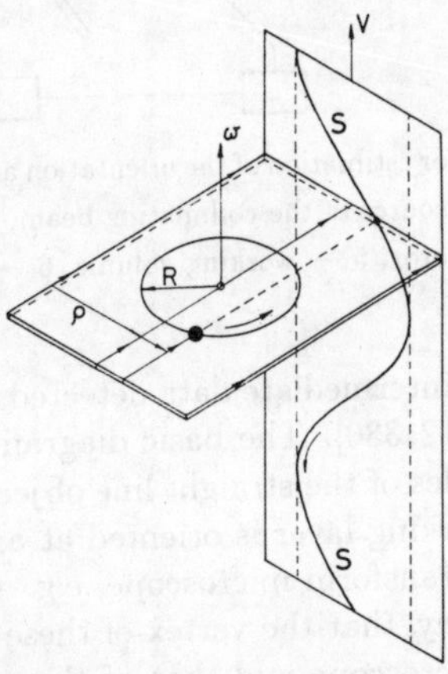

Fig. 8.8. Tomographic illumination of the point-like object and $2D$ detected signals: ω — rotating axis, R — radius of the ring, ρ — impact parameter, V — velocity vector of the screen movement, S — sinogram of the point-like object on the $2D$ screen.

The basic diagram of the tomographic illumination of the point-like object on the stage of detection of the intermediate date in the form of the ray-sums is shown in Fig. 8.8. For a point-like object at distance ρ with respect to the rotation axis, the ray-sums has the form of the sinusoid ("sinogram") in the frame of the (ρ, θ) coordinate system:

$$g(\rho, \theta) = \rho \sin(\theta - \theta_\circ), \tag{8.27}$$

where θ is the position angle of the position vector $\vec{\rho}(\rho, \theta)$ with respect to the orientation of the illuminating light rays, and $\theta_\circ$ is the angle θ, at which the first ray-sum is located. To reconstruct this point-like object in the (ρ, θ) plane, it is sufficient to estimate the amplitude of the sinogram and the position of the extremum of the ray-sums $g(\rho, \theta)$ along θ-angle. If the object in this tomographic slice consists of several point-like elements, there are the same number of sinograms [328] with different ρ and θ. These parameters are reconstructed by the computer. The array of many tomographic slices represents the internal structure of the $3D$ object.

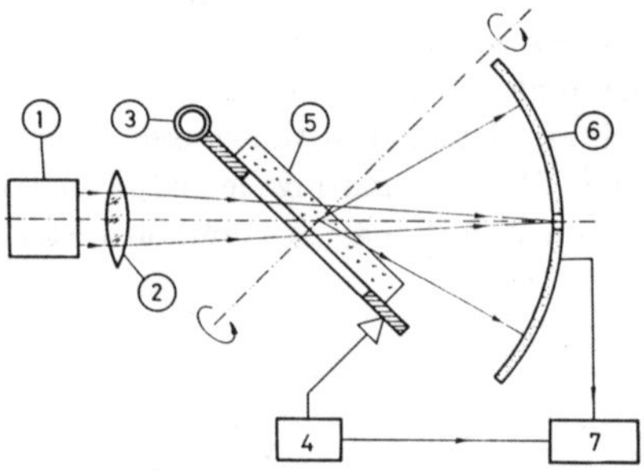

Fig. 8.9. Fourier transform microscope for estimation of the orientation angles θ and φ of two straight line objects with common vertex: 1 — source of the collimating beam, 2 — Fourier transform lens, 3 — rotating stage, 4 — angular indicator, 5 — working volume, 6 — spherical read-out system, 7 — computer memory block.

Consider the structure of the intermediate data detected as output signals in the Fourier transform microscope [312, 330]. The basic diagram of this microscope for estimation of the orientation angles of the straight line objects with common vertex is presented in Fig. 8.9. The working layer is oriented at an angle with respect to the optical axis of this Fourier transform microscope, e.g. at 45°, and is attached to the rotating stage in such a way, that the vertex of these objects is in the point, where the optical axis of the microscope and that of the rotating stage intersect. In the course of the tomographic measurements the working layer is rotating over 360° around this rotating axis with angular increment $\Delta\varphi$. The read-out system is located on the sphere of the spatial frequencies. The photos of the Fourier transform of two straight line objects in the form of a cross is presented in Fig. 3.13 [330]. One of the objects is tilted with respect to the optical axis at the angle 90°, the second one at the angle 90°, 45° and 10°.

An example of two straight line objects 1 and 2 with common vertex and the corresponding sinograms $\underline{1}$ and $\underline{2}$ are shown in Fig. 8.10. The phase of the sinogram is defined by the position of the rotating stage of the Fourier transform microscope, where the angle between the object projection on the plane perpendicular to the optical axis and the rotating axis assumes its maximum value. Only one complete revolution is needed to detect all sinograms of any complex configuration of straight line objects. The reconstruction stage is performed by the traditional methods.

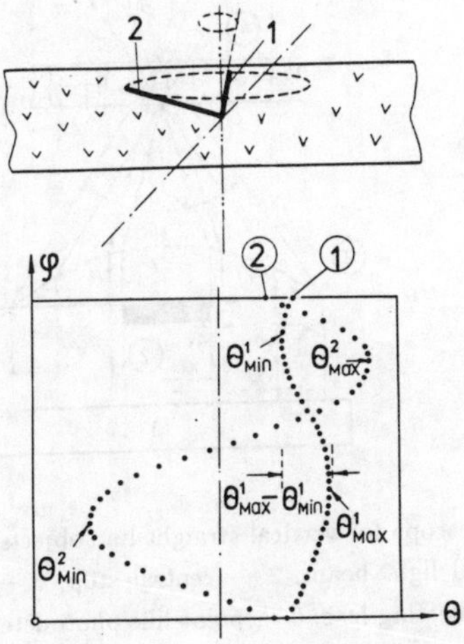

Fig. 8.10. Two sinograms of the line objects 1 and 2 with common vertex in the coordinate frame (θ, φ).

A tomogram constructed in the MFTM for straight line objects has its distinct features. Firstly, the MFTM [2, 3] sees only the straight line objects with small dip angle, that is, perpendicular to the tomographic slice. Secondly, the intermediate information provided by the MFTM comprises **two** meso–optical images. Finally, the reconstructed data in this tomographic slice have the form of the vector values function unlike the data in the traditional tomograms, which represent scalar functions. The properties of the MFTM and its performance are presented in Chapter 3.

To perform fast searching of the microscopic straight line objects oriented parallel to the optical axis another meso–optical microscope has been suggested [313, 329]. The key part of this microscope is the meso–optical condenser (Fig. 8.11). Collimated light from the light source 1 is converted by the meso–optical cylindrical lens 3 with central stop 2 into the cylindrical waves which produce an illuminated region in the form of a short "fence" inside the working layer 4. The generating line of the meso–optical cylindrical lens 3 is a polygon of many line elements being the tangents to the generating line of the cylindrical lens of the conventional condenser. The length of each line element of this polygon Δ is a linear function of the distance x to the symmetry plane of the meso–optical lens 3:

$$\Delta = h\,\frac{x}{L}, \tag{8.28}$$

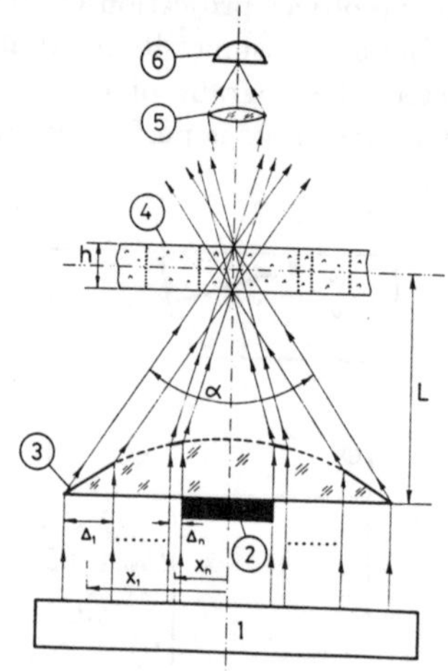

Fig. 8.11. Meso–optical microscope for vertical straight line objects in $3D$ volume of the height h: 1 — source of the collimated light beam, 2 — central stop, 3 — meso–optical cylindrical lens, 4 — working volume, 5 — collecting lens, 6 — point-like photodetector.

where h is the thickness of the working layer and L is a distance between the meso–optical lens and the working layer. To suppress the longitudinal modulation of the light intensity inside the working layer, the distance L must be chosen very large, for example, $L \geq 100$ mm for $h = 0.2$ mm and $\lambda = 0.5$ μm.

The diffracted light is detected by one photomultiplier. The photocurrent as a function of time is registered when the working layer is moving with constant velocity. Only alternating component of this time signal is directed into the computer memory.

The analysis of the light intensity distribution in the illuminated region of the working layer shows [312] that side lobe intensity increases with increasing of the dimension of the central stop 2. The structure of these side lobes is shown schematically in Fig. 8.12. If line object is oriented at large angle with respect to the optical axis of the microscope, the length of such object inside the central lobe will be smaller, than the thickness of the working layer, and the time duration when this object is in the central lobe will be larger than for objects oriented parallel to the optical axis. Hence the alternating component of the time signal will be very small, and the angular selectivity of the microscope will be very high. Thus, the longitudinal dimension of the illuminated region can be made as large as 5–10 mm instead ~ 0.5 mm in the conventional system.

The scanning operation for searching the vertical line objects is performed twice with different orientation angle of the illuminated region with respect to x-axis (Fig. 8.13). For example, at +45° during the first scanning and at −45° during

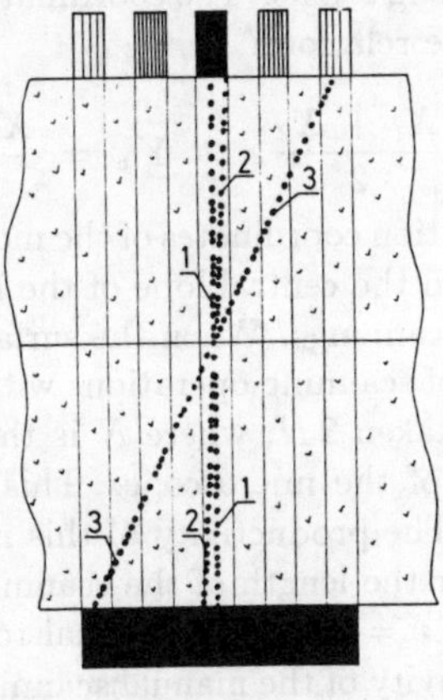

Fig. 8.12. Arrangement of three different straight line objects relative to the central illuminating region (black above) with its two side lobes (hatched above). The black region (bottom) defines the width of the central region in the conventional condenser.

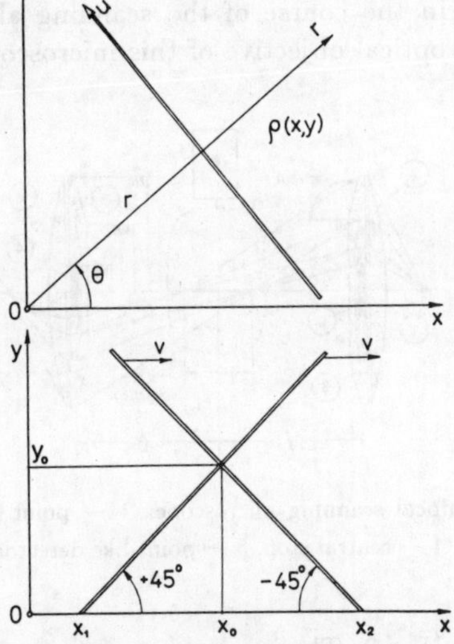

Fig. 8.13. The orientation of the summing lines along "u" axis on the detection stage for one aspect (above) and the position of the illuminating narrow "fence" region relative to the working volume on the detection stage of the vertical straight line object (x_o, y_o) during two successive scanning runs with two different orientations of the fence (below).

the second scanning, both along x-axis. The coordinates $X_\circ$, $Y_\circ$ of the found vertical line object are defined by the relations:

$$X_\circ = \frac{X_1 + X_2}{2}, \qquad Y_\circ = \frac{X_1 - X_2}{2}, \tag{8.29}$$

where X_1 and X_2 are the position coordinates of the moving stage along x-axis, when the vertical line object was in the central lobe of the illuminated region during the first and during the second scannings. When the surface density of the vertical line objects is high, the number of scanning operations with different orientations of the illuminated region must be taken $2N$, where N is the number of the vertical line objects in the field of view of the microscope. This is a general rule typical for reconstructed tomography. The productivity of this microscope with tomographic algorithm of the scanning for the length of the scanning region $D = 2$ mm and for velocity of the moving stage $v = 50$ mm/s is equal to 50 mm^2/s. This is $1.4 \cdot 10^4$ times more than the productivity of the manual scanning, equal to $3.5 \cdot 10^{-3}$ mm^2/s.

8.6. Meso–optical Confocal Microscopes

In the meso–optical confocal microscope, shown in Fig. 8.14, the scanning along z-axis typical for conventional and confocal scanning microscopes is absent. Only ray-sums are detected in the course of the scanning along x and y axes [331]. The illuminating meso–optical objective of this microscope consists of the optical

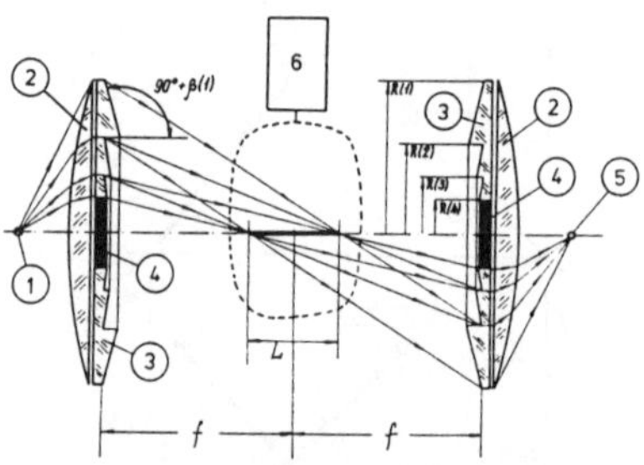

Fig. 8.14. Meso–optical confocal scanning microscope: 1 — point-like light source, 2 — optical collimator, 3 — multiaxicon, 4 — central stop, 5 — point-like detector, 6 — $(X - Y)$ moving stage with position indicator.

collimator 2 and multiaxicon 3. The imaging meso–optical objective consists of the multiaxicon 3 and of the optical collimator 2, places before the point-like photodetector 5. To hold the topology of light rays typical for confocal microscope, the light rays in the object space of the depth L should be parallel in the meridional cross-section of this meso–optical confocal microscope.

The conical surfaces of the multiaxicon 3 are restricted by the circles with radii defined by the equation (confocal condition):

$$\frac{R(i+1)}{R(i)} = \frac{(f-\frac{L}{2})}{(f+\frac{L}{2})}, \qquad \begin{matrix} i = 1,2,...,N, \\ N \geq 3, \end{matrix} \tag{8.30}$$

where R_i is the outer radius of the i-th conical surface, $2f$ is the distance between objectives, L is the depth of the $3D$ space in which a $3D$ object is placed, and the outer radius of the first conical surface is equal to

$$R_{max} = R(1) = \frac{1}{2}(f+\frac{L}{2})\frac{\lambda}{\Delta x}, \tag{8.31}$$

where Δx is the lateral resolution of the microscope. The radius of the central stop 8 is the internal radius of the N-th conical surface. The angle $90° + \beta(i)$ between the generating line of the i-th conical surface and the optical axis of the microscope is defined by the equation

$$\tan\beta(i) = \frac{R(i)}{[n\sqrt{(f+\frac{L}{2})^2 + R^2(i)} - (f+\frac{L}{2})]}, \tag{8.32}$$

where n is the refractive index of the multiaxicon. The depth L does not depend on diffraction of light. To suppress the intensity of the side lobes the number of the conical surfaces in the multiaxicon $N \geq 3$. This meso–optical confocal microscope can be used as tomographic device with extremely high lateral resolution and non-restricted depth of the object space. The absence of the side lobes gives rise to very sparse transformation matrix-operator and to simple inverse calculations.

8.7. New Stereo-microscope

As a starting example of the conventional device we present stereoscopic zoom microscope SMZ-10 with Parallel Optical System of the Nikon company (Fig. 8.15a). The light ray from the object O follows two parallel paths with stereoscopic view angle 14° through the common objective lens L_o, two zooming systems Z_L and Z_R; prism with adjusted distance. With two eyepieces the observer sees two stereoscopic images with total magnification up to 160X. In the stereomicroscope SMZ-1/1B a convergent optical system is used (Fig. 8.15b) with two zoom objectives and with stereoscopic view angle 12°.

The drawback of this stereoscopic microscope is that stereoscopic view angle is small for observation of shallow microscopic relief on the plane backing. The straightforward increasing of the stereoscopic view angle from initial value 14° up to ~ 90° leads to the proportional decreasing of the field of view of the stereo-microscope. The main part of the observed images will be out-of-focus and thus

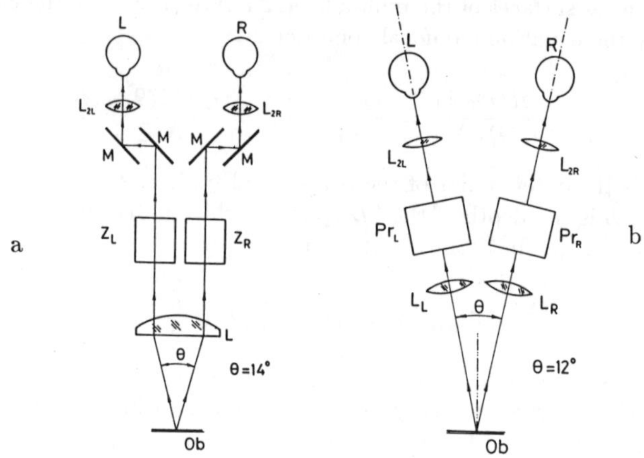

Fig. 8.15. The optical part of the stereoscopic zoom microscopes: *a* — with parallel optical system, *b* — with convergent optical system.

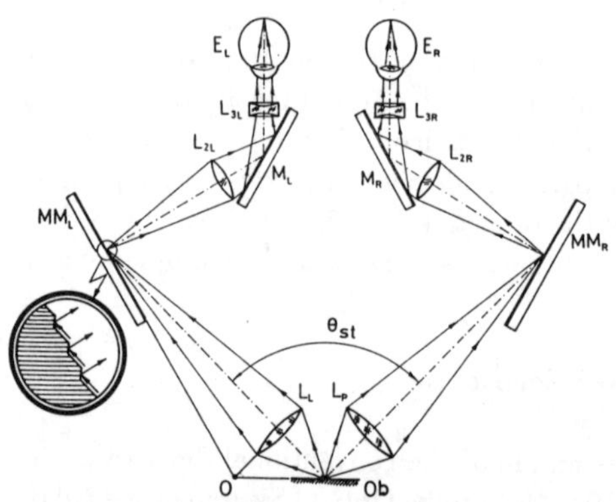

Fig. 8.16. Principal scheme of the new stereoscopic microscope with large field of view: Ob — object, L_L and L_R — left and right microscopic objectives, MM_L and MM_R — left and right multielement mirrors, L_{2L} and L_{2R} — left and right projected objectives, M_L and M_R - left and right plane mirrors, L_{3L} and L_{3R} — left and right eyepieces, E_L and E_R — left and right eyes of the observer.

useless. To eliminate this shortcoming inherent to all conventional stereomicroscopes a new stereo-microscope will be suggested. This device has been developed on the basis of the converted unit used in the meso-optical Fourier transform microscope [162]. In Fig. 3.85 this unit is shown as two conical mirror pieces 6 and 7. Two meso-optical images, projected with sunset light rays on this two units 6 and 7, are observed with imaging lens 8. The latter transforms these two meso–optical images on the screen 9. The same conical mirror pieces 6 and 7 can be used in the proposed new stereo-microscope as reflected intermediate screen to fulfil the Scheimplug condition [94]. The principle scheme of the new stereo-microscope, is presented in Fig. 8.16. The object Ob is imaged twice by the left L_L and by the right L_R microscopic objectives, which are oriented symmetrically at the stereoscopic view angle θ_{st}, on two multielement mirrors MM_L and MM_R. These light rays are going at sunset angles relative to the plane of these two multielement mirrors, provided with the blazed grooves. The grooves of the multielement mirrors MM_L and MM_R are oriented perpendicular to the plane of the stereoscopic angle θ_{st}. The light rays are reflected from the working elements of these multielement mirrors at the angle $\approx 90°$ and are directed to the projected objectives, to the left L_{2L} and to the right L_{2R}. With two systems, composed of the plane mirror M_L (M_R) and the eyepiece L_{3L} (L_{3R}), the observer sees two stereoscopic images. The magnification of the projected objective L_{2L} (L_{2R}) and the groove period of the multielement mirror MM_L (MM_R) are chosen so that the observer eye could not see the microscopic structure of the multielement mirrors MM_L (MM_R). The depth resolution of this new stereo-microscope is equal to

$$\Delta z = \frac{\lambda}{2 \sin \alpha} \sin \frac{\theta_{st}}{2}, \tag{8.33}$$

where α is the angular aperture of the imaging objective L_L (L_R), and θ_{st} is the stereoscopic view angle. For $\theta_{st} = 90°$, $\alpha = 30°$, $\lambda = 0.5$ μm, $\Delta z \cong 0.3$ μm. The diameter of the field of view D of this microscope is equal to

$$D = \frac{D_\circ}{\cos \theta_{st}/2}, \tag{8.34}$$

where $D_\circ$ is the diameter of field of view of the imaging objective L_L in the conventional observation mode. For $\theta_{st} = 90°$, $D = D_\circ\sqrt{2}$. To get large field of view we must use the imaging objective L_L (L_R) with high working distance.

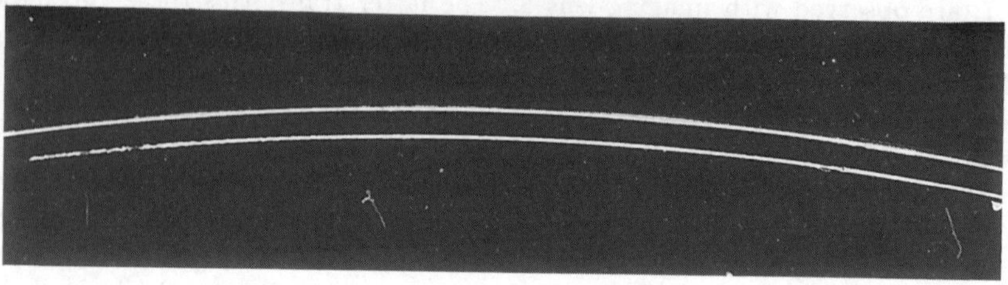

Fragment of the meso-optical image of the open circle of the diameter equal to the distance between two circles shown in Figure.

Far field diffraction picture produced by the coherent system with spatial frequency filter composed of two ortogonal Hilbert filters with phase jumps on the optical axis of the system. The light intensity on the optical axis in the center of the Figure is equal practically to zero.

Chapter 9

MESO–OPTICAL METROLOGICAL DEVICES

9.1. Traditional Testing of the Conical Surfaces

We begin with conventional techniques of the control of the external conical surfaces [332 – 335]. For example, in the device [332] the cone is mounted on the rotating platform. The conical axis is oriented parallel to the plane of the platform (Fig. 9.1) and perpendicular to the rotating axis of the platform. The observation is performed in optical autocollimator with its axis perpendicular to the rotating axis of the platform. The mirror of the autocollimator is in mechanical contact with the cone to be tested. The rotating platform with this cone is mounted in two positions with azimuthal angles φ_1 and φ_2, at which the autocollimator provides zero samplings. The conical angle α is equal to

$$\alpha = 360^\circ - (\varphi_2 - \varphi_1). \tag{9.1}$$

The analogous angular samplings can be made in the arbitrary position of the cone on the rotating platform, in particular, with generating line of the cone parallel to the platform.

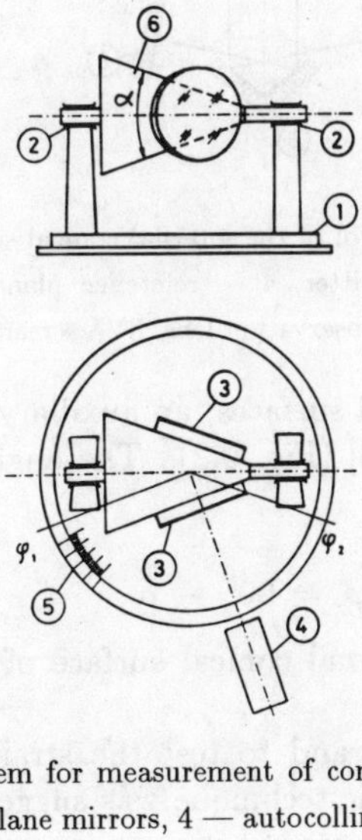

Fig. 9.1. A conventional system for measurement of conical angle α of the cone: 1 — rotating platform, 2 — bearings, 3 — plane mirrors, 4 — autocollimator, 5 — reading vernier, 6 — the cone to be tested.

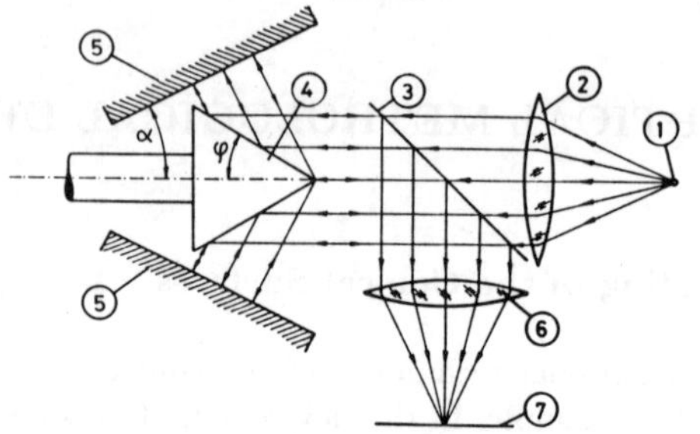

Fig. 9.2. A control system for the internal conical surfaces: 1 — point light source, 2 — optical collimator, 3 — beam-splitter, 4 — auxilliary conical mirror, 5 — object, 6 — observation lens, 7 — screen.

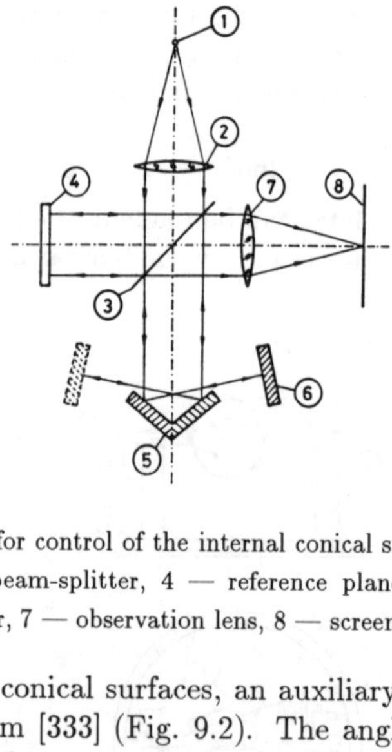

Fig. 9.3. Interferometric device for control of the internal conical surfaces: 1 — point light source, 2 — optical collimator, 3 — beam-splitter, 4 — reference plane mirror, 5 — conical surface, 6 — rotating plane etalon mirror, 7 — observation lens, 8 — screen.

To control the internal conical surfaces, an auxiliary conical mirror is inserted in the collimated light beam [333] (Fig. 9.2). The angle of this auxiliary conical mirror must be chosen equal to

$$\varphi = 90^\circ - \alpha, \tag{9.2}$$

where α is the angle of the internal conical surface of the industrial part to be tested.

To control the conical object and to test the straightness of the generating line of the cone, an interferometric technique was suggested [334] (Fig. 9.3). The internal conical surface 5 is illuminated at various azimuthal angles successively by the collimated beam from the rotating etalon mirror 6 in the autocollimator mode. The ring interference pattern, generated by the light rays, reflected twice from the testing conical surface 5 and by the plane reference beam, is detected.

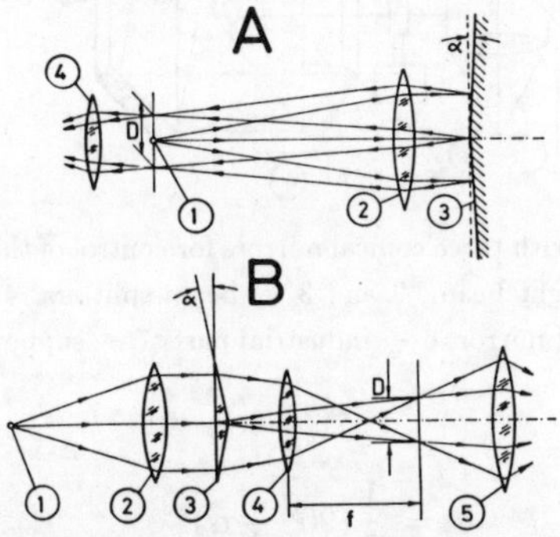

Fig. 9.4. Devices for measurement of small conical angles. "A" — for reflection surfaces: 1 — point light source, 2 — lens, 3 — tested cone, 4 — observation lens. "B" — for refracted axicons: 1 — point light source, 2 — collimator lens, 3 — axicon to be tested, 4 — lens, 5 — observation lens.

For testing conical mirror or axicons with small conical angles, a device [335] has been suggested (Fig. 9.4). The spherical light wave with center in the focal plane of the lens 2 is converted by the lens 2 into the collimated light beam. After the reflection from the conical surface this beam is converted by the same lens 2 into the meso–optical ring image in the focal plane of the lens 2. The diameter of this ring image 2 defines the conical angle α of the test object:

$$\tan\alpha = \frac{D}{2f}, \tag{9.3}$$

where f is the focal length of the lens 2.

9.2. Meso–optical Conicometers

To solve the problem of the fast precise control of the industrial parts with external conical surfaces an autocollimator device with three conical mirrors was suggested [336] (Fig. 9.5). The collimator light beam from the source 1 traverses two beam-splitters, 2 and 3, experiences reflections on the central conical mirror and on two ring-shaped conical mirrors. The external conical surface of the industrial part 6 is illuminated in an autocollimator mode from all directions simultaneously, if the conical angle of the accessory conical mirror 5 is chosen equal to

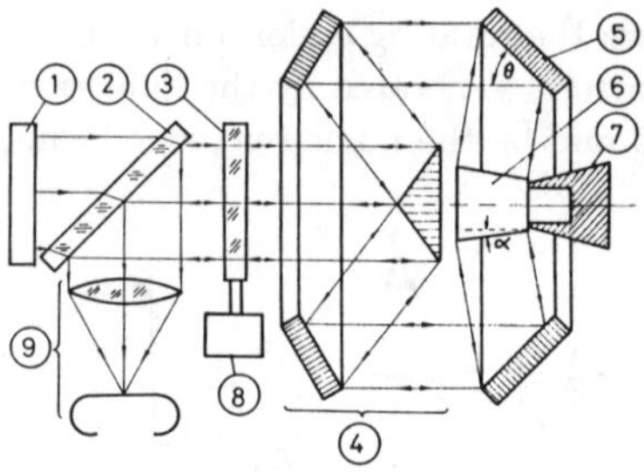

Fig. 9.5. A meso–optical device with three conical mirrors for control of the external conical surfaces: 1 — source of the collimated light beam, 2 and 3 — beam-splitters, 4 — system of two conical lenses, 5 — the accessory conical mirror, 6 — industrial part, 7 — support, 8 — micrometer screw, 9 — photoregistrator.

$$\theta = \frac{1}{2}(90^\circ + \alpha), \tag{9.4}$$

where α is the conical angle of the etalon object. The reflected light rays faithfully copy the traces of the illuminated light rays. The device is provided with micrometer screw 8 of the second beam-splitter 3, and with photoregistrator 9.

The rays reflected from the conical surface of the etalon of the industrial part 6 produce the first point-like image on the observation screen. The second point-like image is produced by the light rays reflected from the front surface of the second beam-splitter 3 and then from the back surface of the first beam-splitter 2. In the more general case the first image assumes the form of a ring. Its diameter is a measure of the deviation of the tested conical surface from the perfect one. When this deviation is very small, the diameter of this ring will be smaller than the diffraction deblouring and both point-like spots can be brought into coincidence. With micrometer screw 8 the operator tunes the destructive mode of the interference picture between these two spots up to n-th diffraction ring. With this superresolution technique the sensitivity of this device can be increased n-times.

The measurement of the internal conical surfaces presents a real challenge for meso–optics. We describe below a meso–optical device by means of which this difficult problem can be overcome. The problem is that the autocollimated mode of the illumination of the internal conical surface cannot be accomplished in the direct fashion for a long conical object. The solution of this problem was found in the sophisticated meso–optical device [337] (Fig. 9.6). The optical collimator 2 with central stop 3 produces ring-shaped light beam which is converted by the circular diffraction grating 4 into the convergent conical wave. After reflection from the circular blazed reflecting grating 5 on the conical backing, the light rays are split into two components. The first component as zero diffraction order of the grating 5 illuminates the tested internal conical surface 6 at small angle and will be out. The second component as plus first diffraction order of the grating 5 is reflected from the internal conical surface 6 in the autocollimator mode back to the grating 5. This time the zero diffraction order will be out as it goes parallel to the optical axis. The plus first diffraction order is propagating to the circular diffraction grating 4. The plus first diffraction order on the circular diffraction grating 4 is propagating to the lens 2 and will be out. The zero diffraction order traverses the circular diffraction grating 4 and illuminates the meso–optical element with ring response 8, which converts the conical wavefront into the converging sphero-conical wavefront with focal ring image in the plane of spatial frequencies. The spatial filter 9 in this plane represents a narrow opaque ring on the transparent region.

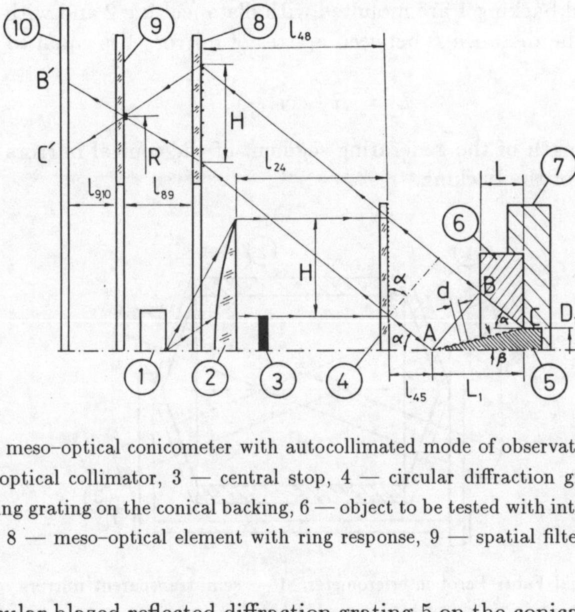

Fig. 9.6. The meso–optical conicometer with autocollimated mode of observation: 1 — point light source, 2 — optical collimator, 3 — central stop, 4 — circular diffraction grating, 5 — circular blazed reflecting grating on the conical backing, 6 — object to be tested with internal conical surface, 7 — support, 8 — meso-optical element with ring response, 9 — spatial filter, 10 — observation screen.

The circular blazed reflected diffraction grating 5 on the conical backing is dedicated to provide an autocollimator mode of illumination of the tested conical surface. The reflected light rays are focused into the focal ring. An operator easily sees any azimuthal deviations of the generating line of the tested object from the etalon. It is instructive to present an example of this device for control of the industrial parts with conical angle of the internal cone $\alpha = 15.6°$ and of length $L = 88$ mm. The axial dimension of grating 5 is equal to $L_1 = 100$ mm, and the external diameter $D = D_1 = 40$ mm. The conical angle of the conical backing of the grating 5 is equal to $\beta = 11.3°$, and the grating period $d = 1.44\ \mu$m. The period of the circular diffraction grating 4 is equal to $d_4 = 2.33\ \mu$m. The distances are: $l_{24} = 158$ mm and $l_{48} = 168$ mm. The width of the working part of the meso–optical element with ring response $H = 47$ mm. The local period of the meso–optical element 6 with ring response 8 is changing monotonously from $d_{max} = 5.38\,\mu$m up to $d_{min} = 1.51\,\mu$m. The local period of the grating 8 on the central circle is equal to $d_{\circ} = 2.33\,\mu$m. The distance $l_{89} = 150$ mm. The effective width of the focal ring is equal to $\Delta r = \frac{\lambda}{H} l_{89} = 2\,\mu$m for $\lambda = 0.628\,\mu$m. This corresponds to the measurement error $\delta\alpha \approx (0.1 - 0.2)''$ (arc sec).

9.3. Meso–optical Interferometer

For very precise measurements of the conical wavefronts, a multipass interference is generated in the meso–optical device [338] (Fig. 9.7). Two semi-transparent mirrors on the conical backing 1 are mounted with distance ring 2 and with mechanical turning unit 3. The distance L between apexes of mirrors 1 is equal to

$$L = R / \cos(\frac{\alpha}{2}), \tag{9.5}$$

where R is the length of the generating segment of the conical mirrors 1, and α is the conical angle of the backing.

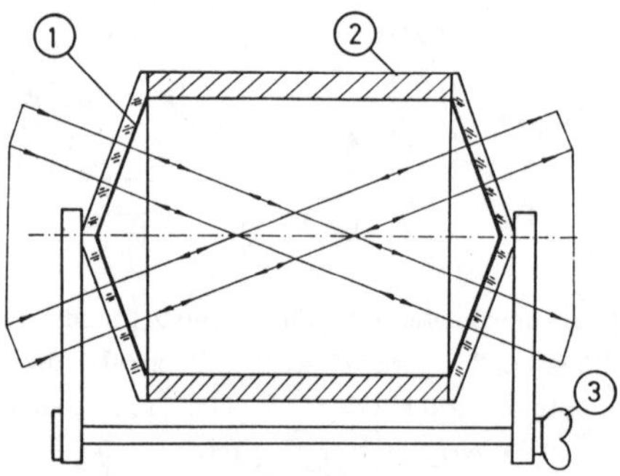

Fig. 9.7. Meso–optical Fabri–Perot interferometer: 1 — semi-transparent mirrors on the conical backing, 2 — distance ring, 3 — mechanical turning unit.

When the interferometer is illuminated by the tested convergent conical wavefront through the semi-transparent conical mirror 1, a linear meso–optical crossover is generated on the axis of the interferometer. The divergent conical wavefront is falling on the semi-transparent conical mirror and is reflecting from this mirror. The process of the multiple walking of the light rays between two conical mirrors is repeated many times. If the generating line of the conical wavefront is parallel to the generating line of the semi-transparent conical mirrors 1, the pure multiple interference is taking place. The effective number of such passages is equal approximately to $(1 - \eta)^{-1}$, where η is the reflectivity of the semi-transparent conical mirrors 1.

The angular interval $\Delta\varphi_k$ between two adjacent angular directions φ_k and φ_{k-1} with maximal transmittances is defined by the equation

$$\Delta\varphi_k = \varphi_k - \varphi_{k-1} = \frac{\lambda}{2L \sin\varphi_k}. \tag{9.6}$$

The effective angular width of such a maximum of the transmittance is defined by the condition

$$\frac{2\pi L\,[\varphi_k^2 - (\varphi_k + \delta_\varphi)^2]}{\lambda} = \frac{1-\eta}{2\sqrt{\eta}} \tag{9.7}$$

and is equal to

$$\delta_\varphi = \frac{(1-\eta)\,\lambda}{2\sqrt{\eta}\cdot 2\pi L\cdot\varphi_k}. \tag{9.8}$$

The operation of this meso–optical interferometer is performed as follows. At first, the axis of the tested conical wavefront is set in coincidence with the axis of the interferometer. Then with mechanical turning unit 4 an adjustment operation is made with the aim to get the coincidence of the angle φ with one of the angular direction φ_k with maximal transmittance. After that an interference pattern produced by auxiliary lens is observed with the purpose to detect the azimuthal deviation in the incident conical wavefront.

9.4. Meso–optical Profilometers

A focusing lens with monotonous longitudinal chromatic aberration is used in the device [339 – 342] for monitoring the surface height of an object by scanning. The principle scheme of the focus-wavelength-encoded profilometer is presented in Fig. 9.8. A white light source illuminates a pinhole placed in the front focal plane of an achromatic collimating lens. The collimated light beam traverses the beam-splitter and the lens with longitudinal chromatic aberration, focuses this beam on the tested surface. The back focal length of this lens varies monotonously with wavelength. A some kind of on-axis rainbow is observed. The light rays reflected from the tested surface of a sample is directed by the same lens and the beam-splitter to be analyzed by a spectroscopic unit which is working as an angular dispersing element. At the back focal plane of the second achromatic lens a linear array of photodetectors is located. Thus only in focus color component is imaged onto this array. The position of the intensity maximum at definite wavelength corresponds to the surface height.

The lens with monotonous longitudinal chromatic aberration consists of five elements and provides a color interval from 400 to 1000 nm over the depth range 1.45 mm. The increment dispersion $dz/d\lambda = 4\,\mu\text{m/nm}$ over the visible spectrum.

The vertical height resolution is equal to 0.5 μm and the lateral surface resolution 20 μm. The depth of focus for a single wavelength is of the order of $\pm 6\ \lambda$. The data acquisition time of the whole device is equal to 100 ms.

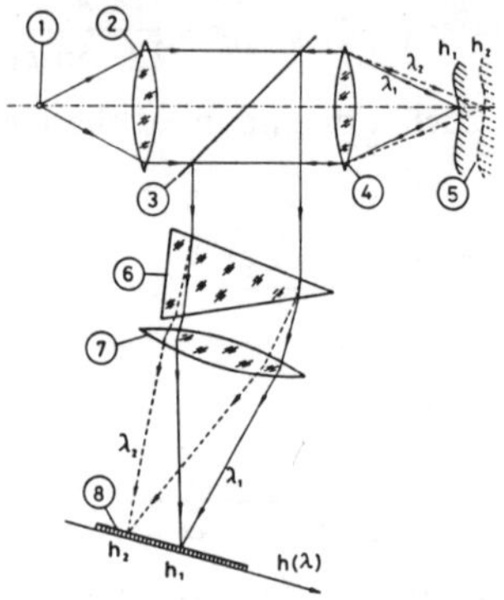

Fig. 9.8. Focus-wavelength-encoded profilometer: 1 — point source of white light, 2 — achromatic collimating lens, 3 — beam-splitter, 4 — chromatically aberrated lens, 5 — profile to be tested, 6 — dispersing prism, 7 — spectroscopic lens, 8 — linear array of micro-photodetectors.

In this device the **refracted** dispersive elements were used. The better performance can be obtained with rainbow profilometer provided with holographic lens in the form of a ring with a hole and with achromatic lens in this hole as a collecting unit of the light reflected from the tested surface. As was shown in Section 7.6, Eq. 7.20 and Eq. 7.21, the relative chromatism of the diffraction grating is 20 times more than that of the spherical refracting surface.

9.5. Meso–optical Keratometer

Now we describe some meso–optical devices for fast estimation of the eye profile [343 – 345]. Recall at first the design of the simplest and rather old traditional optical keratometer [189]. It consists of the coordinate marker of many ($\sim$ 19) coaxial rings 2 (Fig. 9.9) on the spherical surface at distance L = 200 mm from the eye. It is provided with light source 1, objective lens 3, aperture diaphragm 4, located in the focal plane of the objective lens 3, and observation screen 5. The angle γ_i between the straight line directed at the i-th ring and the optical axis of the system assumes discrete set of values from 10° to 80° with angular increment 5°. In the course of measurement the coordinate marker is illuminated by light from the source 1 and the image of the coaxial rings on the surface of the eye is observed on the screen 5 in the condition when the aperture diaphragm 4 transmits only that light rays which are propagating parallel to the optical axis of the system. Owing to this telecentric feature of the optics, the image of the coordinate rings 2 on the eye surface consists of partially distorted circles. The input information comprises the radii R_i of the rings. On the reconstruction stage a local curvature radii of the eye surface for various distances from the optical axis are to be estimated. In some improvement versions of the traditional keratometer the surface of the eye is illuminated by the collimated light beam.

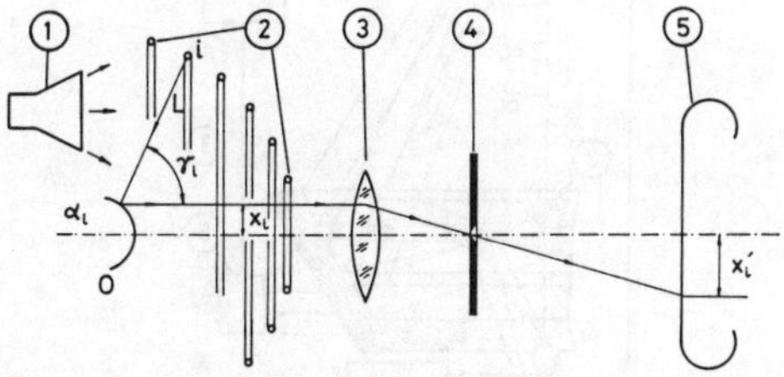

Fig. 9.9. Traditional optical keratometer: 1 — light source, 2 — coaxial rings, 3 — objective lens, 4 — aperture diaphragm, 5 — observation screen.

The main source of measurement errors in this device is the uncertainty in the longitudinal adjustment which gives an error $\Delta R \cong 19\,\mu\text{m}$ in curvature radius of the eye $R = 8\,\text{mm}$ for longitudinal shift $\Delta z = 0.5\,\text{mm}$. This sensitivity of the results to the longitudinal misalignment is a main drawback of the traditional optical keratometer. The another drawback is of pure mathematical nature. To receive the desired information, an inverse calculations must be accomplished.

To eliminate these drawbacks, a new keratometer has been suggested [343]. A narrow light beam with longitudinal light intensity modulation is used in this keratometer which is called a meso–optical one as conical wavefronts produce this longitudinal modulation. The objective of this keratometer sees only that part of the eye surface which is on the optical axis of the system. Therefore this keratometer is a confocal device like a confocal scanning microscope [301].

The principle scheme of the confocal meso–optical keratometer is presented in Fig. 9.10: 1 — point light source, 2 — concave meso–optical mirror with three elements, 3 — convex meso–optical mirror with three elements, 4 — point photodetector, 5 — moving stage with position indicator, 6 — computer memory block. The generating line of the meso–optical mirror 2 involves three straight line segments. The generating line of the meso–optical mirror 3 consists of three arcs of the eccentric parabolas. The external and internal segments of the generating line of the conical mirror 2 are located on the common straight line. The central segment of this generating line is oriented at angle Φ with respect to the common straight line, where

$$\Phi = \sin^{-1}\left[\frac{r_2}{L_1 + L_2 + r_2\cos\gamma}\right] - \sin^{-1}\left[\frac{r_2}{L_1 + r_2\cos\gamma}\right], \tag{9.9}$$

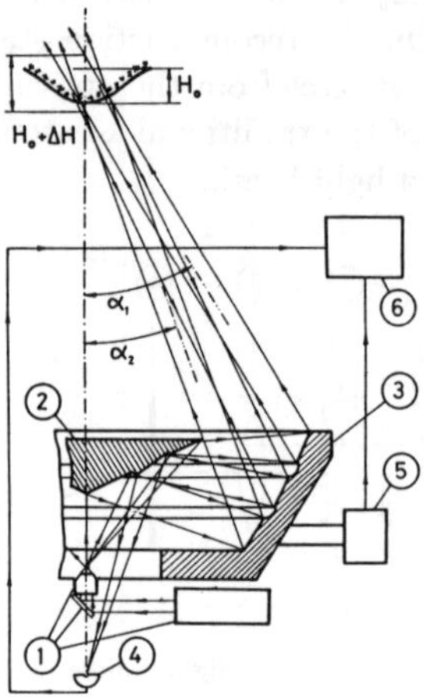

Fig. 9.10. Meso–optical confocal keratometer: 1 — point light source, 2 — concave conical mirror of three elements, 3 — convex conical mirror of three elements, 4 — point photodetector, 5 — moving stage with position encoder, 6 — computer memory block.

where r_2 is the distance from the central segment of the generating line of the conical mirror 2 to the optical axis of the keratometer, 2γ is the vertex angle of the conical mirror 2, L_1 is the distance from this vertex to the point light source 1, and L_2 is the distance from the point light source 1 to the point photodetector 4.

The external part of the generating line of the conical mirror 3 is the arc of the first eccentric parabola with focus in the point where the virtual image of the point light source 1 is produced by the conical mirror 2 and with parabola axis parallel to the central light ray going from the external part of the generating line of the conical mirror 3 to the eye surface. The internal part of the generating line of the mirror 3 is the arc of the second eccentric parabola with parabola focus in the point where the virtual image of the point light source 1 is generated by the conical mirror 2, and with parabola axis parallel to the central light ray going from the internal part of the generating line of the conical mirror 3 to the eye surface. The central part of the generating line of the conical mirror 3 is the arc of the third eccentric parabola with parabola focus in the point where the virtual image of the point photodetector 4 is formed by the conical mirror 2 and with parabola axis parallel to the central light ray going from the illuminated part of the eye surface to the central part of the generating line of the conical mirror 3.

The illuminated part of this keratometer forms the external parts of the conical mirror 2 and of the conical mirror 3. The function of the objective lens is performed by the internal parts of these mirrors.

Two coaxial conical wavefronts produce the longitudinal modulation of the on-axis light intensity. The fringes of this interference pattern are shown in Fig. 9.11 in one of the meridional cross-sections. The period A of this longitudinal modulation is given by the equation

$$A = \frac{\lambda}{\sin(\alpha_2 - \alpha_1)\sin((\alpha_1 + \alpha_2)/2)}, \tag{9.10}$$

where α_1 and α_2 are the angles between the central light rays propagating from the external and from the internal parts of the conical mirror 3, respectively, and the optical axis of the keratometer. For $\lambda = 0.5\,\mu$m, $\alpha_2 = 35°$, $\alpha_1 = 15°$, $A = 3.5\,\mu$m. For $\alpha_2 = 30°$, $\alpha_1 = 20°$, $A = 7\,\mu$m.

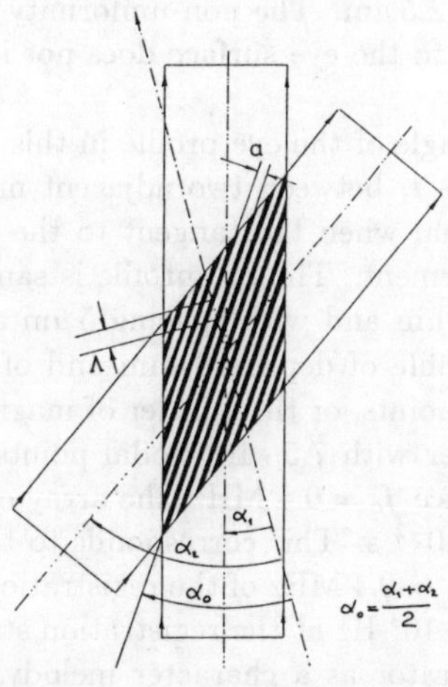

Fig. 9.11. The generation of the longitudinal on-axis interference pattern from volume interference fringes produced by two coaxial conical wave fronts: "a" is a period of these volume interference fringes, and A is a period of the on-axis intensity light modulation.

When the point of the eye surface on the optical axis is in the position with maximal light intensity, the reflected light induces photocurrent in the photodetector 4 of maximal amplitude. When the point of the eye surface on the optical axis is in the position with minimal light intensity, the photocurrent will be minimal. As the objective lens unit of this device sees only the points on the optical axis of the system, its intensity point spread function is equal to

$$I(\rho) = \text{const}\,|\,J_o(\alpha\,\rho)\,|^4\,. \tag{9.11}$$

The intensity of the first side lobe will be equal to 0.016 of the intensity of the central maximum. Recall that the intensity of the first side lobe of an axicon is equal to 0.13 of the central maximum. Due to this confocal feature, the confocal meso–optical keratometer offers very high noise immunity, and this device does not see practically off-axis parts of the eye surface. Only objects located inside the micro-tube with axis on the optical axis of this keratometer and of diameter

$$d = \frac{\lambda}{\sin((\alpha_1 + \alpha_2)/2)} \tag{9.12}$$

are seen in this device. For $\lambda = 0.5\,\mu$m, $\alpha_2 = 30°$, $\alpha_1 = 20°$, $d \approx 1\,\mu$m.

The output data entered into the computer memory block 6 consist of time moments t_i, at which the photocurrent assumes the maximal amplitude and of time moments τ_i, when the meso–optical part of the whole device is in the set of points with increment step equal to $2.5\,\mu$m. The non-uniformity of the lateral moving of the keratometer with respect to the eye surface does not induce any measurement errors.

The tangent inclination angle of the eye profile in this cross-section defines the time increments $\Delta t_i = t_{i+1} - t_i$ between two adjacent moments of time t_i. The time increment Δt_i is maximal when the tangent to the eye profile is parallel to the vector of the lateral movement. The eye profile is sampled over $2D$ grid with movement vector spacing $2.5\,\mu$m and with spacing $5\,\mu$m along the optical axis of the keratometer. For eye profile of depth 10 mm and of length 10 mm we have $2 \cdot 10^4 \cdot 4 \cdot 10^3 = 9 \cdot 10^6$ nodal points, or three order of magnitude grater than in the traditional optical keratometer with $7.5 \cdot 10^3$ nodal points. At clock frequency of the electronic part of the device $f_\circ = 0.4$ MHz, the array of $4 \cdot 10^3$ photopulses can be detected in time interval 10^{-2} s. This corresponds to the lateral linear velocity 1 m/s. The clock frequency $f_\circ = 0.4$ MHz of the registration stage can be converted into the audio frequency $\nu_\circ = 10^3$ Hz at the registration stage. These audio signals can be perceived by the operator as a character melody, which has a high tone at the beginning, then low tone and again high tone to the end of scanning. By adding some reference signal which corresponds to the nominal eye profile we can hear the interference between two signals, one the nominal one and another the distorted one. At normal musical ability the operator can monitor continuously the form of the eye profile during the surgeon operation. Hence we may conclude that measurements in the confocal meso–optical keratometer are direct ones unlike from the traditional optical keratometer which involves the reconstruction stage for solving inverse problem.

One disadvantage of the confocal meso–optical keratometer described above is that the object profile must be smooth. Meanwhile in the mechanical engineering the object profile has sudden changes and kinks, which induce potential ambiguities in the interpretation of the output signals. The source of this disadvantage is that the period A of the longitudinal interference pattern is constant. To meet these demands a new version of the confocal meso–optical keratometer was suggested [344] (Fig. 9.12). The conical mirror 3 generates the interference picture with variable period $A(z)$. Here the generating line of the external (or internal) part of the conical mirror 3 represents an arc of the ellipse with first focus in the point where the virtual

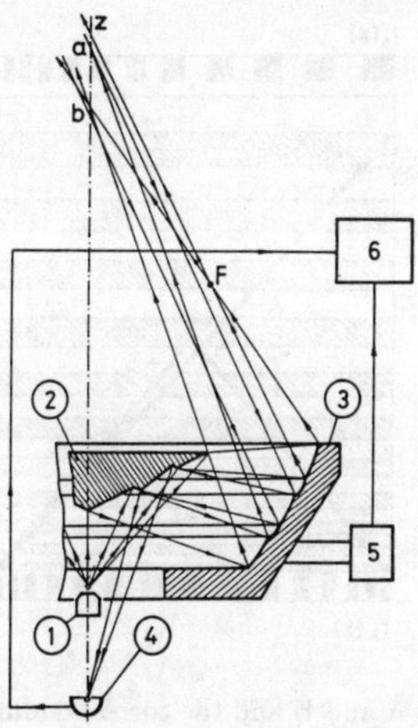

Fig. 9.12. The second version of the meso–optical confocal keratometer: 1 — point light source, 2 — concave conical mirror of three elements, 3 — convex conical mirror of three elements, 4 — point photodetector, 5 — moving stage with position encoder, 6 — computer memory block.

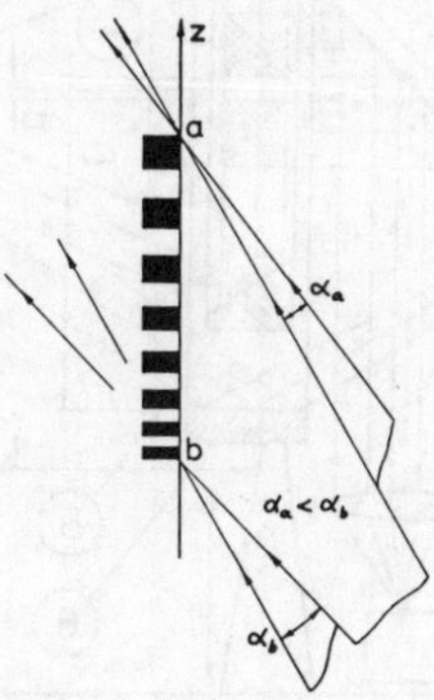

Fig. 9.13. The structure of the on-axis longitudinal modulation of the light intensity. The angle between two wavefronts is varying along z-coordinate.

image of the point light source 1 is produced by the external (or internal) part of the conical mirror 2 and with second focus in the points equally removed from the ellipse arc and from the illuminated region. The angle between the wavevectors of two interfering conical wavefronts is now changing along z-axis. The period $A(z)$ is varying along z-axis (Fig. 9.13) like chirp modulated radar signals. Now the $2D$ sample grid has another structure (Fig. 9.14), which permits to distinguish the profiles of two objects A and B presented on the top and on the bottom of Fig. 9.14.

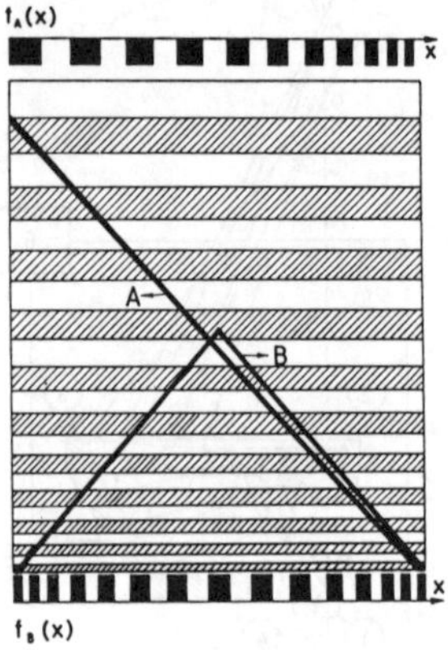

Fig. 9.14. The profiles of two objects A and B and the corresponding output signals in both versions of the meso–optical confocal keratometers: signal $f_A(x)$ is shown on the top and signal $f_B(x)$ is given on the bottom of the figure.

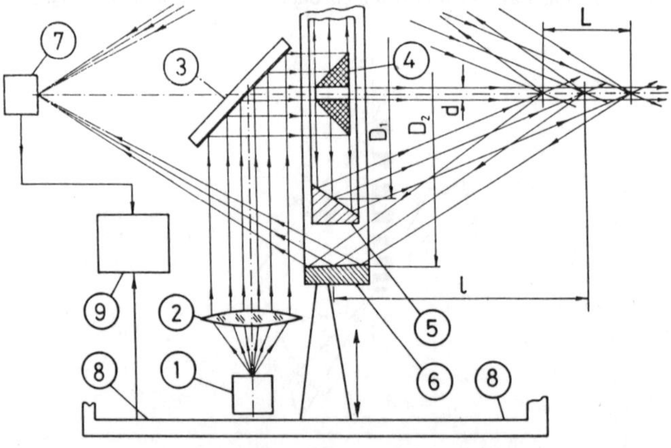

Fig. 9.15. Confocal meso–optical profilometer with central light beam: 1 — point light source, 2 — collimating lens, 3 — plane mirror, 4 — internal conical mirror with central hole, 5 — external conical mirror, 6 — conical (or cylindrical) mirror of the observation system, 7 — point photodetector, 8 — moving stage with position encoder, 9 — computer memory block.

Now we present a design of the confocal meso–optical profilometer, in which a central light beam is used [345] (Fig. 9.15). This device is provided with point light source 1, optical collimator 2, plane mirror 3, internal and external conical mirrors 4 and 5 of the illuminating unit, conical mirror of the observation unit 6, point photodetector 7, moving stage with position encoder 8 and computer memory block 9. The internal mirror 4 has a hole for central light beam. The longitudinal interference is produced by the central plane beam and by the conical wavefront propagating from the external conical mirror 5. The point photodetector 7 sees the points in the vicinity of the optical axis over the range of length L. To get high noise immunity the internal diameter D_2 of the observation conical mirror 6 and the external diameter D_1 of the conical mirror 5 are chosen according to the equation

$$\frac{D_1}{D_2} = \frac{d_1}{d_2} \cong 1.62 ; \qquad d_1 = D_2 . \tag{9.13}$$

In this design the first side lobe of the point spread function of the conical mirror 5 does coincide with first zero of the point spread function of the conical mirror of the observation unit 6. Therefore the overall point spread function of this device does not have practically side lobes with intensity of the order of 10^{-3} relative to the central maximum. The lens equivalent of this device with three coaxial axicons instead of the device with three conical mirrors can be offered.

9.6. Informational Content of the High Order Diffraction Pattern

The analysis of the diffraction pattern of the straight line objects is a typical one in the Fourier optics [54]. This approach gives very large depth of focus both for registration [129] and for processing of the track information [161]. The convergent light beam, which accomplishes the spatial Fourier transformation, is used widely in the various constructions of the meso–optical Fourier transform microscope [2]. In all these designs the object of observation was a diffraction pattern in the region of the first diffraction order. The zero diffraction order is suppressed when we use the convergent illuminated light beam. The high orders of the diffraction pattern were used implicitly in [107] for observation of the meso–optical images of open circles, rectangular slit, shutter or straight line array of many open circles. As was shown in [107], the meso–optical images of these objects had the form of one-dimensional derivative of the input signal. These results were explained in terms of the one-dimensional Fourier–Hilbert transform, accomplished by the meso–optical element with ring response. Also it was proved that the existence of the central dead part in the meso–optical element with ring response is indeed a minor handicap for using this element in the MFTM.

Now we treat the informational content of the diffraction pattern in the region of very high orders [478]. It has been shown that high order diffraction orders allow one to construct a super resolution width indicator of particle track in nuclear emulsion. To detect high order diffraction orders, the backing of the photosensitive layer must be taken in the form of the cylindrical surface with axis which coincides with object axis. We explain some quantitative tolerances to the mutual arrangement of the main parts of the experimental setup.

To treat the informational content of the diffraction pattern in the region of very high diffraction orders, we consider the $1D$ model of the light diffraction on a spectral slit, the amplitude transparency function of which $f(x)$ is equal to

$$f(x) = \begin{cases} 1, |x| \leq d/2, \\ 0, |x| > d/2. \end{cases} \quad (9.14)$$

where d is the width of the spectral slit. The far field diffraction pattern, expressed in terms of spatial frequency ω_x, can be written as one-dimensional Fourier transform

$$f(x) \to F(\omega_z) = \int_{-\infty}^{\infty} f(x) \cdot \exp(-ix\omega_x)\, dx, \quad (9.15)$$

For open spectral slit we get

$$F_s(\omega_x) = \int_{-d/2}^{d/2} \exp(-ix\omega_x)\, dx, = \sin(\frac{\omega_x d}{2})/(\frac{\omega_x d}{2}). \quad (9.16)$$

The intensity distribution of diffraction light has the form

$$I_s(\omega_x) = \sin^2(\frac{\omega_x d}{2})/(\frac{\omega_x d}{2})^2 \quad (9.17)$$

and represents a periodic function $\sin^2(\cdot)$ with envelope $(\omega_x)^{-2}$. The distance between two adjacent minima of the function $I_s(\omega_x)$ is equal to

$$\Omega_s = \frac{2\pi}{d}. \quad (9.18)$$

To estimate this period Ω_s in the region of low spatial frequencies ω_x, an observation plane oriented perpendicular to the optical axis of the convergent illuminated light is usually used. The drop of the intensity distribution $I_s(\omega_x)$ toward high spatial frequencies ω_x is the main limiting factor of this method.

From Eq. (9.17) we see that one single measurement of the first intensity minimum coordinate is enough to estimate the parameter d of the spectral slit. However there is an essential obstacle for precise measurement of high order diffraction mimima. In all cases commonly encountered in practice, the surface of the photosensitive layer backing has the form of a plane. In such a condition we are faced with defocusing effect which reduces drastically the contrast of the detected diffraction pattern in the region of high diffraction orders.

In our experimental setup shown in Fig. 9.16 we use nontraditional form of the observation surface: 1 — point light source, 2 — Fourier transform lens, 3 — aperture diaphragm, 4 — object, 5 — photosensitive layer on the cylindrical backing, 6 — absorbing stop. The object is oriented perpendicular to the optical axis of the convergent light beam and is directed along the axis of the cylindrical surface of the photosensitive layer backing. This configuration is adequate because there is no defocusing effect in the region of high diffraction orders. The fragment of the diffraction pattern produced by the tungsten wire of the diameter $\sim 26\mu$m is presented in Fig. 9.17.

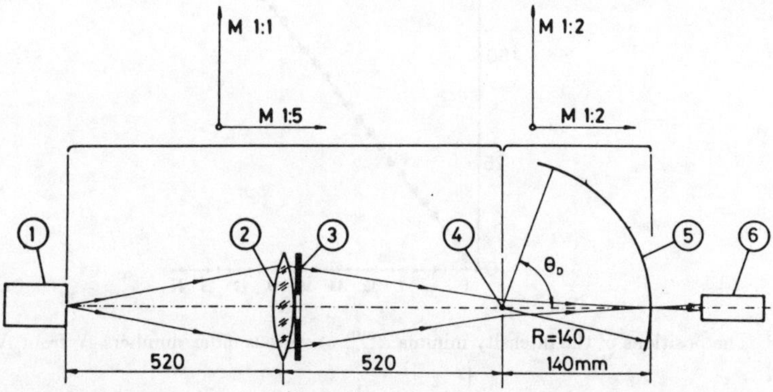

Fig. 9.16. The experimental setup for observation of very high diffraction orders: 1 — point light source, 2 — Fourier transform lens, 3 — aperture diaphragm, 4 — object (tungsten wire), 5 — photosensitive layer, 6 — absorber of the direct light beam.

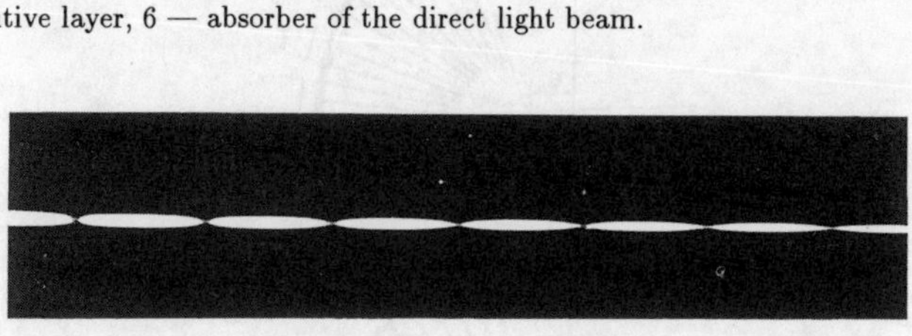

Fig. 9.17. Fragment of the diffraction pattern produced by the tungsten wire of the diameter $\approx 26\mu$m. The distance between two adjacent minima is equal about 4mm on the real photonegative.

This experimental setup is designed for estimation of the width d of the straight line object with error as small as possible from the data on positions of many intensity minima of the diffraction pattern in the region of high orders. The coordinates of these intensity minima $x_{mm}^{\min}$ versus the index number N of the diffraction minimum are shown in Fig. 9.18. This dependence is pure linear one.

Another picture is observed in the case of the accidental eccentricity between the object axis and the axis of the cylindrical backing of the photosensitive layer. A nonlinear effect is shown in Fig. 9.19. The difference between the observed minima positions and the expected linear dependence is shown in the Fig. 9.20.

Another effect observed in this experimental setup is the bending of the diffraction pattern induced by the wrong orientation of the object axis with respect to the

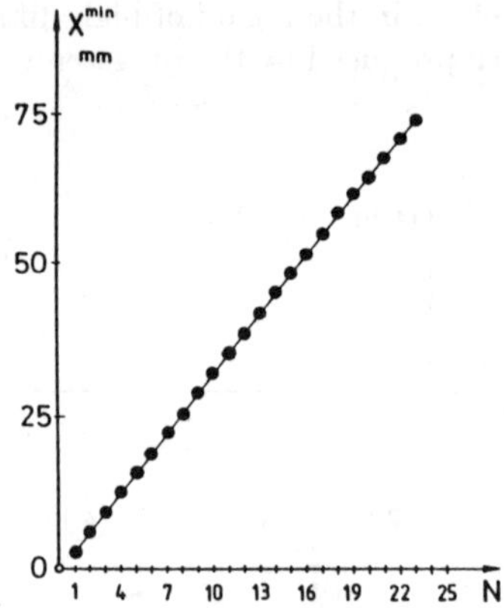

Fig. 9.18. The positions of the intensity minima $x_{mm}^{\min}$ at various index numbers N from $N = 2$ to $N = 24$.

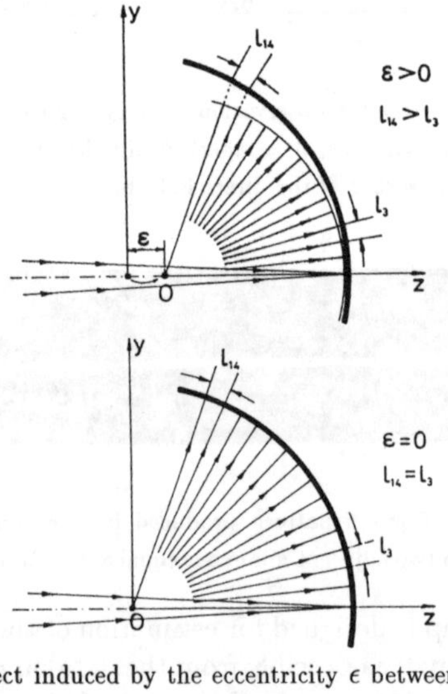

Fig. 9.19. The nonlinear effect induced by the eccentricity ϵ between the object axis and the axis of the cylindrical backing of the photosensitive layer. For $\epsilon > 0$ we have $l_{m+1} > l_m > l_{m-1}$ for every index number m.

axis of the cylindrical backing of the photosensitive layer (Fig. 9.21). The curvature radius ρ of this diffraction pattern is defined by the angle θ_z between the optical axis and the axis of the cylindrical backing. Just this phenomenon was used to measure the dip angle of the straight line object by means of the MFTM [126].

The photo of the diffraction pattern produced by **two** tungsten wires and detected on the cylindrical surface of the radius $R = 140\,\text{mm}$ over the region of the

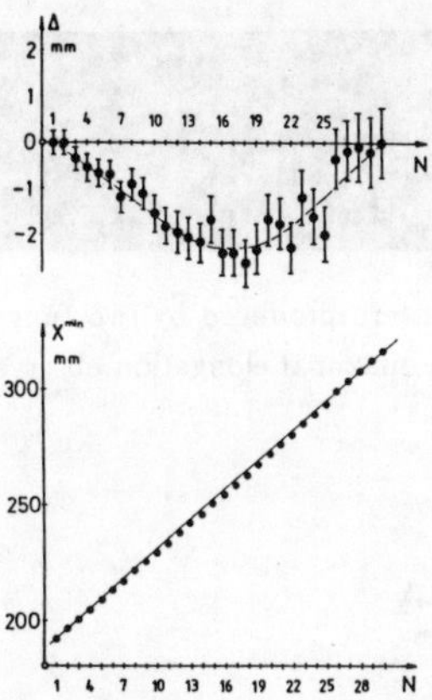

Fig. 9.20. The extra effect of the nonlinearity Δ_{mm} observed in the experiment with $\epsilon > 0$.

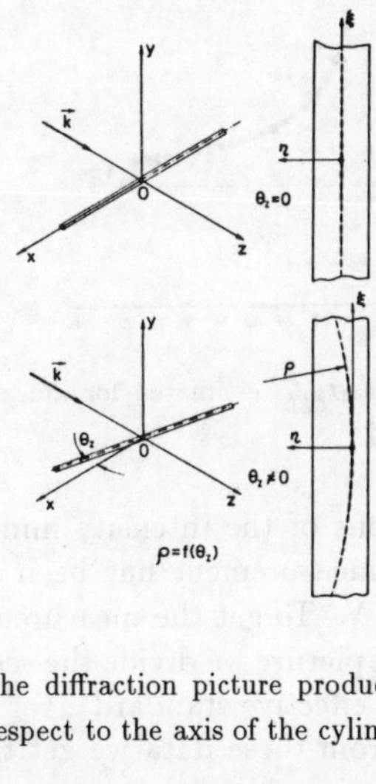

Fig. 9.21. The bending of the diffraction picture produced by the tungsten wire at the wrong orientation of the wire with respect to the axis of the cylindrical backing.

diffraction orders for diffraction angles from $\theta = 0$ to $\theta_\circ \approx 1\,\text{rad}$, is given in Fig. 9.22. The central star-like picture in the region of zero diffraction order is produced by the Fourier transform lens 2 (see Fig. 9.16) with aperture diaphragm 3 composed of six iris stops.

We see that the diffraction picture, ranged from $\theta = 0$ up to $\theta_\circ \approx 1\text{rad}$, goes into the reflection component at very large angles θ. A local interference between diffraction and reflection components can be seen as well.

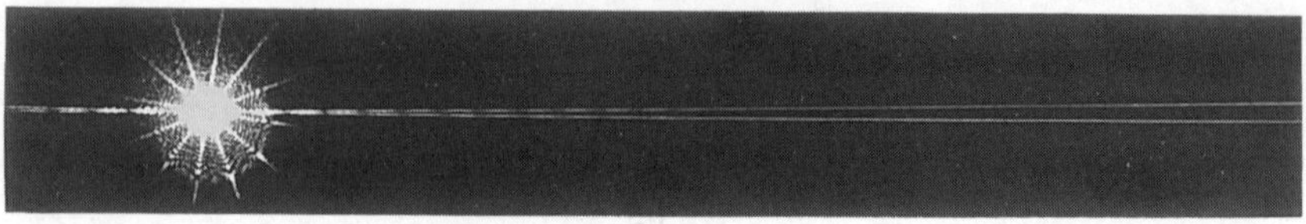

Fig. 9.22. The photo of the diffraction pattern produced by two tungsten wires. The radius of the cylindrical backing is $R = 140$mm. The maximal elongation observed experimentally is equal to 170mm ($\theta > 1$rad).

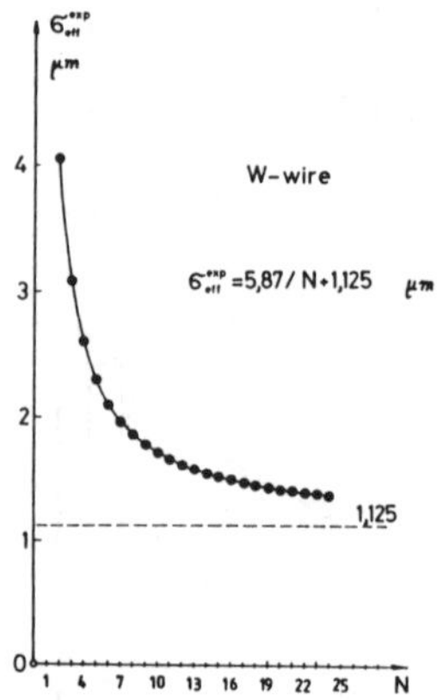

Fig. 9.23. The effective standard errors σ_{eff}^{exp}, estimated for index numbers from $N_1 = 2$ to $N_2 = 24$ on the photo shown in Fig.9.22.

We have estimated the positions of the intensity minima at index number N from $N = 1$ to $N = 24$. Each measurement has been repeated many times to find the real r.m.s. error for each N. To get the measurement error of the average period L of the periodic diffraction picture we divide the received r.m.s. error by the corresponding N. Thus we get the effective standard error σ_{eff}^{exp} of the period L (but not the diameter of the wire!). From these data we get the relative measurement error of the period L, or σ_{eff}^{exp}/L. It is evident that

$$\frac{\Delta L}{L} = \frac{\Delta d}{d}, \tag{9.19}$$

where Δd is the measurement error of the wire diameter.

The effective standard errors σ_{eff}^{exp} are shown in Fig.9.23 for different index numbers N. These data can be approximated for tungsten wire of the diameter $\approx 26\mu$m by the following function

$$\sigma_{eff}^{exp} = (\frac{5.87}{N} + 1.125)\,\mu\text{m}. \tag{9.20}$$

The constant term 1.125 μm is due to the fact that the contrast of the diffraction minima is decreasing at high N's. For $N = 24$ we have $\sigma_{eff}^{exp} = 1.35\mu m = \Delta L$.

The period of the diffraction pattern shown in Fig.9.17 and in Fig.9.22 is equal to $L \approx 4$ mm. Therefore

$$\frac{\Delta L}{L} = \frac{1.35 \cdot 10^{-3}}{4} = 3.4 \cdot 10^{-4} = \frac{\Delta d}{d}. \tag{9.21}$$

From Eq. (9.19) we see that new technique based on the observation of very high diffraction orders is equivalent to superresolution. This technique gives the possibility to estimate the diameter of the tungsten wire of the order 26μm with error as small as 0.03 % [478].

9.7. Meso–optical Microscope of High Productivity for Vertical Particle Tracks

This microscope is designed for selective observation of the vertical particle tracks in the nuclear photoemulsion [483]. To ameliorate the signal-to-noise ratio some new variants of the meso–optical condenser [313] described in Chapter 8, were proposed [312, 329]. The illuminating system with two convergent cylindrical waves is shown in Fig. 9.24. This system is provided with two cylindrical **half-lenses** 1 shifted at the distances $\pm\epsilon$ with respect to the symmetry axis of the whole system. The main feature of the illuminated region is that the period of the interference

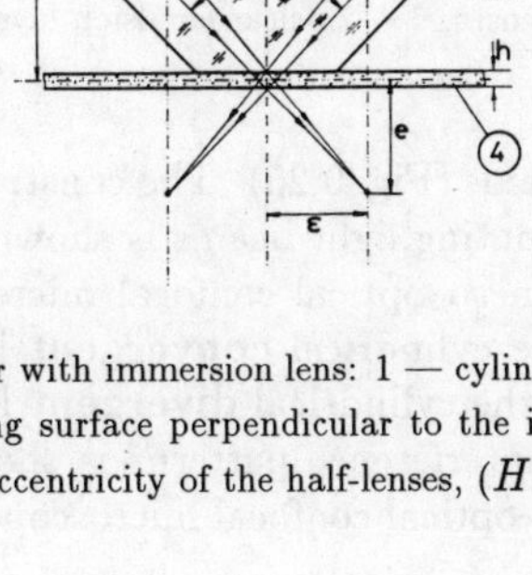

Fig. 9.24. Meso–optical condenser with immersion lens: 1 — cylindrical half-lens, 2 — central stop, 3 — immersion lens with entering surface perpendicular to the incoming light rays, 4 — nuclear emulsion layer, ϵ — transversal eccentricity of the half-lenses, $(H + e)$ — its focal length.

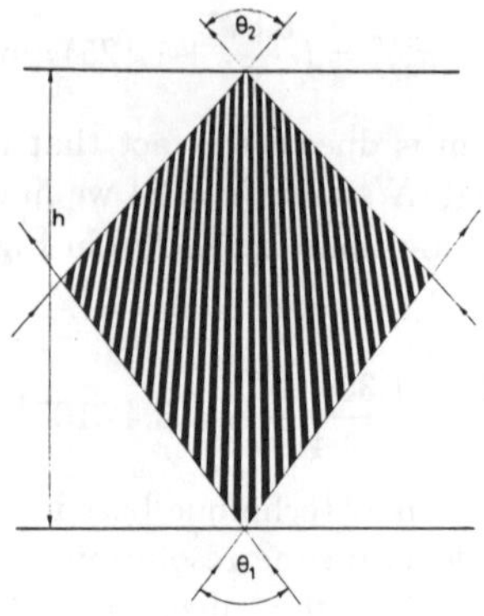

Fig. 9.25. Microscopic view of the interference picture produced by two restricted convergent cylindrical waves: θ_1 and θ_2 are angles between two wavevectors at the upper and at the bottom surfaces of the nuclear emulsion layer, respectively.

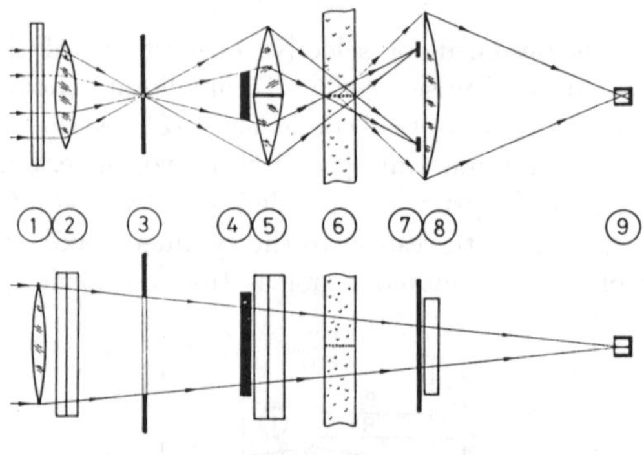

Fig. 9.26. Microscope for selective observation of the vertical particle tracks with convergent cylindrical illuminating light beams: 1 — first cylindrical lens, 2 — second cylindrical lens, 3 — $1D$ pinhole, 4 — stop, 5 — cylindrical half-lenses, 6 — nuclear emulsion layer, 7 — $1D$ stops, 8 — cylindrical lens, 9 — photodetector.

fringes is changing along z-axis (Fig. 9.25). The construction of the proposed device with two convergent illuminating light beams is shown in Fig. 9.26.

Now we consider new meso–optical confocal microscope with illuminating objective, which generates the cylindrical **convergent** light rays, and with imaging objective, which generates the cylindrical **divergent** light rays. A typical example of the overlapping of two interference patterns is shown in Fig. 9.27. The tabulated response of this meso–optical confocal microscope is presented in Fig. 9.28 for

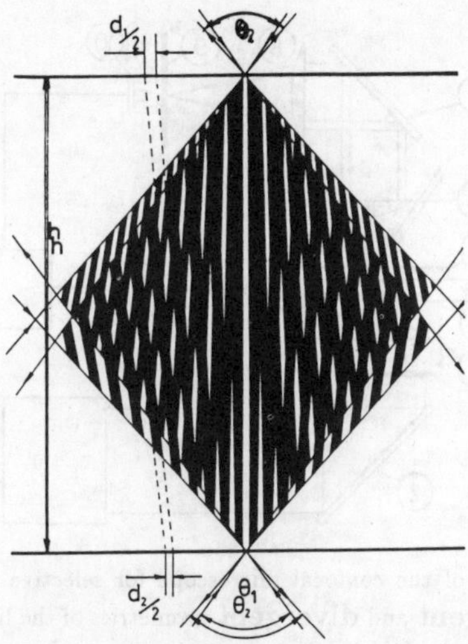

Fig. 9.27. An example of the overlapping of two patterns observed in the device with convergent cylindrical illuminating objective and with divergent cylindrical imaging objective.

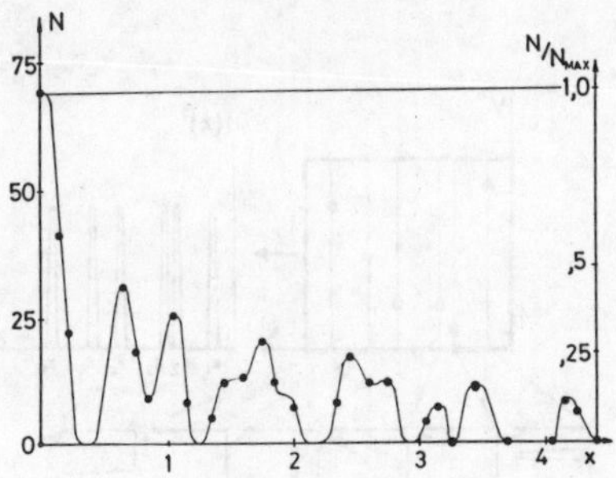

Fig. 9.28. The tabulated response of the confocal system with convergent and divergent objectives, which has been estimated for vertical particle track with 75 silver grains in the nuclear emulsion layer of the depth 200μm.

vertical particle track with 75 silver grains in the nuclear photoemulsion layer of depth 200μm. We see that at comparator level of 37.5 silver grains there are no side lobes at all.

The confocal microscope with convergent light rays geometry of the illuminating objective and with divergent light rays geometry of the imaging objective is shown in Fig. 9.29. The laser 1 generates a collimated light beam, which is reflected from the plane mirror 2 and is directed into the cylindrical interference illuminating

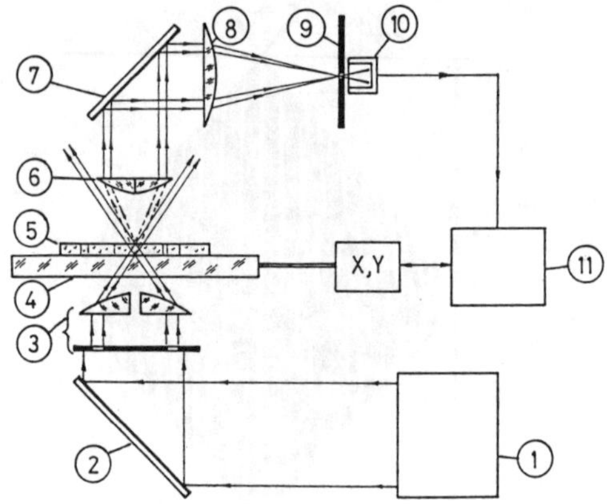

Fig. 9.29. Complete scheme of the confocal microscope for selective observation of the vertical particle tracks with **convergent** and **divergent** geometries of the light rays in the illuminating and imaging objectives: 1 — laser, 2 — plane mirror, 3 — cylindrical interference illuminated system, 4 — $(X - Y)$ stage, 5 — nuclear emulsion layer with vertical particle tracks, 6 — cylindrical interference imaging system, 7 — plane mirror, 8 — lens, 9 — pinhole, 10 — photodetector, 11 — computer.

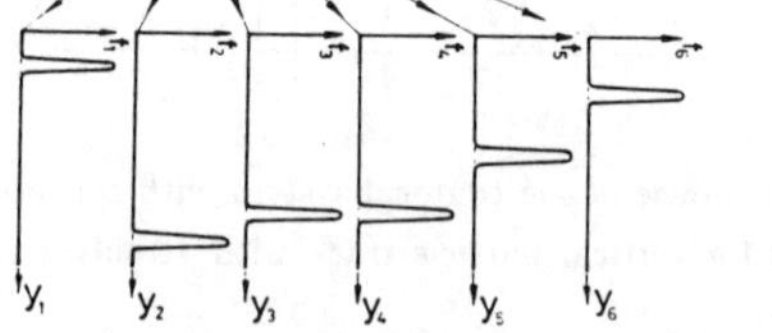

Fig. 9.30. Example of six vertical particle tracks marked as $\odot$, inside the square to be scanned.

system 3 provided with two eccentric cylindrical lenses. The diffracted light is picked up by the imaging objective 8 with focal length which is chosen to be 1.62 times more than the focal length of the illuminating objective 3.

Now we consider the scanning operations in this confocal meso–optical microscope [482]. The $(X - Y)$ stage of this microscope accomplishes the scanning operations prescribed by the following program in the computer 11. We consider an example of six vertical particle tracks inside the square to be scanned (Fig. 9.30). The first scanning operation is performed along x-axis by the illuminated region having narrow "fence" at the regime of the wide-row drill. After this first scanning operation we get one $1D$-function $f(x)$, with six picks. The coordinates of these peaks $x_1, ..., x_6$ of the function $f(x)$ define the positions of the $(X - Y)$ stage of our microscope, at which the vertical particle tracks $1, ..., 6$ are successively in the center of the illuminating region.

Before the second scanning operations, the imaging interference system 7 is to be rotated through the angle 90° in its plane [482]. The response of our device in this configuration has a point-like structure, as two line responses of the systems 3 and 7 are now mutually perpendicular. The second scanning operations are performed sixfold along y-axis at x-coordinates $x_1, ..., x_6$ stored in the computer 11 after the first scanning operation. Now we get six functions $f_1(y), ..., f_6(y)$ versus y-coordinate. The positions of six vertical particle tracks to be searched for are defined by six pairs of coordinates $(x_1, y_1), ..., (x_6, y_6)$, where $y_1, ..., y_6$ are the positions of the maxima of six functions $f_1(y), ..., f_6(y)$.

In the general case of N vertical particle tracks inside the region to be scanned we must accomplish **one** $1D$-scanning with N picks in the function $f(x)$, and then **N** $1D$-scannings with one pick in each $1D$-plot. Thus the total number of $1D$-scannings is equal to $(N + 1)$. In the case of quadrate region with sides of 5mm length, we must make $5 \cdot 10^3 \cdot (N + 1)$ digital measurements with spatial resolution of the order of 1μm. Meanwhile in the traditional microscope the number of the digital measurements will be equal to $(5 \cdot 10^3)^2 \cdot 30$. Here we accept that the number of digital measurements along z-axis in depth is equal to 30. Thus the factor of merit is equal to

$$\kappa = \frac{(5 \cdot 10^3)^2 \cdot 30}{5 \cdot 10^3 (N + 1)} = \frac{15 \cdot 10^4}{(N + 1)}. \tag{9.22}$$

For $N = 25$ we have $\kappa = 5.8 \cdot 10^3 : 1$.

The interference pattern produced by two cylindrical waves has been observed experimentally in the conditions presented in Fig. 9.31. Two cylindrical half-lenses L_L and L_R are illuminated by the plane wave from laser 1. The stop S restricts two external parts of this plane wave, which are focused independently each at eccentricities $\pm\epsilon$. The photoplate 2 can be located in two various positions: in the plane A in the first exposure or in the plane B in the second exposure. In the position A two cylindrical waves are overlapped, but in position B they are separated by distance which is more than the transversal width of the cylindrical waves. The microphotograph in the position A is shown in Fig. 9.32. The number of the contrast interference fringes is equal to 7.

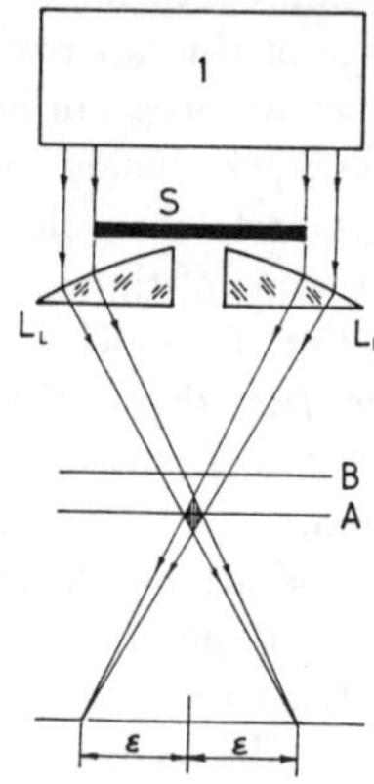

Fig. 9.31. Experimental setup used for observation of the interference picture produced by two cylindrical plane wavelets.

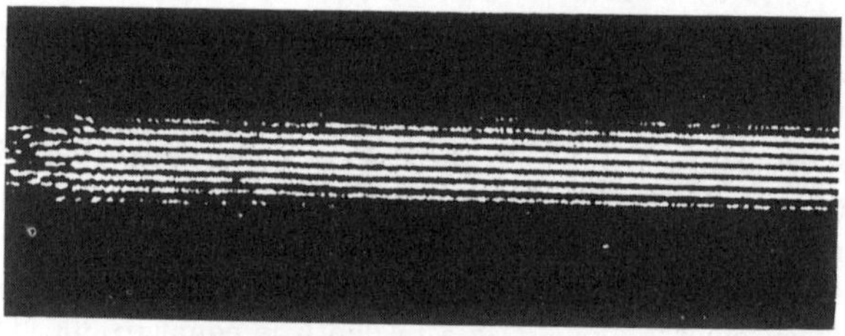

Fig. 9.32. The microphoto taken in the plane A of the experimental setup shown in Fig.9.31.

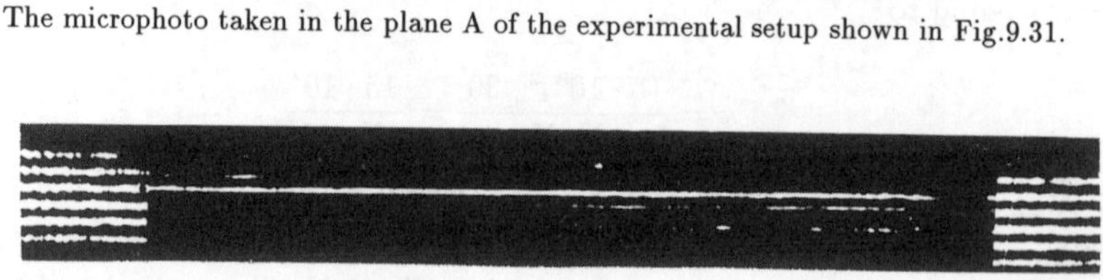

Fig. 9.33. The transparency mounting of two interference patterns with fringe periods ratio equal to 1.62 : 1.

A simulated picture of two different fringe periods in the ratio 1.62 : 1 has been prepared. The mechanical mounting of two interference pictures with this period ratio is shown in Fig. 9.33. We see that the number of contrast maxima in this combined mounting of two tranparencies is equal to **one** instead seven, observed in each interference pattern. To get such amelioration in the total point spread function of the real confocal system, it is sufficient to chose the focal lengths f_1 and f_2 of two half-lenses L_1 and L_2 in the ratio $f_1 : f_2 = 1.62 : 1$.

9.8. Caustic Meso–optical Confocal Microscope for vertical particle tracks

The well known caustic phenomenon [484] can be used to make another one nontraditional microscope for selective observation of the vertical particle tracks in the nuclear photoemulsion [483]. To demonstrate the caustic phenomenon, an experimental setup, shown in Fig 9.34, has been used. A cylindrical half-lens L is illuminated by the collimated light beam. The monochromatic pattern of the cylindrical caustic is detected on the photoplate HPh with high spatial resolution

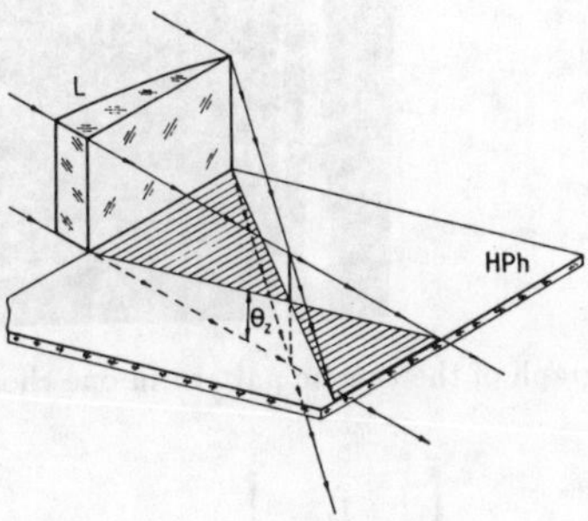

Fig. 9.34. The geometry of the experiment for observation of the caustic pattern with very high spatial resolution: L — cylindrical half-lens, HPh — holographic photoplate, θ_z — inclination angle of the photoplate.

Fig. 9.35. The view of the caustic pattern before and after the focus region.

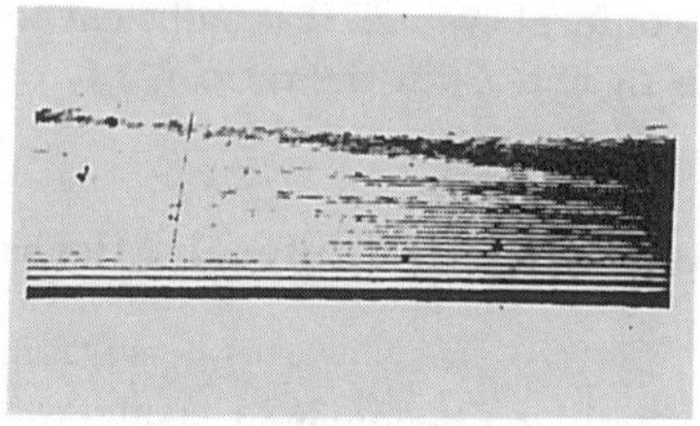

Fig. 9.36. The microscopic structure of the edge of the caustic pattern, shown in Fig.9.35.

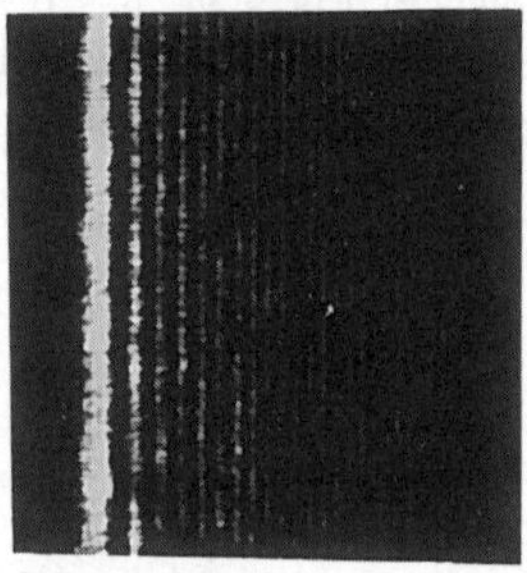

Fig. 9.37. Direct photomicrograph of the caustic pattern in one chosen position of the photofilm.

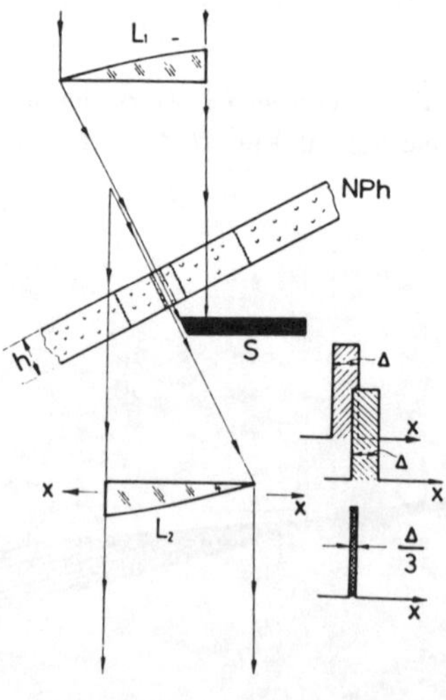

Fig. 9.38. Confocal caustic meso–optical microscope for selective observation of the vertical particle tracks: L_1 — illuminating cylindrical half-lens, NPh — nuclear photoemulsion, S — stop, L_2 — imaging cylindrical half-lens. The moving of the half-lens L_2 is performed along x-axis.

of the order of 10^3 lines per mm. The photoplate HPh is inclined at the angle θ_z with respect to the light beam axis. The monochromatic caustic pattern before and after the focus region is shown in Fig. 9.35. Its microscopic structure is given in Fig. 9.36. In Fig. 9.37 we see the caustic pattern taken by camera with optical axis parallel to the illuminating light beam.

The principal scheme of the confocal caustic meso–optical microscope for selective observation of the vertical particle tracks is shown in Fig. 9.38. The illuminating cylindrical half-lens L_1 produces the caustic interference pattern, shown in Fig. 9.36. The nuclear photoemulsion layer is placed in the vicinity of the focus of the half-lens L_1 and is oriented perpendicular to the external light rays of the caustic pattern. The optical stop S absorbs all internal light rays. The imaging cylindrical half-lens L_2 focuses the light rays of the external part of the caustic interference fringes. The region where two systems of the interference fringes — one from the illuminating half-lens L_1 and another from the imaging half-lens L_2 — are overlapped, can be controlled by transversal moving of these two half-lenses. A typical example of such an overlapping is shown in Fig. 9.38 for the case of the total response width equal to $\Delta/3$.

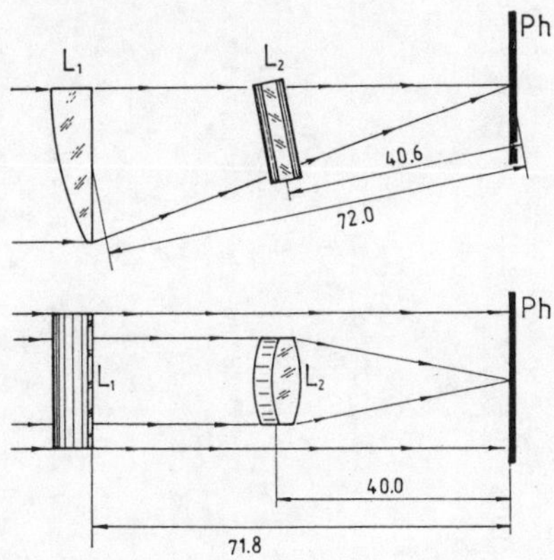

Fig. 9.39. The experimental setup used for observation of the point-spread function in the system with two crossed cylindrical half-lenses.

By means of the experimental setup shown in Fig. 9.39, we have observed the point-spread function of this system which involves two crossed cylindrical half-lenses. The first cylindrical half-lens L_1 focuses the light rays onto the photo-plate Ph. The second cylindrical half-lens L_2, placed after the first cylindrical half-lens L_1 and oriented perpendicular to the half-lens L_1, focuses the light rays onto the same photo-plate Ph. In such a configuration we have detected the point-spread function shown in Fig. 9.40 at two different exposure times. We see that side lobes of this point-spread function are present only in one quadrant with respect to the central maximum and therefore induce only minor effect on the signal-to-noise ratio of the whole confocal system.

The configuration shown in Fig. 9.39, must be used on the second stage of the scanning operations described above [482] (see p.305 of this book).

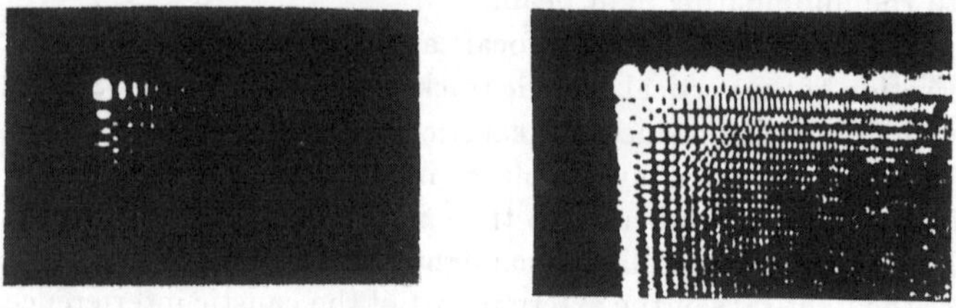

Fig. 9.40. The photo of the point-spread function of the system, shown in Fig. 9.39, with two crossed half-lenses, at two exposure times.

This caustic confocal meso–optical microscope sees selectively only vertical particle tracks and has high productivity explained above [482]. No depth scanning is needed in such a microscope.

Chapter 10

HISTORY OF MESO–OPTICS

10.1. From Big Bang to the Gravitational Lens

According to generally accepted model of the development in time of our universe, an extremelly dense initial state, named "Big Bang", had to be exist about $2 \cdot 10^{11}$ years ago. After very short period of the quantum era of duration $\sim 10^{-43}$ s, a lepton era of the length $\sim$ 25 s followed. The temperature of the universe was $\sim 10^{12}$ K at the age $\sim 2 \cdot 10^{-4}$ s, $\sim 10^{10}$ K at the time moment 1 s and $\sim 5 \cdot 10^{9}$ K to the end of this period.

The matter density ρ_m of the expanding universe of radius $R(t)$ was decreasing as

$$\rho_m(t) \sim R^{-3}(t). \tag{10.1}$$

In the isotropic radiation model, the density of radiation was decreasing as

$$\rho_{rad}(t) \sim R^{-4}(t). \tag{10.2}$$

Due to this difference of the time behaviours of $\rho_m(t)$ and $\rho_{rad}(t)$, the early stage of the universe was radiation-dominated and lasted about $\sim 2 \cdot 10^{13} s \approx 0.6 \cdot 10^{6}$ years. At this age the radiation density and the matter density were 10^{-21} g/cm^3 and 10^{-20} g/cm^3, respectively, whereas at the time moment $2 \cdot 10^{-4}$ s, the radiation density ρ_{max} was $5 \cdot 10^{13}$ g/cm^3, and the matter density ρ_m was only $3 \cdot 10^{5}$ g/cm^3 [346,347].

From the age of 10^6 years, the light rays were propagating along the straight lines, and the phenomenon of the gravitational lensing begun to occur in our universe. Therefore this phenomenon existed well before the creation of our Earth. The author of this book suppose, that this earlier phase of the gravitational lensing was not yet treated in the scientific literature. Of a particular interest is the evolution in time of this phenomenon from the universe age of 10^5 years up to the age of $\sim 10^9$ years.

10.2. Ancient History — Middle Ages — XX Century

The cylindrical and saddle-shaped mirrors were used already in the ancient times. It has been found that in most cases these mirrors transform the object into its image. These observations were described by Ptolemeus (90–168) [348].

The first glasses came into use in the 13 century. Already in 1352 the glasses were seen on the art paintings. To correct the astigmatic eyes, the lenses with nonspherical surfaces, such as with cylindrical ones, have been invented. The first investigation with these surfaces was made by G.B. Airy in 1826, although three new machines for manufacturing of cylindrical, conical and plane optical surfaces have been described by J. Leupold as early as in 1713 [351, 349]. J Leupold had mentioned that astigmatism and distortion would occur when spherical surfaces were used with mutual eccentricity.

The reflection of light from a conical mirror with apex angle $< 45°$ was treated in detail by H. Emsman [352]. He calculated the position of the naturally distorted images for on-axis observer. The theory of thin cylindrical lenses dedicated to focus the optical spectral lines has been developed by E. Reusch in 1868 [353]. The first experiments with lenses having toric surfaces were performed by Poullains in 1870 [349]. There were two types of these lenses: plane-toroidal and sphero-toroidal ones. The meridional and equatorial focus lengths were different in these lenses. The astigmatism was as large as +4 dptr. The anamorphic lens system has been patented by C. Zeiss in 1897 [354]. G.J. Burch [355] described the application of the cylindrical lens for rotation of the images.

10.3. Diastoloscope

The first application of the conical lens as an imaging element in microscopy has been described by C. Chabrié [356 – 360] in 1904. This was the "big bang" of the meso–optics, because from this time conical elements were used for imaging of the microscopic objects. The aim of C. Chabrié [356] was to produce magnified image with tolerated strong geometrical distortions. For this purpose a conical refracting element was chosen. The structure of the real image of the object in the form of a circular disc, illuminated by a collimated light beam, has been investigated. The image had the form of a ring with center on the symmetry axis and of the width defined by the diameter of the disc. For a circular disc of small radius the image assumed the form of a circle. The point-like object was converted into a circle of any desired radius. To produce the secondary image, analogous to the object, a second conical element had been inserted.

In the paper [357] it was shown that the linear magnification is the function of the object dimension x:

$$M(x) = \frac{R}{x} - \frac{R-r}{\rho}, \tag{10.3}$$

where R is the radius of the exterior circle of the ring, r is that of the inner circle, and ρ is the radius of the disc, with $0 \leq |x| \leq \rho$. For $x = 0$, $M \to 0$, and for $x = \rho$, $M = r/\rho$.

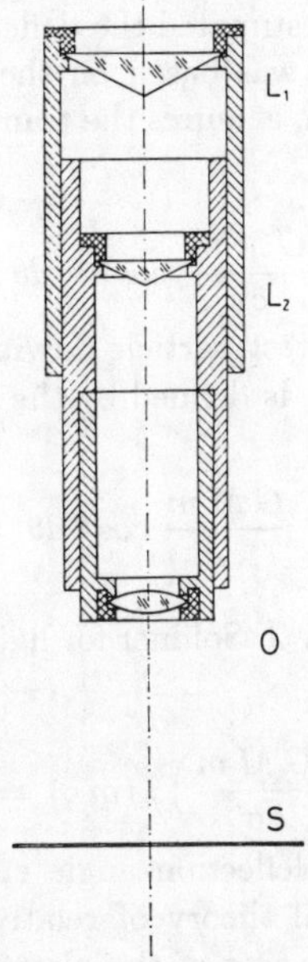

Fig. 10.1. Principal scheme of the diastoloscope of C. Chabrié: L_1 and L_2 are conical lenses, O — microscopic objective, and S is the observation screen.

The diastoloscope (microscope which produces the dilation effect) shown in Fig. 10.1 [358] has been provided with two coaxial conical lenses mounted in the first tube holder. The second tube holder, equipped with microscopic objective, was moving with respect to the first holder. The image was projected on the screen. With this diastoloscope the central part of the object disc has been dilated with high linear magnification $M \sim 6 \cdot 10^3$. The image of the object produced by diastoloscope was subjected to very strong deformations which were analyzed in [359]. The main feature of these deformations was drastic discontinuity of the multiple images of the object. This property of the diastoloscope could be used for precise position measurements of the illuminated microscopic objects.

10.4. Theory of the Gravitational Lens

Exciting history of this theory was started as early as in 1801(!). That was J. von Soldner, German mathematician and geodesist, who has derived in March 1801 the equation for calculation of the angular deflection of the light rays in the gravitational field. J. van Soldner has used the Newtonian theory of gravitation [361]. With this equation he estimated the deflection angle for Earth, Moon, and Sun. The Soldner's derivation was based on the fact, that any body of the mass m, including the light quantum, acquires the transverse momentum $dP_\perp$ equal to

$$dP_\perp = \frac{G\,M\,m}{c\,a}\cos\theta\cdot d\theta\,, \tag{10.4}$$

where θ is the position angle of the test particle moving with impact parameter a. The total transferred momentum $P_\perp$ is defined by the equation

$$P_\perp = \int dP_\perp = \int_{-\pi/2}^{+\pi/2}\frac{G\,M\,m}{c\,a}\cos\theta\,d\theta = \frac{2\,G\,M\,m}{c\,a}\,. \tag{10.5}$$

The deflection angle ε_s, calculated by J. Soldner for light quantum at very small ε_s, is equal to

$$\varepsilon_s = \frac{P_\perp}{P} = \left(\frac{2\,G\,M\,m}{c\,a}\right)/(m\,c) = \frac{2\,G\,M}{c^2\,a}\,. \tag{10.6}$$

The mass m is cancelled, but the deflection angle ε_s is only half of the correct value of ε, predicted by the general theory of relativity. Of course, J. Soldner has neglected the effect of the increasing of the photon mass in the gravitational field. However, being introduced into the calculation, this effect does not alter the value of the deflection angle. The staggering fact was that the prediction of the special theory of relativity, derived by A. Einstein in 1911 [362], does not differ from the J. Soldner's prediction, made more than hundred years before. This dramatic situation has not been discussed in detail in the papers on gravitational lenses, such as [234, 235, 236]. Only in 1916 [363] A. Einstein arrived at the correct value of the deflection angle ε, drawn from general theory of relativity. The astronomical observations performed during the total sun eclipses have proved without any doubt that only general theory of relativity, developed by A. Einstein, gives the correct deflection angle, 1.75″ [364].

The idea to observe the phenomenon now called "gravitational lens" has been suggested by R.W. Mandle, a Czech electrical engineer, in 1935. He asked A. Einstein to publish his idea. In the paper [238] A. Einstein reported that the lens-like action of a star could induce the light intensity intensification and had derived an equation for point-like stars configurations.

Meanwhile, the splitting of the star's image in the gravitational field of the nearest star has been predicted by O.D. Chvolson in 1924 [365]. He showed that this splitting of the star's image could not be detected by optical telescope and that for this purpose more sensitive techniques, spectral and interferometric ones, are to be used. He explained that in the case of on-axis configuration the nearest star would be surrounded by a ring.

Important corrections to the generally adopted history of the gravitational lensing have been made by Jeno M. Barnothy. In his paper [367] he announced that Oliver Lodge was indeed the first who predicted in 1919 the focusing effect of the gravitational lens. E.B. Frost, the director of Yerkes observatory, put forward a program in 1923 to search for this effect for stars. J.M. Barnoty pointed out that now popular term "Einstein ring" must be changed by the correct term "Chvolson ring", because the paper of A. Einstein was published 12 years later than the paper of O.D. Chvolson.

F. Zwicky in 1937 [239] has proposed to observe the effect of the gravitational lens for galaxies (nebulae) rather than for stars. He proved that the mass of the galaxy must be actually about $4 \cdot 10^{11}\, M_\odot$ and the amplification gain can be as high as 100.

A comprehensive theoretical treatment of the gravitational lensing for stars has been made by G.A. Tikhov in 1938 [366]. He calculated the angular splitting of the star's image, the expected intensity gain and the influence of the finite dimensions of the source object on the results of the calculations. He introduced the concept of the "light strikes". These calculations were made both for Newtonian classical theory and for general theory of relativity.

The detailed analysis of the gravitational lensing phenomenon was made in papers [234, 235, 236]. Already in 1965 J.M. Barnothy conjectured that quasars are indeed the nuclei of the Soyfert I galaxies, the visible intensity of which was magnified by the gravitational lensing action of the foreground galaxies. This idea was rekindled in 1979, when D. Walsh et al. discovered the first gravitational lens[242]. The second robust idea expressed by J.M. Barnothy was his interpretation of the neutron star binary system. One such star can act as an ocular lens and the other one as an objective lens, thus forming a cosmic scale gravitational telescope. The expected duration of two light flashes from this binary neutron star was about 100 μs. The effect of the light diffraction on the intensity gain in the gravitational lensing has been considered by B.V. Bliokh and A.A. Minakov [368] in 1975. The effect of gravitational microlensing was treated by K. Chang and S. Refsdal in 1979 [259]. The effects of the moving gravitational lens were explained in 1989 by S.M. Chitre and W.C. Saslaw [240] and by M. Birkinshaw [241].

The first proposal to observe the interference pattern produced by the gravitational lens was made in 1991 by J.B. Peterson and T. Falk [256]. Independently the idea to consider the gravitational lens as a beam-splitter of the cosmic scale and to observe the gravitational lensing in the narrow spectral window has been expressed in 1992 by L.M. Soroko on the seminar of the Laboratory of Theoretical Physics, Joint Institute for Nuclear Research, Dubna, on 2 April.

The observation in 1993–1995 of a dozen formidable microlensing events [436, 437, 453, 455, 464] was indeed the epoch-making discovery. These gravitational microlensing events were induced by single or binary lensing stars. The preliminary dark matter interpretation of these events was analyzed anew, and it was shown, that all detected events were produced by the ordinary stars, not by dark matter objects. The most power forthcoming technique in the gravitational microlensing investigations would be parallax method [445, 460, 463, 466], which will become the real technique just after the lunching of the special satellite with Earth or Jupiter-like orbits. We are in the excitation stage to hear the expected announcement of the discovery of the event induced by the "second Jupiter" in our Galaxy [452, 456, 458, 465, 468].

10.5. Meso–optics

Why the annular aperture of the imaging element, such as lens, mirror, objective, is of great importance in the meso–optics? The reason is that the radial distribution of the light intensity in the focus of the objective with narrow annular aperture is described by the square of the Bessel function $J_{\circ}(\alpha r)$ just in the same way as that in the focus of an axicon. There is no problem with depth-of-focus in such a system. The detailed treatment of the lens with narrow annular aperture has been made by G.C. Steward in 1928 [36].

However only in 1954 an axicon (a conical lens) has been rediscovered by J.H. Mc Leod [10, 11] after a lapse of 50 years after the publications of C. Chabrié. A comprehensive investigation of some meso–optical elements such as axicon, circular diffraction grating and Archimedean spiral grating has been made in his classical paper by J. Dyson [38]. Many results of this investigation were rediscovered later by other authors.

New scopes for the coherent optics were opened by A. Lohmann after the advent of the computer synthesized holographic translator in 1967 [63]. Since that time any input $2D$ signal can be converted by this element into any output $2D$ image. Recall that in traditional "physical" holographic elements only spherical or plane waves were used. Now with this holographic translator of A. Lohmann, the nonspherical wave fronts can be generated by the simple manner. In 1974 O. Bryngdahl [50] demonstrated that with optical devices which are working as a parallel processing system we can perform general geometrical transformations, such as local translation, inversion, reflection and rotation, by using spatial-variant optical filter. In 1975 O. Bryngdahl [369] showed that with computer generated hologram we can construct or store an arbitrarily shaped wavefronts and perform geometrical and phase transformations. He introduced azimuthal phase modulation of the wavefront, typical for helical wavefronts, and also radial phase modulation of the wave amplitude, typical for conical waves.

Confocal scanning microscope, described in 1977 by C.J.R.Sheppard and A.Choudhury [370], opened new approach to $3D$ imaging and to effective depth sectioning. The proposal to use the meso–optical confocal scanning microscope in the microtomography of high spatial resolution was made in 1985. The first observation of the interference of two conical wavefronts of synchrotron radiation was made in 1980 by M.M. Nikitin et al. [285]. The synthesis of optical elements which generate the focal line of the arbitrary shape by means of the focusators working as one-shot computer generated hologram was accomplished by I.N. Sisakyan et al. in 1981–1984 [61, 62, 64 – 68].

The idea that meso–optical element with ring response can be used only in metrology and for beam transformation was destroyed in 1982 [371] by L.M. Soroko, who proposed meso–optical Fourier transform microscope. In this microscope the line object was viewed at two camera-angles and the problem of the depth-of-focus has been resolved without making a sacrifice of the spatial resolution. The term "MESO–OPTICS" was legitimated in 1984 at the 13 Congress of the International Commission of Optics (ICO), in Sapporo [372]. A kinoform-axicon and a kinoform with ring response were fabricated with diffraction limited resolution by V.P. Koronkevich et al. [52, 107]. The meso–optical mirror with ring response of diameter 160 mm was fabricated in 1986 [108]. The computer generated hologram for reconstruction of $3D$ objects composed of line segments (conical waves!) was made and investigated in 1986 by Ch. Frère, D. Leseberg and O. Bryngdahl [373]. The meso–optical Fourier transform microscope with double focusing has been made in 1992 [134]. With this microscope the information about position in space of the microscopic line object was compressed at highest degree, restricted only by diffraction of light.

10.6. Diffraction Free Wavefields

The paper of J. Durnin [75] is doubtless of the highest citation index. New term "diffraction free wavefields" introduced in this paper proved to be concise, useful and become very popular. Soon after this paper, it was understood that for practical realization of the pure Bessel beams we must consider Bessel–Gauss beams [168, 172]. The helical wavefields were introduced as well [167, 176]. With these waves the meso–optical darkness and phase singularities were investigated [177 – 182]. The new phenomenon, the longitudinal interference of coaxial light beams, can be observed only with conical wavefields [187 – 189]. The production of the diffraction free wavefields was considered in many papers [190, 192 – 195]. The generation of the steady-state radiation from laser with phase singularities and vortices was treated in [179, 183 – 186].

10.7. Localized Wave Fields

The history of this branch of meso–optics goes back to the book of R. Courant and D. Hilbert **Methods of Mathematical Physics** (Interscience, New York, 1962), in the second volume of which, p.746, the solutions called "almost undistorted waves" were treated. According to Courant and Hilbert [374] the homogeneous wave equation

$$\partial_x^2 \Psi + \partial_y^2 \Psi + \partial_z^2 \Psi - \partial_{x_\circ}^2 \Psi = 0, \quad \text{with } x_\circ = ct \tag{10.7}$$

has nondispersive solutions of the form

$$\Psi = g(\vec{x}, x_\circ) F(u), \tag{10.8}$$

where u is a solution, called "phase", of the corresponding characteristic equation

$$(\partial_x u)^2 + (\partial_y u)^2 + (\partial_z u)^2 - (\partial_{x_\circ} u)^2 = 0, \tag{10.9}$$

and F is an arbitrary function with continuous partial derivatives. The function $g(\vec{x}, x_\circ)$ can be considered as an attenuation factor. This fact was discovered by P. Hillion in 1992 [375] and 1993 [376]. The important innovation was made by G.A. Deschamps in 1971 [201] who introduced the notion of "complex source" for description of the propagation of Gaussian beam. This concept will be used widely in the theory of the localized wave fields, in particular in [209].

The focus wave modes were first described by J.N. Brittingham [203] and by A.P. Kiselev [377]. In the modern classification presented by P. Hillion [375] we may distinguish three types of electromagnetic inhomogeneous pulses with phase depending on space and time: the focus wave modes [205], the Gaussian pulses and the Bessel pulses.

It is worthwhile to add that H.F. Harmuth in 1977 [379] gave the analysis of the non-sinusoidal electromagnetic waves **without** sinusoidal carrier. He has shown that angular resolution of the non-sinusoidal electromagnetic waves was better than for classical system with narrow frequency modulated sinusoidal carrier. H.F. Harmuth also noticed [380], that group velocity is useless to describe the energy propagation of non-sinusoidal signals. In 1990 P. Hillion proved [381] that focus wave modes and some other inhomogeneous scalar waves can be interpreted as nondispersive plane waves with a non-uniform transverse structure. The main point of this approach was that the problem of the localized wave fields can be treated as the Goursat boundary value problem, for which the boundary data are given on the characteristics of the initial wave equation. P. Hillion has shown [207], that focus wave modes can be generated by plane waves diffracted by a pinhole moving with a velocity of light. The remarkable hit of the theory of the localized wave fields was its experimental verification on acoustical waves [213, 214].

APPENDIX

INTEGRAL TRANSFORMS

A.1. Fourier Transform

Let $\vec{x}\,(x,\,y)$ be $2D$ position vector in the input plane and $\vec{\omega}\,(\omega_x,\,\omega_y)$ be $2D$ vector of spatial frequencies ω_x and ω_y, corresponding to x and y coordinates, respectively. The Fourier transform $F(\vec{\omega})$ of the original $2D$ function $f(\vec{x})$ is defined by the equation [382, 383]

$$F(\vec{\omega}) = \iint\limits_{-\infty}^{+\infty} f(\vec{x}) \exp(-i\,\vec{\omega}\,\vec{x})\, d\vec{x} \equiv \hat{\mathcal{F}}[f(\vec{x})]. \tag{A.1}$$

The inverse Fourier transformation is defined by the equation [382]

$$f(\vec{x}) = \hat{\mathcal{F}}^{-1}[F(\vec{\omega})] = \frac{1}{2\pi} \iint\limits_{-\infty}^{+\infty} F(\vec{\omega}) \exp(i\,\vec{\omega}\,\vec{x})\, d\vec{\omega}\,. \tag{A.2}$$

In the polar coordinates $\vec{x}\,(\rho,\,\Phi)$ and $\vec{\omega}\,(\sigma,\,\chi)$ the equations (A.1) and (A.2) assume the forms [383]:

$$F(\vec{\omega}) \equiv F(\sigma,\,\chi) = \int\limits_0^{\infty} \int\limits_{\alpha}^{\alpha+2\pi} \rho\, f(\rho,\,\Phi) \exp[-i\,\sigma\,\rho\,\cos(\Phi-\chi)]\, d\Phi\, d\rho \tag{A.3}$$

and

$$f(\vec{x}) \equiv f(\rho,\,\Phi) = \frac{1}{2\pi} \int\limits_0^{\infty} \int\limits_{\beta}^{\beta+2\pi} \sigma\, F(\sigma,\,\chi) \exp[i\,\sigma\,\rho\,\cos(\Phi-\chi)]\, d\chi\, d\sigma\,, \tag{A.4}$$

where α and β are the arbitrary angles. The functions of polar coordinates $f(\rho,\,\Phi)$ and $F(\sigma,\,\chi)$ can be written in the forms:

$$\left.\begin{aligned} f(\rho,\,\Phi) &= \sum_{m=-\infty}^{\infty} c_m(\rho) \exp(i\,m\,\Phi)\,, \\ F(\sigma,\,\chi) &= \sum_{m=-\infty}^{\infty} \gamma_m(\sigma) \exp(i\,m\,\chi)\,, \end{aligned}\right\} \tag{A.5}$$

where

$$c_m(\rho) = i^m \int_0^\infty \gamma_m(\sigma)\, J_m(\sigma\rho)\, \sigma\, d\sigma\,, \tag{A.6}$$

and $J_m(\bullet)$ is the Bessel function of the m-order.

For $2D$ functions with circular symmetry $f(\vec{x}) \equiv f(r)$ we have

$$F(\omega) = \int_0^\infty r\, f(r)\, J_\circ(\omega r)\, dr\,, \tag{A.7}$$

where $F(\omega)$ is called the Hankel transform of radial function $f(r)$. The inverse transformation is described as

$$f(r) = \int_0^\infty \omega\, F(\omega)\, J_\circ(\omega r)\, d\omega\,, \tag{A.8}$$

For discrete sequence $p(x)$, $x = 0, 1, ..., N-1$ the forward discrete Fourier transform of the function $p(x)$ is given by [383, 384]

$$P_F(u) = \sum_{x=0}^{N-1} p(x) \cdot \exp(-i\, 2\pi\, u\, x / N)\,, \tag{A.9}$$

and the inverse discrete Fourier transform is defined as

$$p(x) = \frac{1}{N} \sum_{u=0}^{N-1} P_F(u)\, \exp(i\, 2\pi\, u\, x / N)\,. \tag{A.10}$$

A.2. Hartley Transform

For real valued input discrete sequence $p(x)$ its discrete Fourier transform $P_F(u)$ is defined by N complex numbers or $2N$ real numbers, but only N of them are independent ones. This redundancy can be used to construct the discrete Hartley transform [385, 386, 388] of the input real valued function $p(x)$:

$$P_H(u) = \sum_{x=0}^{N-1} p(x)\, \mathrm{cas}(2\pi\, u\, x / N)\,, \tag{A.11}$$

where

$$\mathrm{cas}(2\pi\, u\, x / N) = \cos(2\pi\, u\, x / N) + \sin(2\pi\, u\, x / N)\,. \tag{A.12}$$

Hence

$$P_H(u) = Re\{P_F(u)\} - Im\{P_F(u)\}\,. \tag{A.13}$$

As the summands of the new kernel function *cas* (•) have different parities, we may separate directly the real and the imaginary parts of the discrete Fourier transform $P_F(u)$. For example, if

$$P_H(u) = P_H^{EVEN}(u) + P_H^{ODD}(u), \tag{A.14}$$

then

$$\left.\begin{aligned} P_H^{EVEN} &= \tfrac{1}{2}[P_H(u) + P_H(-u)] \\ P_H^{ODD} &= \tfrac{1}{2}[P_H(u) - P_H(-u)] \end{aligned}\right\} \tag{A.15}$$

and

$$P_F(u) = P_H^{EVEN}(u) - i\, P_H^{ODD}(u). \tag{A.16}$$

The inverse discrete Hartley transform is defined by the equation

$$p(x) = \frac{1}{N} \sum_{u=0}^{N-1} P_H(u) \operatorname{cas}(2\pi u x / N). \tag{A.17}$$

The main advantage of the discrete Hartley transform over the discrete Fourier transform is that the real valued input function goes into the real valued transform function [389].

However we meet with difficulties to construct the discrete Hartley transform in two dimensions. There are at least, two ways to perform the extension of the discrete Hartley transform into two dimensions [390]. The first one is to introduce the kernel functions of the type

$$\operatorname{cas}\left[2\pi\left(\frac{u\,x}{N} + \frac{v\,y}{M}\right)\right]. \tag{A.18}$$

The kernel functions in the second approach are the separable ones [390]

$$\operatorname{cas}(2\pi x u / N) \cdot \operatorname{cas}(2\pi v y / M). \tag{A.19}$$

Only the first one, Eq. (A.18) manifests the main feature of the discrete Hartley transform, namely,

$$P_H(u, v) = Re\{P_F(u, v)\} - Im\{P_F(u, v)\}, \tag{A.20}$$

where $P_F(u, v)$ is $2D$ discrete Fourier transform of the $2D$ input function $p(x, y)$. Hence

$$P_H(u, v) = \sum_{x=0}^{N-1} \sum_{y=0}^{M-1} p(x, y) \operatorname{cas}\left(2\pi \frac{u\,x}{N} + 2\pi \frac{v\,y}{M}\right). \tag{A.21}$$

The absence of separability in this definition of $2D$ Hartley transform does not permit to use the fast algorithms of calculation [386, 387].

The separable $2D$ Hartley transform is defined by the equation

$$P_{H,S}(u, v) = \sum_{x=0}^{N-1} \sum_{y=0}^{M-1} p(x, y) \operatorname{cas} \frac{2\pi u x}{N} \cdot \operatorname{cas} \frac{2\pi v y}{M}, \tag{A.22}$$

for forward transform, and

$$p(x, y) = \frac{1}{NM} \sum_{u=0}^{N-1} \sum_{v=0}^{M-1} P_{H,S}(u, v) \operatorname{cas} \frac{2\pi u x}{N} \cdot \operatorname{cas} \frac{2\pi v y}{M}, \tag{A.23}$$

for inverse transform.

The relation between separable $2D$ discrete Hartley transform and $2D$ discrete Fourier transform is defined by the equation

$$P_{H,S}(u, v) = Re\{P_F(u, -v)\} - Im\{P_F(u, v)\}. \tag{A.24}$$

The inverse relation has the form

$$P_F(u, v) = P_{H,S}(u, -v) - i\, P_{H,S}(u, v). \tag{A.25}$$

If $p(x, y)$ is the real and even function, then

$$P_{H,S}(u, v) = P_F(u, -v). \tag{A.26}$$

The convolution theorems for discrete Fourier and discrete Hartley transforms are given in [386, 387].

A.3. Hilbert Transform

The Hilbert transform of $1D$ function $f(x)$ is defined by the equation

$$\hat{\mathcal{H}}[f(x)] = \frac{1}{\pi} P \int \frac{f(\xi)}{(x-\xi)} d\xi = f(x) \otimes \frac{1}{\pi x}, \tag{A.27}$$

where the integral is taken in the sense of the Cauchy principal value. The Hilbert transformation, Eq. (A.27), is the convolution of the function $f(x)$ with the kernel $1/\pi x$. To find the Hilbert transform $\hat{\mathcal{H}}[f(x)]$ we must transmit the Fourier transform $F(\omega)$ of this function $f(x)$ through the spatial frequency filter of the type

$$G(\omega) = -i \operatorname{sgn} \omega = \begin{cases} -i, & \omega > 0, \\ +i, & \omega < 0, \end{cases} \tag{A.28}$$

and then perform the inverse Fourier transformation [383]. This spatial frequency filter $G(\omega)$ is called the "Hilbert knife" [157].

A.4. Abbe Transform

With Abbe transform [391, 392] we may converts the double diffraction integral over the optical aperture in the opaque screen to the line integral along the boundary of this aperture. Very simple relations are deduced for boundary composed of the straight line segments.

The Abbe transform is valid only for light amplitude $u(\vec{x})$ which satisfies the Helmholtz equation

$$\nabla^2 u(\vec{x}) + k^2 u(\vec{x}) = 0, \tag{A.29}$$

for example, for plane wave with amplitude

$$u_\circ(\vec{x}) = \exp(i\vec{k}\,\vec{x}), \tag{A.30}$$

To prove the Abbe transform, it is sufficient to use the Gauss theorem, according to which

$$\iint_S \nabla \vec{A}(\vec{x})\, d\vec{x} = \oint_l \vec{A}(\vec{x})\,\vec{n}\, dl, \tag{A.31}$$

where $\vec{n}$ is a unit vector directed along the outer normal to the curve l and $\vec{A}(\vec{x})$ is an arbitrary vector valued function. By choosing

$$\vec{A}(\vec{x}) = \nabla u_\circ(\vec{x}) = -i\vec{k}\, u_\circ(\vec{x}), \tag{A.32}$$

where $u_\circ(\vec{x})$ is given by Eq. (A.30), we get

$$\nabla \vec{A}(\vec{x}) = \nabla^2 u_\circ(\vec{x}) = -k^2 u_\circ(\vec{x}), \tag{A.33}$$

and

$$-k^2 \iint_S u_\circ(\vec{x})\, d\vec{x} = \oint_l (-i\, k\, u_\circ(\vec{x})\,\vec{n})\, dl. \tag{A.34}$$

The Abbe transform [391, 393] follows from Eq. (A.34):

$$\iint_S u_\circ(\vec{x})\, d\vec{x} = \frac{i}{k}\oint_l u_\circ(\vec{x})\,\vec{n}\, dl. \tag{A.35}$$

The left part of Eq. (A.35) represents the wave amplitude of the Fraunhofer diffraction field. Now it may be evaluated as a line integral. The proof of the Abbe transform can be accomplished as well in the reverse direction, by going from the line integral to a surface integral [402].

Several approaches have been developed for direct calculation of the Fraunhofer diffraction pattern for regular aperture such as polygon [396, 397, 404], but the approach emerged from the Abbe transform turns to be the most simple and elegant. For polygon aperture the line integral in Eq. (A.35) can be represented as a sum of the contributions from all sides of the polygon [393]. The contribution to the line integral from the j-th side of the polygon, with end points defined by the radius vectors $\vec{x}_{j+1}$ and $\vec{x}_j$, $j = 1, ..., n$, is given by the equation [393]

$$\Psi_j(\vec{k}) = \frac{i}{k} \vec{k} \cdot \vec{n}_j \exp(-\vec{k}\,\vec{x}_j) \times \int_0^{L_j} \exp(-i\,\vec{k} \cdot \vec{t}_j\, l)\, dl =$$

$$= -\frac{1}{k^2} \frac{\vec{k} \cdot \vec{n}_j}{\vec{k} \cdot \vec{t}_j} \exp(-i\,\vec{k} \cdot \vec{x}_j) \times [\exp(-i\,\vec{k} \cdot \vec{t}_j \cdot L_j) - 1], \tag{A.36}$$

where

$$\vec{x}_{j+1} - \vec{x}_j = \vec{t}_j \cdot L_j, \tag{A.37}$$

and L_j is the length of the j-th side of the polygon.

For regular n-polygon inscribed into the circle of radius R, the length L_j of the j-th side is equal to

$$L_j = L = 2R \sin \frac{\pi}{n}, \tag{A.38}$$

and the area of the polygon

$$D = n R^2 \cos(\frac{\pi}{n}) \sin(\frac{\pi}{n}). \tag{A.39}$$

In Eq. (A.35) the area of the aperture was accepted as unit.

In the polar coordinate system over the observation plane in the far field region (ρ, χ) with $\rho = k\,R$ and azimuthal angle χ, the amplitude of the Fraunhofer diffraction field is given by the equation [393]:

$$\begin{aligned} u(\rho, \chi) = {} & \frac{2i}{n\,\rho \cos\frac{\pi}{n}} \sum_{j=1}^{n} \cos[\chi - \frac{\pi}{n}(2j+1)] \times \\ & \times \operatorname{sinc}\{\frac{\rho}{\pi} \sin\frac{\pi}{n} \cdot \sin[\chi - \frac{\pi}{n}(2j+1)]\} \times \\ & \times \exp\{-i\,\rho \cos\frac{\pi}{n} \cdot \cos[\chi - \frac{\pi}{n}(2j+1)]\}. \end{aligned} \tag{A.40}$$

At the directions in the (ρ, χ) plane, which are perpendicular to the sides of the polygon aperture and for which

$$\chi_j = \frac{\pi}{n}(2j+1), \tag{A.41}$$

the intensity of the diffraction pattern decreases with ρ more slowly, with envelope ρ^{-1}, than along other directions, at which the envelope decreases as ρ^{-2}. Due to this we observe fans in the diffraction picture, which are oriented along the radial straight lines going perpendicular to the sides of the regular polygon.

The Abbe transform was mentioned by R. Straubel in 1888 [392]. The properties of the far field diffraction picture explained above were known since 1835 from the work of F.M. Schwerd [100]. Two theorems attributed to E. Abbe were proved by R. Straubel in 1895 [399]: a) if the aperture has an axial symmetry of the n-th order, then the far field diffraction picture has the axial symmetry of the $2n$-th order, b) when the number n is even, the symmetry axes coincide in pairs, and the null intensity lines are observed along n fans; when the number n is odd, the null intensity fans are absent. To suppress the null intensity lines in the case of even n, the π-phase-shifting plate must be inserted in the half-plane of the aperture [396]. The experiments with such phase plate were made in 1895 [399].

A generalization of the Abbe transform, written by Eq. (A.35), has been found by N. Saga and K. Tanaka [398]. To explain this new form of the Abbe transform, let us rewrite Eq. (A.35) in the form

$$\begin{aligned} S(\vec{k}) \equiv S(u,v) &= \frac{-i}{|\vec{k}|^2}\oint_l \exp(i\vec{k}\,\vec{r})\,(\hat{n}\cdot\vec{k})\,dl = \\ &= \frac{-i}{u^2+v^2}\oint_l \exp\left[i\,(u\,x + v\,y)\right](u\,dy - v\,dx), \end{aligned} \tag{A.42}$$

where u and v are Cartesian components of the wave vector $\vec{k}\,(u, v)$.

Let us chose the vector valued function

$$\vec{b} = \left[i\,\frac{\vec{k}}{|\vec{k}|^2} - i\,\frac{\vec{x}}{\vec{k}\cdot\vec{x}} + \frac{\vec{x}}{(\vec{k}\cdot\vec{x})^2} \right] \exp(i\,\vec{k}\,\vec{x}), \tag{A.43}$$

for which

$$\nabla\vec{b} = 0 \tag{A.44}$$

and

$$\iint_S \nabla\vec{b}\,d\vec{x} = \oint_l \vec{b}\cdot\hat{n}\,dl = 0. \tag{A.45}$$

We may add the line integral from Eq. (A.45) to Eq. (A.42). Now we have

$$S(\vec{k}) \equiv S(u,v) = \oint_l \frac{1 - i\,(\vec{k}\,\vec{x})}{(\vec{k}\,\vec{x})^2}\exp(i\,\vec{k}\,\vec{x})\,(\hat{n}\cdot\vec{x})\,dl =$$

$$= \oint_l \frac{1 - i(u\,x + v\,y)}{(u\,x + v\,y)^2} \exp[\,i\,(u\,x + v\,y)]\,(x\,dy - y\,dx)\,. \tag{A.46}$$

By using the identity

$$\oint_l \frac{x\,dy - y\,dx}{(u\,x + v\,y)^2} = \oint_l d\left[\frac{-\frac{x}{v}}{u\,x + v\,y}\right], \tag{A.47}$$

we get the equations:

$$S(u, v) = \tfrac{i}{v} \oint_l \exp[\,i\,(u\,x + v\,y)]\,dl \qquad \text{with} \qquad y = y(x) \tag{A.48}$$

or

$$S(u, v) = -\tfrac{i}{u} \oint_l \exp[\,i\,(u\,x + v\,y)]\,dl \qquad \text{with} \qquad x = x(y)\,. \tag{A.49}$$

It is evident, that Eq. (A.48) or Eq. (A.49) represents more simple expression of the Abbe theorem, than Eq. (A.42).

The Abbe transform is in a direct relationship with Maggi–Rubinowicz transformation [398]. By using the divergence theorem, we may convert the surface integral into a line integral [141]. The multiple application of the line integral, Eq. (A.46), has been used in [403] for evaluation of the diffraction of light beam at a series of boundaries, coaxially distributed along a laser beam axis.

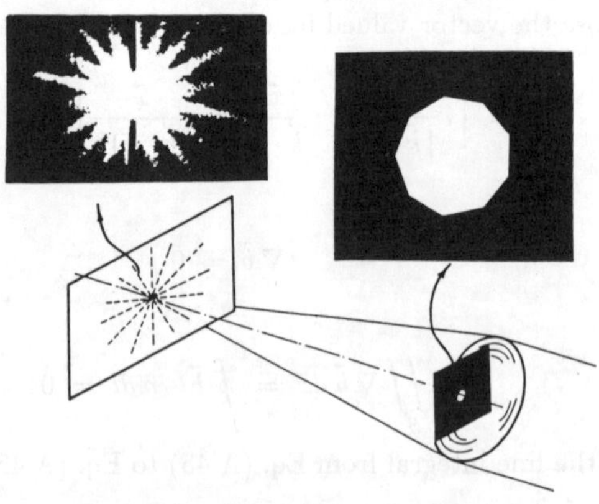

Fig. A.1. Experimental setup for observation of the diffraction pattern from aperture in the form of polygon (top). The structure of the polygon aperture (bottom, left), and the diffraction pattern observed at high overexposure time (bottom, right).

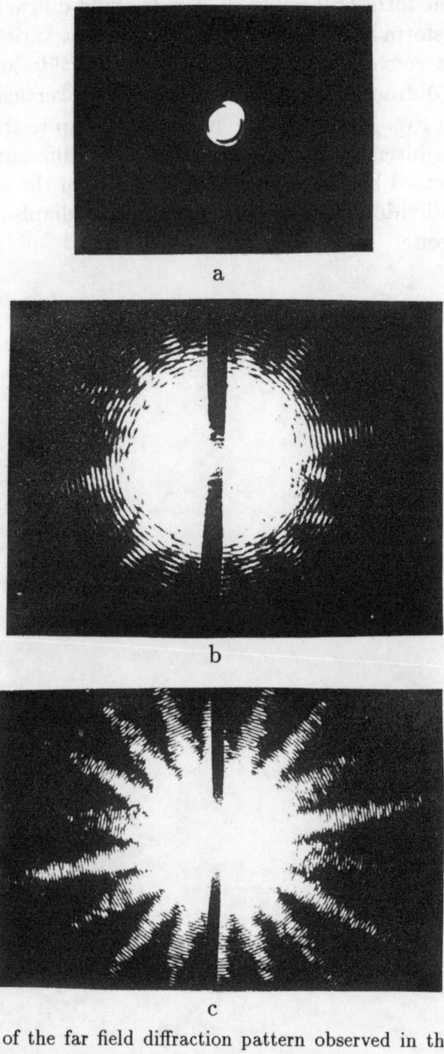

a

b

c

Fig. A.2. The structure of the far field diffraction pattern observed in the experiment shown in Figure A.1: a — conventional Airy disk with one side lobe detected at short time exposure, b — the diffraction pattern with absorbed central part which is observed at small overexposure, c — the same as in "b" at very high overexposure.

By closing the Appendix we present diffraction patterns of the polygon aperture which are in line with Abbe theorem, considered above. The experimental layout is shown in Fig. A.1. The lens objective used in this experiment was provided with iris diaphragm in the form of 9-polygon. The far field diffraction patterns observed in the Fourier transform plane is shown in Fig. A.2 at various exposure times. In Fig. A.2.a we see a conventional Airy disk with one side lobe. In Fig. A.2.b the central part of the diffraction pattern is blocked by a vertical opaque strip. There are no star lines. In Fig. A.2.c the star lines are seen up to the 78-th ring side lobe. The length and the intensity of these star lines are different. This spread of star lines structure is caused by the spread of the length, of the straightness and of the roughness of the individual sides of the aperture iris diaphragm of the lens in the form of the 9-polygon.

REFERENCES

1. S.Kawata, O.Nakamura and S.Minami, "Optical microscope tomography. I. Support constraint," *J. Opt. Soc. Am.* **A 4** (1987), 292–297.
2. A.Ya.Astakhov, Yu.A.Batusov, Gy.L.Bencze, I.Farago, A.Kisváradi, L.Molnar, L.M.Soroko and J.Vegh, "Meso–optical Fourier transform microscope — a new device for high energy physics," *Nucl. Instrum. Meth.* **A 283** (1989), 13–23.
3. L.M.Soroko, "Axicons and meso–optical imaging devices," in *Progress in Optics*, edited by Emil Wolf (Elsevier, Amsterdam, 1989) **27**, pp. 109–160.
4. L.M.Soroko, "Optics, holography and meso–optics in bubble chamber of vertex detector," *Commun. JINR* **D1–82–642** (Dubna, 1982).
5. J.B.Goodell, "Eccentric lenses for producing ring images," *Appl. Opt.* **8** (1969), 2566.
6. Gy.L.Bencze and L.M.Soroko, "Meso–optics and high energy physics," *Commun. JINR* **E13–84–310** (Dubna, 1984), submitted to **ICO–13** Congress, Sapporo, 1984. The 13-th Congress of the International Comission for Optics.
7. A.Kolodziejczyk, S.Bará, Z.Jaroszewicz and M.Sypek, "The light sword optical element — a new diffraction structure with extended depth of focus," *J. Mod. Opt.* **37** (1990), 1283–1286.
8. S.Bará, C.Frère, Z.Jaroszewicz, A.Kolodziejczyk and D.Leseberg, "Modulated on-axis circular zone plates for a generation of three-dimensional focal curves," *J. Mod. Opt.* **37** (1990), 1287–1295.
9. D.Leseberg, "Computer generated holograms: cylindrical, conical and helical waves," *Appl. Opt.* **26** (1987), 4385–4390.
10. J.H.Mc Leod, "The axicon: a new type of optical element," *J. Opt. Soc. Am.* **44** (1954), 592–597.
11. J.H.Mc Leod, "Axicons and their uses," *J. Opt. Soc. Am.* **50** (1960), 166–169.
12. F.P.Schäfer, "On some properties of axicons," *Appl. Phis.* **B 39** (1986), 1–8.
13. G.Roy and Trembley, "Influence of the divergence of a laser beam in the axial intencity distributions of an axicon," *Opt. Commun.* **34** (1980), 1.
14. N.A.Volkova, V.V.Korobkin, E.Yu.Maljisheva, L.Ya.Polonsky, V.P.Poponin and L.N.Pyatnitsky, *"Focusing of the laser radiation by axicons"*, Preprint of the Institute of High Temperature of the Acad. Sci. USSR **No 5–126** (1983).

15. O.Bryngdahl, "Shearing interferometry with constant radial displacement," *J.Opt. Soc.Am.* **61** (1971), 169–172.

16. M.Rioux, R.Tremblay and P.-A.Belanger, "Linear, annular and radial focusing with axicons and applications to laser machining," *Appl.Opt.* **17** (1978), 1532–1536.

17. O.Bryngdahl, "Reversed-radial-shearing interferometry," *J. Opt. Soc. Am.* **60** (1970), 915–917.

18. L.Lavoie, "Conical axicons used as optical polar evertors," *Appl. Opt.* **14** (1975), 1482–1484.

19. W.S.Bichel, "The spectroscopic properties of the ionic beam source," *Appl. Opt.* **6** (1967), 1309–1317.

20. J.O.Stoner, "Properties of axicon systems for collecting foil-excited accelarator beam spectra," *Appl. Opt.* **9** (1970), 53–57.

21. M.Platiša, "Azimuthal variations in the brightness of Fabry-Perot axicon fringes," *Optics and Laser Technology* **6** (1974), 114–116.

22. M.Platiša and J.Puric, "Analysis of the instrumental profile of an axicon-scanned Fabry-Perot spectrometer," *Fizika* **3** (1971), 175–187.

23. R.M.A.Azzam and M.E.R.Khan, "Polarization-preserving single-layer-coated beam displacers and axicon," *Appl. Opt.* **21** (1982), 3314–3322.

24. W.H.Southwell, "Multilayer coatings producing 90° phase change," *Appl. Opt.* **18** (1979), 1875.

25. L.Ya.Polonsky and L.N.Pyatnitsky, "Continuons extended laser sparks in air," *Sov. J. Opt. Atmosph.* **1** (1988), 86–92.

26. D.Fink, "Polarization effects of axicons," *Appl.Opt.* **18** (1979), 581–582.

27. G.C.Dente, "Polarization effects in resonators," *Appl.Opt.* **18** (1979), 2911-2912.

28. R.A.Chodzko, S.B.Mason, E.B.Turner and W.W.Plummer, "Annular (HSURIA) resonators: some experimental studies including polarization effects," *Appl. Opt.* **19** (1980), 778–789.

29. T.R.Ferguson, "Polarization effects in interferograms of conical optical elements," *Appl. Opt.* **21** (1982), 514–517.

30. R.M.A.Azzam, "Scheme to polarization-correct a waxicon," *Appl. Opt.* **21** (1983), 3067–3068.

31. S.I.Zowgorodneva, V.I.Kuprenyuk, V.V.Sergeev and V.E.Sherstobitov, "On the convertion of the unstable-resonator laser radiation by means of the conical optics," *Sov. J. Quantum Electron.* **7** (1980), 142–146.

32. V.F.Gavricov, A.K.Piskunov and V.A.Shcheglov, "On the use of conical mirrors in laser units," *Sov. J. Quantum Electron.* **13** (1986), 2135–2137.

33. J.W.Ogland, "Mirror system for uniform beam transformation in high-power annular lasers," *Appl. Opt.* **17** (1978), 2917–2923.

34. G.S.Bakken, "The parabolic axicon," *Appl. Opt.* **13** (1974), 1291–1292.

35. J.W.Y.Lit and E.Brannen, "Optical properties of a reflecting cone," *J. Opt. Soc. Am.* **60** (1970), 370–171.

36. G.C.Steward, *The Symmetrical Optical System* (Cambridge University Press, Cambridge, 1928).

37. P.-A.Bèlanger and M.Rioux, "Ring pattern of a lens-axicon doublet illuminated by a Gaussian beam," *Appl. Opt.* **17** (1978), 1080–1086.

38. J.Dyson, "Circular and spiral diffraction gratings," *Proc. R. Soc. London, Ser.* **A 248** (1958), 93–106.

39. L.R.Staronski, J.Sochacki, Z.Jaroszewicz and A.Kolodziejczyk, "Design of uniform-intensity refractive axicons," *Opt. Eng.* **31** (1992), 516–521.

40. J.Sochacki, A.Kolodziejczyk, Z.Jaroszewicz and S.Bará, "Nonparaxial design of generalized axicons," *Appl. Opt.* **31** (1992), 5326–5330.

41. J.Sochacki, S.Bará, Z.Jaroszewicz and A.Kolodziejczyk, "Phase retardation of the uniform-intensity axilens," *Opt. Lett.* **17** (1992), 7–9.

42. L.R.Staronski, J.Sochacki, , Z.Jaroszewicz and A.Kolodziejczyk, "Lateral distribution and flow of energy in uniform-intensity axicons," *J. Opt. Soc. Am.* **A 9** (1992), 2091–2094.

43. V.P.Koronkevitch, V.P.Korolkov, A.G.Poleschuk, I.G.Palchikova, Yu.I.Yurlov, I.A.Mikhaltsova, E.G.Churin, A.P.Sokolov and A.G.Sedukhin, "Kinoforms: technologies, new elements and optical systems," Preprint IAE SB (USSR Academy of Science) No **421** (Novosibirsk, 1989), p. 15.

44. H.L.Kazansky, "Computer experiment-bases procedure for correction of focuser phase function," in *Computer Optics* (Intern. Center for Scientific and Technical Inform., Moscow, 1987), Vol. **1**, pp. 90–96.

45. O.Bryngdahl and A.W.Lohmann, "Interferograms are image holograms," *J. Opt. Soc. Am.* **58** (1968), 141–142.

46. T.S.Huang, "Digital holography," *Proc. IEEE* **59** (1971), 1335–1346.

47. O.Bryngdahl, "Computer-generated holograms as generalized optical components," *Opt. Eng.* **14** (1975), 426–435.

48. L.P.Yaroslavskii and N.S.Merzlyakov, *Methods of digital holography* (Plenum, New-York, 1980).

49. O.Bryngdahl, "Radial- and circular-fringe interferogram," *J. Opt. Soc. Am.* **63** (1973), 1098–1104.

50. O.Bryngdahl, "Geometrical tranformations in optics," *J. Opt. Soc. Am.* **64** (1974), 1092–1099.

51. I.B.Lesem, P.M.Hirsch and J.A.Jordan, "The kinoform: a new wavefront reconstruction device," *IBM J. Res. Dev.* **13** (1969), 150–155.

52. I.A.Mikhaltsova, V.I.Nalivaiko and I.S.Soldatenkov, "Kinoform axicons," *Optik* **67** (1984), 267–278.

53. V.P.Koronkevitch, G.A.Lenkova, F.I.Kokoulin, I.G.Palchicova, A.G.Poleshchuk, A.G.Sedukhin, E.G.Churin, A.M.Shcherbachenko and Yu.I.Yurlov, "Fabrication of kinoform optical elements," *Optic* **67** (1984), 257–266.

54. V.P.Korolkov, V.P.Koronkevitch, I.A.Mikhaltsova, I.G.Palchicova, A.G.Poleschuk, A.G.Sedukhin, A.P.Sokolov, E.G.Churin and Yu.I.Yurlov, "Kinoforms: phototechnology, new elements and optical systems," *Sov. J. Autometria, No* **3** (1989), 95–102.

55. V.P.Korolkov,V.P.Koronkevitch, I.A.Mikhaltsova, I.G.Palchicova, A.G.Poleshchuk, A.G.Sedukhin, A.P.Sokolov, E.G.Churin and Yu.I.Yurlov, "Kinoforms: phototechnology, new elements and optical systems," *Sov. J. Autometria, No* **4** (1989), 47–64.

56. I.G.Palchicova, "Kinoform conoid axicons," *Sov. J. Autometria, No* **6** (1988), 78–82.

57. B.I.Spector, "About one technique of the phase kinoform structure synthesis," *Sov. J. Autometria, No* **6** (1985), 34–38.

58. I.G.Palchicova, "The action of the kinoform lens fabrication errors on the image quality," *Sov. J. Autometria, No* **6** (1984), 98–99.

59. I.G.Palchicova and A.G.Ryabchuck, "About the action of the kinoform lens fabrication errors on the pupil function," *Sov. J. Autometria, No* **6** (1985), 38–42.

60. L.D'Auria, J.P.Huignard, A.M.Roy and E.Spitz, "Photolithographic fabrication of thin film lenses," *Opt. Comm.* **5** (1972), 232–235.

61. M.A.Golub, S.V.Karneev, A.M.Prokhorov, I.N.Sisakyan and V.A.Soifer, "Focusing of radiation into the prescribed space region with computer generated holograms," *Sov. Tech. Phys. Lett.* **7** (1981), 618–623.

62. V.A.Danilov, V.V.Popov, A.M.Prokhorov, D.M.Sagatelyan, I.N.Sisakyan and V.A.Soifer, "Synthese of the optical elements, which produce the focal line of the arbitrary shape," *Sov. Tech. Phys. Lett.* **8** (1982), 810–815.

63. A.W.Lohmann, D.P.Paris and H.W.Werlich, "A computer generation spatial filter, applied to code translation," *Appl. Opt.* **6** (1967), 1139–1140.

64. A.V.Goncharskii,V.A.Danilov,V.V.Popov, A.M.Prokhorov, I.N.Sisakyan, V.A.Soifer and V.V.Stepanov, "Solution of the inverse problem of the focusing radiation

into the arbitrary line," *Dok. Akad. Nauk SSSR [Sov. Phys. Dokl.* **273** (1983), 605–608.

65. I.N.Sisakyan and V.A.Soifer , "Thin optics sythesized by computer," in *Fizicheskie osnovi i prikladnie voprosi holografii* (LIJAPH, Leningrad, 1984), pp. 142–164.

66. A.V.Goncharskii, I.N.Sisakyan and V.V.Stepanov, "About solution of some inverse problems of the focusing of the laser radiation," *Dok. Akad. Nauk SSSR [Sov. Phys Dokl.]* **279** (1984), 68–71.

67. A.V.Goncharskii,V.A.Danilov,V.V.Popov, A.M.Prokhorov, I.N.Sisakyan,V.A.Soifer and V.V.Stepanov, "Focusators of an obliquely incident laser radiation," *Sov. J. Quantum Electron.* **11** (1984), 166–168.

68. A.V.Goncharskii, V.A.Danilov, V.V.Popov, I.N.Sisakyan, V.A.Soifer and V.V.Stepanov, "Flat focusing elements for the visible light," *Sov. J. Quantum Electron* **13** (1986), 660–662.

69. W.R.Edmonds, "The reflaxicon, a new reflective optical element and some applications," *Appl. Opt.* **12** (1978), 1940–1945.

70. J.Turunen and A.T.Friberg, "Electromagnetic theory of reflaxicon fields," *JEOS Part A: Pure and Applied Optics,* **2** (1993), 539–548.

71. S.R.Mishra, "A vetor wave analysis of a Bessel beam," *Opt. Commun.* **85** (1991), 159–161.

72. J.Durnin, "Exact solutions for nondiffracting beams. I. The scalar theory," *J. Opt. Soc. Am.* **A 4** (1987), 651–654.

73. J.Durnin, J.J.Miceli and J.H.Eberly, "Diffraction-free beams," *Phys. Rev. Lett.* **58** (1987), 1499–1501.

74. E.H.Linfoot and E.Wolf, "Diffraction images in systems with an annular aperture," *Proc. Phys. Soc.* **B 66** (1953), 145–149.

75. C.Rivolta, "Multiple annular circular aperture: maxima and minima of the diffraction pattern," *Appl. Opt.* **26** (1987), 3119–3122.

76. J.Ojeda-Castañeda, L.R.Berriel-Valdos and E.Montes, "Bessel-annular apodizers: imaging characteristics," *Appl. Opt.* **26** (1987), 2770–2772.

77. J.Ojeda-Castañeda, E.Tepichin and A.Diaz, "Arbitrary high focal depth with a quasioptimum real and positive transmittance," *Appl. Opt.* **28** (1989), 2666–2669.

78. J.Ojeda-Castañeda and A.Diaz, "High focal lenth by quasibifocal apodizer," *Appl Opt.* **27** (1988), 4163–4165.

79. J.Ojeda-Castañeda, R.Ramos and A.Noyola-Isgleas, "High focal depth by apodization and digital restoration," *Appl. Opt.* **27** (1988), 2583–2586.

80. H.F.A.Tschunko, "Imaging performance of annular aperture. 6: Limitations by optical surface deviations," *Appl. Opt.* **26** (1987), 5333–5336.

81. H.F.A.Tschunko, "Imaging performance of annular apertures," *Appl. Opt.* **13** (1974), 1820–1823.

82. V.V.Korobkin, L.Ya.Polonskii, V.P.Poponin and L.N.Pjatnitsky, "Focusing of the Gaussian and hyper-Gaussian laser beams by axicons for production of countinious laser sparks," *Sov. J. Quantum Electron.* **13** (1986), 265–270.

83. J.Sochacki, Z.Jaroszewicz, L.R.Staronshi and A.Kolodziejczyk, "Annular aperture logarithmic axicon," *J. Opt. Soc. Am.* **A 888** (1993).

84. J.W.Goodman, *Introduction to Fourier Optics* (Mc Graw-Hill, New York, 1968).

85. A.J.Cox and J.D'Anna, "Constant-axial-intensity nondiffacting beam," *Opt. Lett.* **17** (1992), 232–234.

86. Ch.Frère and O.Bryngdahl, "Computer-generated holograms: reconstruction of curves in $(3 - D)$," *Opt. Commun.* **60** (1986), 369–372.

87. L.M.Soroko, *Optical microscope*, USSR patent No 1,278,773 (29 July 1985), Bull. No **47** (1986), p.186.

88. B.Ya.Zeldovitch, B.F.Mulchenko and N.F.Pilipetsky, "Observation of an extended light spark," *Sov. Phys. JETP* **58** (1970), 794–796.

89. R.Tremblay, Y.D'Astons, G.Roy and M.Blanchard, "Laser plasmas optically pumped by focusing with an axicon a $CO_2 - TEA$ laser beam in a high pressure gas," *Opt. Commun.* **28** (1979), 193–196.

90. F.V.Bunkin, V.V.Korobkin, Yu.A.Kurinii, L.Ya.Polonskii and L.N.Pjatnitskii, "Laser spark with continious channel in air," *Sov. J. Quantum Electron.* **10** (1983), 443–444.

91. V.V.Korobkin, L.Ya.Polonskii, V.P.Poponin and L. N. Pjatnitskii, "Focusing of the Gaussian laser beam by axicons for production of continious laser sparks," *Sov. J. Quantum Electron.* **13** (1986), 265–270.

92. G.Häusler and W.Hechel, "Light sectioning with large depth and high resolution," *Appl. Opt.* **27** (1988), 5165–5169.

93. G.Bickel, G.Häusler and M.Maul, "Triangulation with extended range of depth," *Opt. Eng.* **24** (1985), 975–977.

94. H.Slevogt, *Technische Optik* (Berlin, 1974).

95. J.Katzenstein, "Axicon-scanned Fabri–Perot spectrometer," *Appl. Opt.* **4** (1965), 263–266.

96. M.Platiša, "The dispersion of an axicon-scanned Fabry–Perot spectrometer," *Fizika* **5** (1973), 83–91.

97. L.M.Soroko, *Objective of the track chamber photoregistrator*, USSR Patent No 1,597,812 (15 July 1988), Bull. No **37** (1990).

98. L.M.Soroko, "Isoplanatic systems in meso–optics," *Exper. Techn. of Phys.* **38** (1990), 401–410, *JINR Preprint* **D13–89–549** (Dubna, 1989).

99. R.Mavaddat, "Development of optically excited soft X-ray laser techniques: a travelling wave lens," *Opt. Commun.* **25** (1978), 215–220.

100. A.Ya.Astakhov, G.M.Komov, V.I.Sidorova, I.I.Skril and L.M.Soroko, "Design description of Fourier transform microscope for nuclear emulsion," *Commun. JINR* **P13–83–119** (Dubna, 1983).

101. A.Ya.Astakhov and L.M.Soroko, "Metric characteristics of the Fourier transform microscope for nuclear emulsion," *Commun. JINR* **P13–83–120** (Dubna, 1983).

102. Gy.L.Bencze and L.M.Soroko, "Imaging properties of the meso–optical Fourier transform microscope for nuclear research emulsion," *Commun. JINR,* **P13-85-136** (Dubna, 1985).

103. A.Ya.Astakhov, Gy.L.Bencze, A.Kisváradi, T.P.Nidermaier, G.Nitrai and L.M.Soroko, "Semiautomatic measurements of particle tracks by the meso–optical Foureir transform microscope for nuclear research emulsion," *Commun. JINR*, **P13-85-378** (Dubna, 1985).

104. H.Maitre, "Matched spatial filtering and the Hough transform," **ICO-13** Congress (Sapporo, 1984). The 13-th Congress of the International Commision for Optics.

105. S.Aoki, K.Hoshino, M.Nakamura, K.Niu, K.Niwa and N.Torii, "Fully automated emulsion analysis system," *Nucl. Instrum. Methods* **B 51** (1990), 466–472.

106. Gy.L.Bencze and L.M.Soroko, "Tangent algorithm for event searching in meso–optical Fourier transform microscope for nuclear emulsion," *Commun. JINR,* **P13-85-502** (Dubna, 1985).

107. Gy.L.Bencze, I.G.Palchikova, A.G.Poleshchuk and L.M.Soroko, "Investigation of the imaging properties of the kinoform with transverse meso–optisity," *Commun. JINR,* **P13-86-240** (Dubna, 1986).

108. Gy.L.Bencze, A.Kisváradi, G.Nitrai and L.M.Soroko, "Observation of relativistic proton tracks by means of meso–optical Fourier transform microscope for nuclear emulsion," *Commun. JINR,* **P13-86-630** (Dubna, 1986).

109. Gy.L.Bencze and L.M.Soroko, "Geometro-optical performances of the meso–optical Fourier transform microscope in the meridional cross-section," *Commun. JINR,* **P13-86-659** (Dubna, 1986).

110. L.M.Soroko, "Measuring technique of coordinates of the origin and of the end of particle track in the meso–optical Fourier transform microscope," *Commun. JINR,* **P13-87-169** (Dubna, 1987).

111. L.M.Soroko, "Measuring technique of small scattering angles and decay length in the meso–optical Fourier transform microscope," *Commun. JINR*, **P13-87-170** (Dubna, 1987).

112. T.Hara, "Study of the lifetime of the charm-particle (E–531 Collaboration)," Ph. D. thesis, Osaka-City Univ. (1982). (Chamū ryūshi no jumpyō no kenkyū), (in Japanese).

113. L.M.Soroko, "Dirac delta-plus (or minus) function in optics and meso–optics," *Preprint JINR*, **E13-87-292** (Dubna, 1987).

114. L.M.Soroko, "A technique of the particle track curvature measurement by means of the meso–optical Fourier transform microscope," *Commun. JINR*, **P13-87-358** (Dubna, 1987).

115. Gy.L.Bencze and L.M.Soroko, "Basic parameters of the meso–optical Fourier transform microscope for nuclear research emulsion," *Commun. JINR*, **E13-87-387** (Dubna, 1987).

116. Gy.L.Bencze and L.M.Soroko, "Imaging properties of the meso–optical Fourier transform microscope for nuclear research emulsion," *Commun. JINR*, **E13-87-388** (Dubna, 1987).

117. Gy.L.Bencze and L.M.Soroko, "Event searching algorithms in meso–optical Fourier transform for nuclear research emulsion," *Commun. JINR*, **E13-87-389** (Dubna, 1987).

118. L.M.Soroko, "Meso–optical Fourier transform microscope for the observation of relativistic neon nuclei," *Commun. JINR*, **P13-87-468** (Dubna, 1987).

119. Gy.L.Bencze, I.G.Paltchikova, A.G.Poleshchuk and L. M. Soroko, "Simulation of the direct technique of "star" registration in nuclear emulsion by means of the kinoform with a ring response," *Commun. JINR*, **P13-87-474** (Dubna, 1987).

120. L.M.Soroko, "The meso–optical Fourier transform microscope: the structure of the signals in the sagittal section," *Commun. JINR*, **P13-87-527** (Dubna, 1987).

121. J.Sklansky, "On the Hough technique for curve detection," *IEEE Trans. Comp.* **C-27** (1978), 923–926.

122. T.Szoplik, "Line detection and directional analysis of images," in *Optical processing and computing*, ed. H. H. Arsenault (1989), pp. 391–419.

123. A.Ya.Astakhov, Gy.L.Bencze, A.Kisváradi, L.Molnar, and L.M.Soroko, "Comparison of the output data in meso–optical Fourier transform microscope and in traditional optical microscope. I. Calibration technique," *Exp. Techn. Phys.* **38** (1990), 425–430.

124. A.Kisváradi, L.Molnar, L.M.Soroko, V.V.Tereshchenko and I.Torma, "Signal-to-noise ratio in the meso–optical Fourier transform microscope," *Exp. Techn. Phys.* **38** (1990), 443–462.

125. W.H.Steier and R.K.Shori, "Optical Hough transform," *Appl. Opt.* **25** (1986), 2734–2738.

126. L.M.Soroko and V.V.Tereshchenko, "The technique for measuring of the dip angle of the straight particle tracks in the nuclear emulsion by means of the meso–optical Fourier transform microscope," *Commun. JINR,* **P13-91-204** (Dubna, 1991).

127. A.Ya.Astakhov, A.Kisváradi, V.I.Krasnoslobodzev, I.Molnar, L.Molnar, L.M.Soroko, V.V.Tereshchenko and I.Torma, "Meso–optical Fourier transform microscope with one-channel photodetectors," *Commun. JINR,* **P13-91-299** (Dubna, 1991).

128. A.Ya.Astakhov, A.Kisváradi, V.I.Krasnoslobodzev, I.Molnar, L.Molnar, L.M.Soroko, V.V.Tereshchenko and I.Torma, "Electronic-computer units of the meso–optical Fourier transform microscope with one-channel photodetectors and with CCD-matrix," *Commun. JINR,* **P13-91-391** (Dubna, 1991).

129. L.M.Soroko, "Meso–optical Fourier transform microscope: principles and main features," *Nucl. Tracks. Radiat. Meas.* **18** (1991), 391–402.

130. L.M.Soroko, "Meso–optical Fourier transform microscope as a new measurement device," *Nucl. Tracks. Radiat. Meas.* **19** (1991), 267–270.

131. Yu.A.Batusov, L.M.Soroko and V.V.Tereshchenko, "Fourier transform microscope of the direct observation for nuclear emulsion," *Commun. JINR,* **E13-91-427** (Dubna, 1991).

132. L.M.Soroko, "Acceptance of the Fourier transform microscopes for nuclear emulsion," *Commun. JINR,* **E13-91-474** (Dubna, 1991).

133. A.Ya.Astakhov, Gy.L.Bencze, A.Kisváradi, L.Molnar and L.M.Soroko, "Comparison of the output data in meso–optical Fourier transform microscope and in the traditional optical microscope. II. Measurement errors," *Exper. Techn. Phys.* **38** (1990), 431–442.

134. Yu.A.Batusov, L.M.Soroko and V.V.Tereshchenko, "Meso–optical Fourier transform microscope with double focusing," *Commun. JINR* **E13-92-179** (Dubna, 1992).

135. P.V.C.Hough, "Method and means for recognizing complex patterns," *U. S. Patent No* **3,069,654** (18 December 1962).

136. R.O.Duda and P.E.Hart, "Use of the Hough transformation to detect lines and curves in pictures," *Commun. ACM* **15** (1972), 11–15.

137. Nevatia and Babu, "Linear feature extraction and description," *Comput. Gr. Image Process.* **13** (1980), 257–269.

138. H.Maitre, "Un panorama sur la transformation de Hough," *Traitement du signal* **2** (1985), 305–317.

139. J.L.Krahe and P.Pousset, "Transformee de Hough discrete et bornee application a la detection de droites paralleles et du reseau routier," in *II Traitement d'Images: du Pixel à l'Interprétation* (Aussois, Savoie, Avril 1988), 19–22.

140. H.Maitre, "Contribution to the prediction of performances of the Hough transform," *IEEE Trans. Pat. Anal. Mach. Inter.*, **PAMI-8** (1986), 669–674.

141. M.Born and E.Wolf, *Principles of optics* (Pergamon Press, Oxford, 1964).

142. A.J.Durelli and V.J.Parks, *Moire analysis of strain* (Prentice-Hall, Englewood Cliffs, New York, 1970).

143. L.M.Soroko, "Device for automatic measurement of the angular distribution of the particle tracks," USSR Patent No 743,424 (12 December 1978), *Bull.No* **21** (1981), p.262.

144. L.M.Soroko, "Device for survey of the nuclear photoemulsion," USSR Patent No 5,007,168 (12 August 1991), in preparation.

145. O.Lummar and F.Reiche, *Die Lehre von der Bildenstehung in Microscop von Ernst Abbe* (Fr.Vieweg and Sohn, Braunschweig, 1910).

146. A.K.Ghatak and K.Thyagarajan, *Contemporary optics* (Plenum Press, New York, 1978).

147. W.T.Welford, "Aplanatism and isoplanatism," in *Progress in Optics,* ed. E.Wolf (North Holland, Amsterdam, 1976), **13**, pp. 287–292.

148. A.E.Conrady, *Applied Optics and Optical Design. Part 1* (Oxford University Press, Oxford, 1929).

149. V.N.Tchurilovsky, *Theory of the optical devices* (Mashinostroenie, Moscow,1966), pp. 116, 484.

150. W.T.Welford, "The most general isoplanatic theorem," *Opt. Commun.* **3** (1971), 1–6.

151. W.T.Welford, "Practical design of an aplanatic hologram lens of focal length 50 mm and numerical aperture 0.5," *Opt. Commun.* **15** (1975), 46–49.

152. S.T.Bobrov, G.I.Groysukh and Yu.G.Turkevich, *The optics of the diffracted elements and systems* (Mashinostroenie, Leningrad, 1986).

153. K.S.Mustaphin and Ph.A.Sattarov, "About aberrations of the on-axis hologramm with spherical backing and advanced pupil," *Sov. J. Opt. and Spectrosc.* **47** (1979), 1204–1206.

154. W.T.Welford, "Aplanatic hologram lenses," *Photogr. Sci.* **23** (1975), 84–87.

155. K.Tatsuno, K.Kataoka and S.Yonezawa, "Holographical lens," *U.S.Patent No* **4,688,880** (25 August 1987).

156. H.M.Nussenzweig, *Causality and dispersion relations* (Academic Press, New York, 1972).

157. L.M.Soroko, *Hilbert optics* (Nauka, Moscow, 1981).

158. Y.Belvaux and J.C.Vareille, "Visualisation d'objects de phase par trasformation de Hilbert," *Nouv. Rev. Opt. Appl.* **2** (1971), 149–162.

159. L.M.Soroko and V.A.Soujetin, "Diffraction grating with phase jump as a new optical element," *Commun. JINR* **P13-65-65** (Dubna, 1972).

160. W.T.Cathey, *Optical information processing and holography* (John Wiley and Sons, New York, 1974).

161. L.M.Soroko, "Meso-optics and the track-detector technique," *Sov. J. Part. Nucl.* **22** (1989), 67–85.

162. L.M.Soroko, "Device for observation of the particle tracks in nuclear photoemulsion," USSR Patent No 1,739,772 (6 August 1990), *Bull. No* **21** (1992).

163. S.Czapski and P.Culmann, in *Geometrical investigations of the formation of images in optical instruments*, ed. M.von Rohr (H. M. Stationery Office, London, 1920), pp. 125–211.

164. G.E.Belovizky, L.N.Korablev, L.V.Sukhov and I.V.Stranich, "Device for automatic measurements of multiple scattering of the particles," *Prib. Tekh. Eksp. [Instrum. Exp. Tech. (USSR)]* **2** (1959), 86–92.

165. M.V.Pyatakhin and A.F.Souchkov, "On possibility of "difractionless" propagation of Bessel beams," Preprint of the Physical Institute of Acd. Sci. USSR No **99** (1988).

166. J.Durnin, J.J.Micali and J.H.Eberly, "Comparison of Beseel and Gaussian beams," *Opt. Lett.* **13** (1988), 79–80.

167. Yu.A.Ananjev, "On "nondiffractive" light waves," *Sov. J. Opt. Spectr.* **64** (1988), 1211–1213.

168. F.Gori, G.Guattari and C.Padovani, "Bessel-Gauss beams," *Opt. Commun.* **64** (1987), 491–495.

169. M.Zahid and M.S.Zubairy, "Directionality of partially coherent Bessel-Gauss beams," *Opt. Commun.* **70** (1989), 361–364.

170. J.K.Jabczynski, "A "diffraction-free" resonator," *Opt. Commun.* **77** (1990), 292–294.

171. B.Hafizi and P.Sprangle, "Diffraction effects in directed radiation beams," *J. Opt. Soc. Am.* **A 8** (1991), 705–717.

172. P.L.Overfelt and C.S.Kenney, "Comparison of the propagation characteristics of Bessel, Bessel-Gauss and Gaussian beams diffracted by a circular aperture," *J. Opt. Soc. Am.* **A 8** (1991), 732–745.

173. C.J.R.Sheppard and M.Hrynevych, "Diffraction be a circular aperture: a generalization of Fresnel diffraction theory," *J. Opt. Soc. Am.* **A 9** (1992), 274–281.

174. V.V.Kotliar and V.A.Soifer, "Helical spatial phase filter," *Sov. J. Computer Optics No* **12** (1992), 61–65.

175. L.M.Soroko, "Device for laser sighting," USSR Patent No 1,651,263 (13 February 1989) *Bull. No* **19** (1991).

176. J.M.Vaughan and D.V.Willetts, "Interference properties of a light beam having a helical wave surface," *Opt. Commun.* **30** (1975), 263–267.

177. N.B.Baranova and B.Ya.Zel'dovich, "Dislocations on the wavefront surfaces and zero-amplitudes," *Sov. Phys. JETP* **80** (1981), 1789–1797.

178. N.B.Baranova, B.Ya.Zel'dovich, A.V.Mamaev, N.F.Pilipetskii and V.V.Shku- kov, "Wavefront dislocations of the speckle in homogeneity (theory and experiment)," *Sov. JETP Lett.* **33** (1981), 206–210.

179. A.G.White, C.P.Smith, N.R.Heckenberg, H.Rubinsztein-Dunlop, R.McDuff, C.O.Weiss and C.Tamm, "Interferometric measurements of phase singularities in the output of a visible laser," *J. Mod. Opt.* **38** (1991), 2531–2541.

180. M.Brambilla, F.Battipede, L.A.Lugiato, V.Penna, F.Prati, C.Tamm and C.O.Weiss, "Transverse laser patterns. I. Phase singularity crystals," *Phys. Rev.* **A 43** (1991), 5090–5113.

181. F.T.Arecchi, G.Giacomelli, P.L.Ramazza and S.Residori, "Vorteces and defect statistics of two-dimensional optical chaos," *Phys. Rev. Lett.* **67** (1991), 3749–3752.

182. N.R.Heckenberg, R.McDuff, C.P.Smith and A.G.White, "Generation of optical phase singularities by computer-generated holograms," *Opt. Lett.* **17** (1992), 221–223.

183. P.Coullet, L.Gil and F.Rocca, "Optical vortices," *Opt. Commun.* **73** (1989), 403–408.

184. G.Indebetouw, "Coherence properties of doughnut modes with and without a vortex," *J. Optics (Paris)* **24** (1993), 3–6.

185. C.Tamm and C.O.Weiss, "Bistability and optical switching of spatial patterns in a laser," *J. Opt. Soc. Am.* **B 7** (1990), 1034–1038.

186. J.M.Vaughan and D.V.Willets, "Temporal and interference fringe analysis of TEM_{01}^* laser modes," *J. Opt. Soc. Am.* **73** (1983), 1018–1021.

187. L.M.Soroko, "Longitudinal interference of the diffraction free wave fields. I. Theory," *Commun. JINR* **E13-90-592** (Dubna, 1990).

188. L.M.Soroko, "Longitudinal interference of the diffraction free wave fields. II. Experiments," *Commun. JINR* **E13-90-593** (Dubna, 1990).

189. L.M.Soroko, "Longitudinal interference of the diffraction free wave fields. III. Applications," *Commun. JINR* **E13-90-594** (Dubna, 1990).

190. R.M.Herman and T.A.Wiggins, "Production and uses of diffractionless beams," *J. Opt. Soc. Am.* **A 8** (1991), 932–942.

191. W.H.Steel, "Axicons with spherical surfaces," Colloquia of the ICO in *Optics in Metrology* , ed. P.Mollet (Pergamon Press, Oxford, 1960), pp. 181–192.

192. A.J.Cox and D.C.Dibble, "Nondiffracting beam front in a spatially filtered Fabry–Perot resonator," *J. Opt. Soc. Am.* **A 9** (1992), 282–286.

193. J.Turunen, A.Vasara and A.T.Friberg, "Holographic generation of diffraction-free beams," *Appl. Opt.* **27** (1988), 3959–3962.

194. A.Vasara,J.Turunen and A.T.Friberg,"Realization of general nondiffracting beams with computer-generated holograms," *J. Opt. Soc. Am.* **A 6** (1989), 1748–1754.

195. A.J.Cox and D.C.Dibbie,"Holographic reproduction of a diffraction-free beam," *Appl. Opt.* **30** (1991), 1330–1332.

196. N.Davidson, A.A. Friesem and E.Hasman, "Holographic axilens: high resolution and long focal depth," *Opt. Lett.* **16** (1991), 523–525.

197. G.Indebetouw, "Nondiffracting optical fields: some remarks on their analysis and synthesis," *J. Opt. Soc. Am.* **A 6** (1989), 150–152.

198. C.W.McCutchen, "Generalized aperture and the three-dimensional diffraction image," *J. Opt. Soc. Am.* **54** (1964), 240–244.

199. G.Inebetouw, "Polychromatic self imaging," *J. Mod. Opt.* **35** (1988), 243–252.

200. W.D.Montgomery, "Self-imaging objects of infinite aperture," *J. Opt. Soc. Am.* **57** (1967), 772–778.

201. G.A.Deschamps, "Gaussian beam as a bundle of complex rays," *Electron. Lett.* **7** (1971), 684–685.

202. L.B.Felsen, "Complex-source-point solutions of the field equations and their relation to the propagation and scattering of Gaussian beams," *Symp. Math.* **18** (1976), 39–56.

203. J.N.Brittingham, "Focus waves modes in homogeneous Maxwell's equations: Transverse electric mode," *J. Appl. Phys.* **54** (1983), 1179–1189.

204. P.A.Belanger, "Packetlike solutions of the homogeneous wave equation," *J. Opt. Soc. Am.* **A 1** (1984), 723–724.

205. A.Sezginer, "A general formulation of focus wave modes," *J. Appl. Phys.* **57** (1985), 678–685.

206. P.Hillion, "Some exotic solutions of the wave equation in unbounded isotropic media," *Wave motion* **10** (1988), 143–147.

207. P.Hillion, "Focus wave modes and diffraction of plane waves," *J. Optics (Paris)* **23** (1992), 233–235.

208. I.M.Besieris, A.M.Shaarawi and R.W.Ziolkowski, "A bidirectional travelling plane wave wave representation of exact solutions of the scalar wave equation," *J. Math. Phys.* **30** (1989), 1254–1269.

209. R.W.Ziolkowski, "Exact solutions of the wave equation with complex source locations," *J. Math. Phys* **26** (1985), 861–863.

210. P.L.Overfelt, "Bessel–Gauss pulses," *Phys. Rev.* **A 44** (1991), 3941–3947.

211. R.W.Ziolkowski, "Localized transmission of electromagnetic energy," *Phys. Rev.* **A 39** (1989), 2005–2033.

212. R.W.Ziolkowski, " Properties of electromagnetic beams generated by ultrawide bandwidth pulse-driven arrays," *IEEE Trans. Antennas Propag.* **40** (1992), 888–905.

213. R.W.Ziolkowski, D.K.Lewis and B.D.Cook, "Experimental verification of the localised wave transmission effect," *Phys. Rev. Lett.* **62** (1989), 147–150.

214. R.W.Ziolkowski and D.K.Lewis, "Verification of the localized wave transmission effect," *J. Appl. Phys.* **68** (1990), 6083–6086.

215. R.W.Ziolkowski, "Localized wave physics and engineering," *Phys. Rev.* **A 44** (1991), 3960–3984.

216. J.Durnin, J.J.Miceli and J.H.Eberly, "Durnin, Miceli and Eberly Reply," *Phys. Rev. Lett.* **66** (1991), 838.

217. R.W.Ziolkowski, "Ziolkowski Replies," *Phys. Rev. Lett.* **66** (1991), 839.

218. A.M.Shaarawi, I.M.Besieris and R.W.Ziolkowski, "A novel approach to the synthesis of nondispersive wave packet solutions to the Klein–Gordon and Dirac equation," *J. Math. Phys.* **31** (1990), 2511–2519.

219. R.W.Ziolkowski, I.M.Besieris and A.M.Shaarawi, "Localized wave representation of acoustic and electromagnetic radiation," *Proc. IEEE* **79** (1991), 1371–1378.

220. R.W.Ziolkowski, A.M.Shaarawi and L.M.Besieris, "A space-time representation of a massive, relativistic, spin zero particle," *Nucl. Phys.* **B** *(Proc. Suppl.)* **6** (1989), 255–258.

221. M.K.Tippett and R.W.Ziolkowski, "A bidirectional wave transformation of the cold plasma equations," *J. Math. Phis.* **32** (1991), 488–492.

222. E.Heyman, "Focus wave modes: a dilemma with causality," *IEEE Trans. Antennas Propag.* **37** (1989), 1604–1608.

223. P.Hillion, "*Approximation of the scalar focus wave modes*," Preprint of the Institute Henri Poincaré (Paris, 1993).

224. E.Heyman, B.Z.Steinberg and L.B.Felson, "Spectral analysis of focus wave modes (FWM)," *J. Opt. Soc. Am.* **A 4** (1987), 2081–2091.

225. P.Hillion, "The Goursat problem for the wave equation," *J. Math. Phis.* **31** (1990), 1939–1942.

226. A.Wünsche, "Embedding of focus wave modes into a wider class of approximate wave equation solutions," *J. Opt. Soc. Am.* **A 6** (1990), 1661–1668.

227. T.T.Wu, "Electromagnetic missiles," *J. Appl. Phys.* **57** (1985), 2370–2373.

228. T.T.Wu and H.Lehman, "Spreading of electromagnetic pulses," *J. Appl. Phys.* **58** (1985), 2064–2065.

229. T.T.Wu, R.W.P.King and H.-M.Shen, "Spherical lens as a launcher of electromagnetic missiles," *J. Appl. Phys.* **62** (1987), 4036–4040.

230. H.E.Moses, "The time-dependent inverse source problem for the acoustic and electromagnetic equations in the one- and three-dimensional cases," *J. Math. Phys.* **25** (1984), 1905–1923.

231. H.E.Moses and R.T.Prosser, "Initial conditions, sources and currents for prescribed time-dependent acoustic and electromagnetic fields in three dimentions. Part I: The inverse initial value problem. Acoustic and electromagnetic "bullets," expanding waves and imploding waves," *IEEE Trans. Antennas Propag.* **34** (1986), 188–196.

232. E.P.Ney, *Electromagnetism and relativity* (Harper and Row, New York and Evanston, 1962).

233. V.A.Fock, *The theory of space, time and gravitation* (Pergamon Press, New York, 1959).

234. Yu.G.Klimov, "Deviation of the light rays in the fields of the graviting galaxes," *Dok. Akad. Nauk SSSR [Sov. Phys. Dokl.]* **148** (1963), 789–792.

235. W.C.Saslaw, D.Narasimha and S.M.Chitre, "The gravitational lens as an asstronomical diagnostic," *Astron. J.* **292** (1985), 348–356.

236. S.Liebes, "Gravitational lenses," *Phys. Rev.* **B 133** (1964), 835–844.

237. W.C.Saslaw, J.Faulkner and P.A.Strittmatter, "Rapidly pulsing radio sources," *Nature (London)* **217** (1968), 1222–1227.

238. A.Einstein, "Lens-like action of a star by the deviation of light in the gravitation field," *Science* **84** (1936), 506–507.

239. F.Zwicky, "Nebulae as gravitational lenses," *Phys. Rev.* **51** (1937), 290.

240. S.M.Chitre and W.C.Saslaw,"Diagnostic properties of moving gravitational lenses," *Nature (London)* **341** (1989), 38–41.

241. M.Birkinshaw, "Moving gravitational lenses," in *Gravitational lenses*, eds. J.M.Moran, J.N.Hewitt and K.Y.Lo (Springer, Berlin, 1989), pp. 59–64.

242. D.Walsh, R.F.Carswell and R.J.Weymann, "0957 + 561 A, B: twin quasistellar objects ar gravitational lens ?" *Nature (London)* **279** (1979), 381–384.

243. D.Walsh, "0957 + 561: The unpublished story," in *Gravitational lenses*, eds. J.M.Moran, J.N.Hewitt and K.Y.Lo (Springer, Berlin, 1989), pp. 11–22.

244. R.J.Weymann, F.H.Chaffe, M.Davis, N.P.Carleton, D.Walsh and R.F.Carswell, "Multiple-mirror telescope observations of the twin QSO 0957 + 561 A, B," *Astrophys. J.* **233** (1979), L43–L46.

245. D.H.Roberts, P.E.Greenfield and B.F.Burke, "The double quasar 0957 + 561: a radio study at 6-centimaters wavelength," *Science* **205** (1979), 894–896.

246. P.Young, J.E.Gunn, J.Kristian, J.B.Oke and J.A.Westphal, "The double quasar Q 0957 + 561 A, B: a gravitational image formed by a galaxy at z = 0.39," *Astrophys. J.* **241** (1980), 507–520.

247. P.Young, J.E.Gunn, J.Kristian, J.B.Oke and J.A.Westphal, "Q 0957 + 561: detailed models of the gravitational lens effect," *Astrophys. J.* **244** (1981), 736–755.

248. R.Narayan, "Gravitational lensing, time delay and angular diameter distance," *Astrophys. J.* **378** (1991), L5–L9.

249. R.J.Weymann, D.Lathan, J.R.P.Angal, R.F.Green, J.W.Liebert, D.A.Turn- shek, D.E.Turnshek and J.A.Tyson, "The triple QSO PG 1115 + 08: another probable gravitational lens," *Nature (London)* **258** (1980), 641–643.

250. C.R.Lawrence, D.P.Schneider, M.Schmidt, C.L.Bennett, J.N.Hewitt, B.F.Bur- ke, E.L.Turner and J.E.Gunn, "Discovery of a new gravitational lens system," *Science* **223** (1984), 46–49.

251. G.I.Langston, D.P.Schneider, S.Conner, C.L.Carilli, J.Lehar, B.F.Burke, E.L.Turner, J.E.Gunn, J.N.Hewitt and M.Schmidt, "MG 1654 + 1346: an Einstein ring image of a quasar radio lobe," *Astron. J.* **97** (1989), 1283–1290.

252. G.I.Laugston, S.R.Conner, J.Lehar, B.F.Burke and K.W.Weiler, "Galaxy mass deduced from the structure of Einstein ring MG 1654 + 1346," *Nature (London)* **344** (1990), 43–45.

253. H.Fizeau, "Recherches sur les modifications que subit la vitesse de la lumière dans le verre et plusieurs autres corps solides sous l'influense de la chaleur," *Ann. Chim. Phys.* **66** (1862), 429.

254. R.J.Bell, *Introduction into Fourier spectroscopy* (Academic Press, New York, 1972).

255. L.M.Soroko, *Multiplex measurements in physics* (Moscow, Atomizdat, 1980) (in Russian).

256. J.B.Peterson and T.Falk, "Gravitational lens interference," *Astrophys. J.* **374** (1991), L5–L8.

257. L.M.Krauss and T.A.Small, "A new approach to gravitational microlensing: time delays and the galactic mass distribution," *Astrophys. J.* **378** (1991), 22–29.

258. K.Griest, "Galactic microlensing as a method of detecting massive compact halo objects," *Astrophys. J.* **366** (1991), 412–421.

259. K.Chang and S.Refsdal, "Flux variations of QSO 0957 + 561 A,B and image splitting by stars near the light path," *Nature (London)* **282** (1979), 561–564.

260. B.Grieger, R.Kayser and S.Refsdal, "A parallax effect due to gravitational microlensing," *Nature (London)* **324** (1986), 126–127.

261. B.Paczynski, "Gravitational microlensing at large optical depth," *Astrophys. J.* **301** (1986), 503–516.

262. R.Kayser, S.Refsdal and R.Stabell, "Astrophysical applications of gravitational microlensing," *Astron. Astrophys.* **166** (1986), 36–52.

263. W.D.Watson, "Gravitational microlensing," in *Gravitational lenses*, eds. J.M.Moran, J.N.Hewitt and K.Y.Lo (Springer, Berlin, 1989), pp. 195–205.

264. Ch.Vanderriest, J.Schneider, G.Herpe, M.Chevreton, G.Wlerick and M.Moles, "Optical determination of the time delay in 0957 + 561," in *Gravitational lenses*, eds. J.M.Moran, J.N.Hewitt and K.Y.Lo (Springer, Berlin, 1989), pp. 11–22.

265. L.M.Soroko, "Inverse meso-optics and its potential applications," *Commun. JINR* **E13-91-175** (Dubna, 1991).

266. W.Peatman, C.Carbone, W.Gudat, W.Heinen, P.Kuske, J.Pflüger, F.Schäfers and T.Schroeter, "The BESSY wiggler/undulator-TGM-5 beamline," *Rev. Sci. Instr.* **60** (1989), 1445–1450.

267. W.Lavander, G.Brown and T.Troxel, "Observation of X-ray undulator radiation on PEP," *Rev. Sci. Instr.* **60** (1989), 1414–1418.

268. D.Attwood, K.Halbach and K.-J.Kim, "Tunable coherent X-rays," *Science* **228** (1985), 1265–1272.

269. D.F.Alferov, Yu.A.Bashmakov and E.G.Bessonov, "On the theory of the undulator radiation," *Sov. Phys. Tech. Phys.* **18** (1974), 1337–1343.

270. H.Kitamura, "On the spectral and polarization characteristics of an undulator," *Nucl. Instrum. Methods* **177** (1980), 235–238.

271. S.Aoki and S.Kikuta, "X-ray holographic microscopy," *Jap. J. Appl. Phys.* **13** (1974), 1385–1392.

272. P.Elleaume, "Generation of various polarization states from insertion devices: a review," *Rev. Sci. Instr.* **60** (1989), 1830–1833.

273. H.Onuki, "Elliptically polarized synchrotron radiation source with crossed and retared magnetic fields," *Nucl. Instrum. Methods* **A 246** (1986), 94–98.

274. S.Yamamoto and H.Kitamura, "Generation of quasi-circularly polarized undulator radiation with higher harmonics," *Jap. J. Appl. Phys.* **26** (1987), L1613–L1615.

275. R.Tatchyn, "Variable-period electrostatic and magnetostatic undulator designs for generating polarized soft X-rays at PEP," *Rev. Sci. Instr.* **60** (1989), 2571–2578.

276. A.I.Chechin and N.V.Smolyakov, "Irregular structure undulator," *Nucl. Instrum. Methods* **A 308** (1991), 86–88.

277. H.Kitamura, "Polarization of undulator radiation," *Jap. J. Appl. Phys.* **19** (1980), L185–L188.

278. K.-J.Kim, "A synchrotron radiation source with arbitrary adjustable elliptical polarization," *Nucl. Instrum. Methods* **219** (1984), 425–429.

279. J.Goulon, P.Elleaume and D.Raoux, "Special multipole wiggler design producing circularly polarized synchrotron radiation," *Nucl. Instrum. Methods* **A 254** (1987), 192–201.

280. T.Miyahara, "Wave-optical property of undulator radiation and its effect on grating monochromators," *Jap. J. Appl. Phys.* **25** (1986), 1672–1676.

281. L.M.Soroko, "Device for production of the undulator radiation," Russia Patent No. 4,934,858 (10 January 1992), in preparation.

282. J.Kirz, "Phase zone plates for X-rays and the extreme UV," *J. Opt. Soc. Am.* **64** (1974), 301–309.

283. L.M.Soroko, "Undulator," Russia Patent No.4,934,526 (29 June 1992), in preparation.

284. L.M.Soroko, "Undulator," Russia Patent No.4,934,495 (29 June 1992), in preparation.

285. M.M.Nikitin, A.F.Medvedyev, M.B.Moiseyev and V.Ya.Epp, "Interference of synchrotron radiation," *Sov. Phys. JETP* **79** (1980), 763–774.

286. M.Benot, J.Litt and R.Meunier, "Cherenkov counters for particle identification at high energies," *Nucl. Instrum. Methods* **105** (1972), 431–444.

287. M.Benot, J.M.Howie, J.Litt and R.Meunier, "A spot-focusing Cherenkov coun- ter for the detection of multiparticle events at high energies," *Nucl. Instrum. Methods* **111** (1973), 397–412.

288. M.Benot, J.C.Bertrand, A.Maurer and R.Meunier, "A spot-focusing Cherenkov detector (SFD)," *Nucl. Instrum. Methods* **165** (1979), 439–461.

289. M.Benot and R.Meunier, "Multiplexed focusing Cherenkov counter," *Nucl. Instrum. Methods* **165** (1980), 445–455.

290. R.Meunier, "Cherenkov imaging of multiparticle events with large solid angle acceptance," *Nucl. Instrum. Methods* **A 235** (1980), 290–295.

291. G.Hatakoshi, M.Kawachi, K.Tereshima, Y.Vematsu, A.Amano and K.Veda, "-Grating radiation waves," *Opt. Lett.* **15** (1990), 1336–1338.

292. L.M.Soroko, "Gas Cherenkov counter," USSR Patent No. 1,695,240, *Bull. No.* **44** (1991), p. 175.

293. B.Gröbler, "Die (3–D)-Aperturen verschiedener Mikroskoptypen als unmittelbare Folge der zugehörigen Übertragungsgleichungen," *Optik* **83** (1989), 139–144.

294. C.W.McCutchen, "Generalized aperture and the three-dimensional diffraction image," *J. Opt.Soc.Am.* **54** (1964), 240–244.

295. G.A.Laub, G.Lenz and E.R.Reinhardt, "Three-dimensional object representation in imaging systems," *Opt. Eng.* **24** (1985), 901–905.

296. M.Francon, *Progress in microscopy* (Pergamon Press, Oxford, 1961).

297. N.Streibl, "Three-dimensional imaging by a microscope," *J. Opt. Soc. Am.* **A 2** (1985), 121–127.

298. K.A.Nugent, "Three-dimensional optical microscopy: a sampling theorem," *Opt. Commun.* **69** (1988), 15–19.

299. S.F.Gibson and F.Lanni, "Diffraction by a circular aperture as a model for three-dimensional optical microscopy," *J. Opt. Soc. Am.* **A 6** (1989), 1357–1367.

300. C.J.R.Sheppard and X.Q.Mao, "Three-dimensional imaging in a microscope," *J. Opt. Soc. Am.* **A 6** (1989), 1260–1269.

301. T.Wilson and C.Sheppard, *Theory and practice of scanning optical microscopy* (Academic Press, London, 1984).

302. C.W.Mc Cutchen, "Convolution relation with the three-dimensional diffraction image," *J. Opt. Soc. Am.* **A 8** (1991), 868–870.

303. D.A.Agard, "Optical sectioning microscopy: cellular architecture in three dimensions," *Ann. Rev. Biophys. Bioeng.* **13** (1983), 191–219.

304. O.Nakamura and S.Kawata, "Three-dimensional transfer-function analysis of the tomographic capability of a confocal fluorescence microscope," *J. Opt. Soc. Am.* **A 7** (1990), 522–526.

305. T.Wilson and S.J.Hewlett, "Superresolution in confocal scanning microscopy," *Opt. Lett.* **16** (1991), 1062–1064.

306. C.J.R.Sheppard and T.Wilson, "Image formation confocal optical systems," *SPIE* **232**, Book II (1980), 197–202.

307. T.Wilson and A.R.Carlini, "Depth discrimination criteria in confocal optical systems," *Optik* **76** (1987), 164–166.

308. M.Gu, X.Gan and C.J.R.Sheppard, "Three-dimensional coherent transfer functions in fiber-optical confocal scanning microscopes," *J. Opt. Soc. Am.* **A 8** (1991), 1019–1025.

309. K.Carlsson, P.E.Danielsson, R.Lenz, A.Liljebord, L.Majlöf and N.Åslund, "Three-dimensional microscopy using a confocal laser scanning microscope," *Opt. Lett.* **10** (1985), 53–55.

310. C.J.R.Sheppard, "Super-resolution in confocal imaging," *Optik* **80** (1988), 53–54.

311. T.Wilson and A.R.Carlini, "The images of thick step objects in the confocal scanning optical microscope," *Optik* **72** (1986), 109–114.

312. L.M.Soroko, "Meso–optical microscope as a tomographic device," *Exper. Techn. der Physik* **38** (1990), 411–424.

313. L.M.Soroko, "Meso–optical microscopes for observation of the vertical particle tracks in nuclear emulsion," *Commun. JINR* **P13-87-576** (Dubna, 1987).

314. R.Frieden, "Optical transfer of the three-dimensional object," *J. Opt. Soc. Am.* **57** (1967), 56–66.

315. R.K.Mueller, M.Kaveh and G.Wade, "Reconstructive tomography and applications to ultrasonics," *Proc. IEEE* **67** (1979), 567–587.

316. E.Abbe, "Beitrage zur Theorie des Mikroskops und der mikroskopischen Wahrnehmung," *Arch. Mikrosk. Anat. Entwicklungsmeh.* **9** (1873), 413–459.

317. T.Wilson, "Confocal light microscopy," in *Annals of the New York Academy of Sciences*, Vol. **483**, Recent Advances in Electron and Light Optical Imaging in Biology and Medicine, ed. A.P.Somlyo (The New York Academy of Sciences, New York, 1986), pp. 416–427.

318. E.Wolf, "Three-dimensional structure determination of semi-transparent objects from holographic data," *Opt. Commun.* **1** (1969), 153–156.

319. A.W.Lohmann, "Three-dimensional properties of wavefields," *Optik* **61** (1978), 105–117.

320. M.Gu and C.J.R.Sheppard, "Three-dimensional imaging in confocal fluorescent microscopy with annular lenses," *J. Mod. Opt.* **38** (1991), 2247–2263.

321. H.Ooki and J.Iwasaki, "A novel type of laser scanning microscope: theoretical considerations," *Opt. Commun.* **85** (1991), 177–182.

322. H.T.M.van der Voort, G.J.Brakenhoff and G.C.A.M.Jabssen, "Determination of the 3-dimensional optical properties of a confocal scanning laser microscope," *Optik* **78** (1988), 48–53.

323. G.J.Brakenhoff, P.Blom and P.Barends, "Confocal scanning light microscopy with high aperture immersion lenses," *J.Microsc.* **117** (1979), 219–232.

324. D.K.Hamilton and C.J.R.Sheppard, "Differential phase contrast in scanning optical microscopy," *J.Microsc.* **133**, Pt. 1, (1984), 27–39.

325. R.W.Gerchberg, "Super-resolution through error energy reduction," *Opt. Acta* **21** (1974), 709–720.

326. K.Minami, S.Kawata and S.Minami, "Superresolution of Fourier transform spectra of autoregressive model fitting with singular value decomposition," *Appl. Opt.* **24** (1985), 162–167.

327. T.Inouye, "Image reconstruction with limited angle projection data," *IEEE Trans. Nucl. Sci.* **5–26** (1979), 2666–2669.

328. G.T.Hermann, *Image reconstruction from projections* (Academic Press,New York, 1980).

329. L.M.Soroko, "Microscope," USSR Patent No. 1,273,861 (1 August 1986), *Bull. No.* **44** (1986), p. 180.

330. L.M.Soroko, "Device for measurement of the events in the nuclear photoemulsion," USSR Patent No. 782,523 (25 June 1979), *Bull. No.* **15** (1982), p. 313.

331. L.M.Soroko, "Microscope," USSR Patent No. 1,183,934 (8 June 1985), *Bull. No.* **37** (1985), p. 191.

332. R.Noch, "Messen von Kegelwinkeln mit Kreisteiltisch und Autokollimationsfernrohr," *Werkstattstechnik* **58** (1968), 163–166.

333. E.M.Morozov and N.S.Soloviev,"Optical device for surface marking-out,"USSR Patent No. 142,037 (2 December 1960), *Bull. No.* **20** (1961).

334. V.G.Zubakov and B.A.Chunin, "Interferometric control technique of the conical surfaces," USSR Patent No. 171,591 (13 April 1964), *Bull. No.* **11** (1965).

335. D.T.Puryaev, "Optical technique of the measurement of the angles," USSR Patent No. 200,165 (19 March 1966), *Bull. No.* **16** (1967).

336. L.M.Soroko, "Measurement device," USSR Patent No. 1,730,534 (15 February 1990), *Bull. No.* **16** (1992).

337. L.M.Soroko, "Device for measurement of the conical angle of the internal conical surfaces," USSR Patent No.1,737,265 (14 February 1990), *Bull. No.* **20** (1992).

338. L.M.Soroko, "Interferometer," USSR Patent No.1,620,817 (17 February 1989), *Bull. No.* **2** (1991).

339. F.Quercioli, B.Tiribilli and G.Molesini, "Optical surface profile transducer," *Opt. Eng.* **27** (1988), 135–142.

340. D.Bertani, M.Cetica and S.Ciliberto, "A fast optical profilometer," *Opt. Commun.* **46** (1983), 1–3.

341. Ch.Gorecki, G.Tribillon and J.Mignot,"Profilomètre optique en lumier blanche," *J. Opt. (Paris)* **14** (1983), 19–23.

342. G.Molesini, G.Perdini, P.Poggi and F.Quercioli, "Focus wavelength encoded optical profilometer," *Opt. Commun.* **49** (1984), 229–233.

343. L.M.Soroko, "Keratometer," USSR Patent No. 1,680,058 (15 December 1988), *Bull. No.* **36** (1991).

344. L.M.Soroko, "Device for measurement of the profile of the object," USSR Patent No. 1,753,260 (9 February 1990), *Bull. No.* **29** (1992).

345. L.M.Soroko, "Keratometer," USSR Patent No. 1,806,588 (12 October 1990), *Bull. No.* **13** (1993).

346. P.T.Landsberg and D.A.Evans, *Mathematical cosmology.An introduction* (Clarendon press, Oxford, 1979).

347. D.W.Sciama, *Modern cosmology* (Cambridge University Press,New York,1971).

348. G.Govi, *L'ottica di Claudio Tolomeo de Eugenio Ammiraglio di Sicilia-scrittore del secolo XII ridotto* 133–135.

349. H.Boegehold, "Die Abbildung durch nichtsphärische Flächen," in S.Czapski and O.Eppenstein: *Gundzüge der Theorie der optischen Instrumente nach Abbe* (J.A. Barth, Leipzig, 1924), Abschniitt IV.

350. M.von Rohr, "Die Brille," in S.Czapski and O.Eppenstein: *Gundzüge der Theorie der optischen Instrumente nach Abbe* (J.A.Barth, Leipzig, 1924), Abschniitt XIII.

351. J.Leupold, *Imm. Tietz* (1713), 6 S, 8°.

352. H.Emsmann, "Über die Construction der Anamorphisen im Kegelspiegel," *Ann. Chim. Phys.* **77** (1849), 571–574 and **85** (1852), 99–106.

353. E.Reusch, *Theorie der Cylinderlinsen* (B.G.Teubner, Leipzig, 1868).

354. C.Zeiss, "Anamorphotisches Linsensystem," D.R.P. 99722 (von 30.11.1897, 25.10.1898).

355. G.J.Burch, "Some uses of cylindrical lens-systems, including rotation of images," *Proc. R. Soc. London* **73** (1904), 281–286.

356. C.Chabrié, "Sur le principe de la construction d'un appareil d'optique destiné à obtenir de tres forts grossissements," *C. R. Acad. Sci.* **138** (1904), 265–268.

357. C.Chabrié, "Sur la function qui représente le grossissement des objets vus à travers un cône de crystal," *C. R. Acad. Sci.* **138** (1904), 349–351.

358. C.Chabrié, "Sur le diastoloscope et les resultants qu'il a permis d'obtenir," *C. R. Acad. Sci.* **138** (1904), 560–563.

359. C.Chabrié, "Sur les applications du diastoloscope à l'étude des déplacements des objets lumineux," *C. R. Acad. Sci.* **138** (1904), 799–802.

360. C.Chabrié, "Sur le diastoloscope, nouvel appareil d'optique destiné à obtenir de trés forts grossissements et à mesurer de trés petits déplacements d'objets lumineux," *Ann. Chim. Phys.* **163** (1904), 449–465.

361. J.Soldner, "Über die Ablenkung eines Lichtstrahls von seiner geradlinigen Bewegung durch die Attraktion eines Weltkorpers, an welchem er nahe vorbeigeht," *Berliner Astron. Jahrb.* **161** (1801).

362. A.Einstein,"Über den Einfluss der Schwerkraft auf die Ausbreitung des Lichtes," *Ann. Phys.* **35** (1911), 898–908.

363. A.Einstein, "Die Grundlage der allgemeinen Relativitätstheorie," *Ann. Phys.* **49** (1916), 769–822.

364. A.Edington, *Space, time and gravitation* (Harper and Brothers, New York, 1920).

365. O.Chwolson, "Über eine mögliche Form fiktiver Doppelsterne," *Astronomische Nachrichten* **221** (1924), 329–330.

366. G.A.Tikhov, "The consequences of the possible deflection of light rays in the gravitational field of the stars," *Priroda (Moscow)* **27**, No. 6 (1938), 7–13.

367. J.M.Barnothy, "History of gravitational lenses," in *Gravitational lenses*, eds. J.M. Moran, J.N.Hewitt and K.Y.Lo (Springer, Berlin, 1989), pp. 23–27.

368. B.V.Bliokh and A.A.Minakov, "Diffraction of light and lens effect of the stellar gravitational field," *Astrophysics and Space Science* **34** (1975), L7–L9.

369. O.Bryngdahl,"Computer-generated holograms as generalized optical components," *Opt. Eng.* **14** (1975), 426–435.

370. C.J.R.Sheppard and A.Choudhury, "Image formation in the scanning microscope," *Opt. Acta* **24** (1977), 1051–1073.

371. L.M.Soroko, "Optics, holography and meso–optics in bubble chamber of vertex detector," *Commun. JINR* **D1-82-642** (Dubna, 1982).

372. Gy.L.Bencze and L.M.Soroko, "Meso–optics and high energy physics," *Preprint JINR* **E13-84-310**. Submitted to ICO–13 Congress, Sapporo, 1984.

373. Ch.Frère, D.Leseberg and O.Bryngdahl, "Computer-generated holograms of three-dimensional objects composed of line segments," *J. Opt. Soc. Am.* **A 3** (1986), 726–730.

374. R.Courant and D.Hilbert, *Methods of mathematical physics* (Interscience, New York, 1962), Vol. 2.

375. P.Hillion, "Nonhomogeneous nondispersive waves: classification," *Wave motion* **16** (1992), 81–87.

376. P.Hillion, "Nondispersive waves: energy," *Int. J. Mod. Phys.* **A 8** (1993), 2507–2515.

377. A.P.Kiselev, "Modulated Gaussian beams," *Radiofisica (Soviet J. Izvestiya Visshikh Uchebnykh Zavedenii)* **26** (1983), 1014–1020.

378. P.Hillion, "Electromagnetic inhomogeneous pulses," *J. Electr. Waves and Appl.* **5** (1991), 959–969.

379. H.F.Harmuth, *Sequency theory* (Academic Press, New York, 1977).

380. H.F.Harmuth, "Propagation velocity of electromagnetic signals," *IEEE Trans. Elec. Comp. E. M. C.* **28** (1986), 267.

381. P.Hillion, "Plane waves with transverse structure," *Rev. Math. Phys.* **3** (1991), 371–378.

382. R.Bracewell, *The Fourier Transform and its Application* (Mc Graw-Hill, New York, 1965).

383. A.Papoulis, *Systems and Transforms with Applications in Optics* (Mc Graw-Hill, New York, 1968).

384. H.J.Nussbaumer,*Fast Fourier Transformation and Convolution Algorithms* (Springer, New York, 1982).

385. R.V.L.Hartley, "A more symmetrical Fourier analysis, applied to transmission problems," *Proc. IRE* **30** (1942), 144–150.

386. R.N.Bracewell, "Discrete Hartley transform," *J. Opt. Soc. Am.* **73** (1983), 1832–1835.

387. R.N.Bracewell, "The fast Hartley transform," *Proc. IEEE* **72** (1984), 1010–1018.

388. R.N.Bracewell, *The Hartley Transform* (Oxford University Press, New York,1986).

389. R.N.Bracewell, H.Bartelt, A.W.Lohmann and N.Streibl, "Optical synthesis of the Hartley transform," *Appl. Opt.* **24** (1985), 1401–1402.

390. A.B.Watson and A.Poirson, "Separable two-dimensional discrete Hartley transform," *J. Opt. Soc. Am.* **A 3** (1986), 2001–2004.

391. J.Komrska, "Abbe transform and Abbe theorem in the theory of Fraunhofer diffraction," *Cesk. Cas. Fys. (A)* **27** (1977), 579–586.

392. R.Straubel, "Über die Berechnung der Fraunhoferschen Beugungserscheinungen durch Randintegrale mit besonderer Berücksichtigung der Theorie der Beugung im Heliometer," Ph. D. thesis, Phil. Fac. zu Jena (Fromannsche Buchdruckerei, Jena, 1888), pp. 6–9.

393. J. Komrska, "Simple derivation of formulas for Fraunhofer diffraction of polygonal apertures," *J. Opt. Soc. Am.* **72** (1982), 1382–1384.

394. J. Komrska, "Fraunhofer diffraction from sector stars," *Opt. Acta* **30** (1983), 887–925.

395. J. Komrska, "Algebraic expressions of shape amplitudes of polygons and polyhedra," *Optik* **80** (1988), 171–183.

396. R.M.Silitto and W.Sillitto, "A simple Fourier approach to Fraunhofer diffraction by triangular apertures," *Opt. Acta* **22** (1988), 999.

397. W.Sillitto, "Fraunhofer diffraction at straight-edged apertures," *J. Opt. Soc. Am.* **69** (1979), 765–770.

398. N.Saga and K.Tanaka, "Relationship between the Abbe transform and the Maggi–Rubinowicz transformation formulas," *J. Opt. Soc. Am.* **A 3** (1986), 1450–1451.

399. R.Straubel, "Zwei allgemeine Sätze über Fraunhoferiśche Beugungserscheinungen," *Ann. Phys. Chemie* **56** (1895), 746–761.

400. F.M.Schwerd, *Die Beugungserscheinungen aus den Fundamentalgesetzen der Undulationstheoria analytisch entwickelt und in Bildern dargestellt* (Schwan und Goetzsche Hofbuchhandlung, Mannheim, Germany, 1835).

401. S.Ganci, "Simple derivation of formulas for Fraunhofer diffraction at polygonal apertures from Maggi–Rubinowicz transformation," *J. Opt. Soc. Am.* **A 1** (1984), 559–561.

402. J.S.Asvestas, "Line integrals and physical optics. Part I. The transformation of the solid-angle surface integral to a line integral," *J. Opt. Soc. Am.* **A 2** (1985), 891–895.

403. G.Otis, J.-L.Lachambre and P.Lavigne, "Focusing of laser beams by a sequence of irises," *Appl. Opt.* **18** (1979), 875–883.

404. R.C.Smith and J.S.March, "Diffraction patterns of simple apertures," *J. Opt. Soc. Am.* **64** (1974), 798–803.

405. Yu.A.Batusov, L.M.Soroko and V.V.Tereshchenko, "Stereoscopic meso–optical Fourier transform microscope with double focusing," *Commun. JINR* **D13-94-478** (Dubna, 1994).

406. Yu.A.Batusov, L.M.Soroko and V.V.Tereshchenko, "Experiments on meso–optical Fourier transform microscope with double focusing," *Rad. Meas.* **25** (1995), 753–756.

407. L.M.Soroko, "Isoplanatic meso–optical Fourier transform microscope for nuclear emulsion," *Commun. JINR* **E13-95-147** (Dubna, 1995).

408. Yu.A.Batusov, V.P.Rumyantseva, L.M.Soroko and V.V.Tereshchenko, "Investigation of the geometrical distortions in the nuclear emulsion," *Commun. JINR* **E13-94-502** (Dubna, 1994).

409. S.Refsdal, "The gravitational lens effect," *Mon. Not. R. Astron. Soc.*, **128** (1964), 295–306.

410. S.Refsdal, "On the possibility of determining Hubble's parameter and the masses of galaxies from the gravitational lens effect," *Mon. Not. R. Astron. Soc.* **128** (1964), 307–310.

411. S.Refsdal, "On the possibility of testing cosmological theories from the gravitational lens effect," *Mon. Not. R. Astron. Soc.* **132** (1966), 101–111.

412. T.W.Noonan, "The effect of the Einstein light deflection on the observed properties of clusters of galaxies," *Astronom. J.* **76** (1971), 765–767.

413. J.H.Cooke and R.Kantowski, "Time delays for multiply imaged quasars," *Astrophys. J.* **195** (1975), L11–L14.

414. U.Borgeest, "The difference in light travel time between gravitational lens images. II. Theoretical foundations," *Astron. Astrophys.* **128** (1983), 162–164.

415. R.Narayan, R.Blandford and R.Nityananda, "Multiple imaging of quasars by galaxies and clusters," *Nature (London)* **310** (1984), 112–115.

416. B.Paczyński, "Gravitational microlensing by the galactic halo," *Astrophys. J.* **304** (1986), 1–5.

417. U.Borgeest, "Determination of galaxy masses by the gravitational lens effect," *Astrophys. J.* **309** (1986), 467–471.

418. M.A.Walker, "Microlensed image motions," *Astrophys. J.* **453** (1995), 37–39.

419. B.Paczyński, "Giant luminous arcs discovered in two clusters of galaxies," *Nature (London)* **325** (1987), 572–573.

420. L.Nottale, "Gravitational lensing in observational cosmology," *Annales de Physique (Ann. Phys. Fr.)* **13** (1988), 223–260.

421. F.Hammer and F.Rigant, "Giant luminous arcs from lensing: determination of the mass distribution inside distant cluster cores," *Astron. Astrophys.* **226** (1989), 45–56.

422. S.Mao and B.Paczyński, "Gravitational microlensing by double stars and planetary systems," *Astrophys. J.* **374** (1991), L37–L40.

423. C.McGill, "The redshift projection. I. Caustics and correlation functions," *Mon. Not. R. Astron. Soc.* **242** (1990), 428–438.

424. J.A.Tyson, F.Valdes and R.A.Wenk, "Detection of systematic gravitational lens galaxy image alignments: mapping dark matter in galaxy clusters," *Astrophys. J.* **349** (1990), L1–L4.

425. B.Paczyński, "Gravitational microlensing of the Galactic bulge stars," *Astrophys. J.* **371** (1991), L63–L67.

426. A.Gould, "Extending the MACHO search to $\sim 10^6 M_\odot$," *Astrophys. J.* **392** (1992), 442–451.

427. R.Pelló, J.F.LeBorgne, B.Sanahuja, G.Mathez and B.Fort, "The system of arcs in the cluster of galaxies Abell 2218: photometry, spectroscopy and geometry," *Astron. Astrophys.* **266** (1992), 6–14.

428. R.Narayan and S.Wallington, "Introduction to basic concepts of gravitational lensing," In: *Gravitational lenses*, eds.R.Kayser, T.Schramm, L.Nieser (Springer, 1992), pp. 12–26.

429. S.Refsdal, "Determining the Hubble parameter," In: *Gravitational lenses*, eds. R.Kayser, T.Schramm, L.Nieser (Springer, 1992), pp. 61–66.

430. R.Kayser, "Gravitational microlensing," In: *Gravitational lenses*, eds.R.Kayser, T.Schramm, L.Nieser (Springer, 1992), pp. 143–155.

431. J.Wambsganss, "Expected color variations of the microlensed QSO 2237 + 0305," In: *Gravitational lenses*, eds. R.Kayser, T.Schramm, L.Nieser (Springer, 1992), pp. 183–188.

432. B.F.Burke, J.Lehàr and S.R.Conner, "Einstein rings and related phenomena," In: *Gravitational lenses*, eds. R.Kayser, T.Schramm, L.Nieser (Springer, 1992), pp. 237–244.

433. M.Bartelmann, "Gravitational lensing by large-scale structures," In: *Gravitational lenses*, eds. R.Kayser, T.Schramm, L.Nieser (Springer, 1992), pp. 345–349.

434. R.D.Blandford, "Concluding summary," In: *Gravitational lenses*, eds.R.Kayser, T.Schramm, L.Nieser (Springer, 1992), pp. 350–361.

435. J.-P.Kneib, Y.Mellier, B.Fort and G.Mathez, "The distribution of dark matter in distant cluster-lenses: Modeling A 370," *Astron. Astrophys.* **273** (1993), 367–376.

436. C.Alcock et al., "Possible gravitational microlensing of a star in the Large Magellanic Cloud," *Nature (London)* **365** (1993), 621–623.

437. E.Aubourg et al., "Evidence for gravitational microlensing by dark objects in the Galactic halo," *Nature (London)* **365** (1993), 623–625.

438. E.Maoz, "Gravitational microlensing by dark clusters in the Galactic halo," *Astrophys. J.* **428** (1994), L5–L8.

439. J.Miralda-Escudè, "Statistics of highly magnified gravitational images in clusters of galaxies: I. Implications for the clusters," *Astrophys. J.* **403** (1993), 497–508.

440. J.Miralda-Escudè, "Statistics of highly magnified gravitational images in clusters of galaxies: II. Implications for the sources," *Astrophys. J.* **403** (1993), 509–516.

441. S.Wallington and R.Narayan, "The influence of core radius on gravitational lensing by elliptical lenses," *Astrophys. J.* **403** (1993), 517–529.

442. H.J.Witt, "An efficient method to compute microlensed light curves for point sources," *Astrophys. J.* **403** (1993), 530–541.

443. K.C.Sahu, "Stars within the Large Magellanic Cloud as potential lenses for observed microlensing events," *Nature (London)* **370** (1994), 275–276.

444. A.Gould, "Proper motions of MACHOs," *Astrophys. J.* **421** (1994), L71–L74.

445. A.Gould, "MACHO velocities from satellite-based parallaxes," *Astrophys. J.* **421** (1994), L75–L78.

446. A.Gould, J.Miralda-Escudè and J.N.Bahcall, "Microlensing events: thin disk, thick disk, or halo?," *Astrophys. J.* **423** (1994), L105–L108.

447. D.Maoz and A.Gould, "A spectroscopic method to measure MACHO proper motions," *Astrophys. J.* **425** (1994), L67–L71.

448. K.Griest, "Galactic microlensing as a method of detecting massive compact halo objects," *Astrophys. J.* **366** (1991), 412–421.

449. S.Mao and B.Paczyński, "Gravitational microlensing by double stars and planetary systems," *Astrophys. J.* **374** (1991), L37–L40.

450. E.Maoz, "Gravitational microlensing by dark clusters in the Galactic halo," *Astrophys. J.* **428** (1994), L5–L8.

451. A.Gould and J.Villumsen, "Weak lensing by nearby structures," *Astrophys. J.* **428** (1994), L45–L48.

452. A.Gould and A.Loeb, "Discovering planetary systems through gravitational microlenses," *Astrophys. J.* **396** (1992), 104–114.

453. B.Paczyński et al., "Are the OGLE microlenses in the Galactic bar?," *Astrophys. J.* **435** (1994), L113–L116.

454. F.K.Baganoff and M.A.Malkan, "Gravitational microlensing is not required to explain quasar variability," *Astrophys. J.* **444** (1996), L13–L15.

455. A.Udalski et al., "The optical gravitational lensing experiment: OGLE No. 7: Binary microlens or a new unusual variable?," *Astrophys. J.* **436** (1994), L103–L106.

456. A.D.Bolatto and E.E.Falco, "The detectability of planetary companions of compact Galactic objects from their effects on microlensed light curves of distant stars," *Astrophys. J.* **436** (1994), 112.

457. N.Kaiser, "Nonlinear clusters lens reconstruction," *Astrophys. J.* **439** (1995), L1–L3.

458. S.Mao and R.DiStefano, "Interpretation of gravitational microlensing by binary systems," *Astrophys. J.* **440** (1995), 22–27.

459. H.S.Zhao, D.N.Spergel and R.M.Rich, "Microlensing by the galactic bar," *Astrophys. J.* **440** (1995), L13–L16.

460. A.Gould, "MACHO parallaxes from a single satellite," *Astrophys. J.* **441** (1995), L21–L24.

461. A.Gould, "Self-lensing by a stellar disk," *Astrophys. J.* **441** (1995), 77–78.

462. M.Kamionkowski, "Microlensing by stars," *Astrophys. J.* **442** (1995), L9–L12.

463. A.Gould, "Transverse velocities of galaxies from microlens parallaxes," *Astrophys. J.* **444** (1995), 556–560.

464. C.Alcock et al. (The MACHO Collaboration), "Probable gravitational microlensing toward the galactic bulge," *Astrophys. J.* **445** (1995), 133–139.

465. A.Gould, "Self-lensing by binaries," *Astrophys. J.* **446** (1995), 541–542.

466. M.Bartelmann and R.Narayan, "The lens parallax method: determining redshifts of faint blue galaxies through gravitational lensing," *Astrophys. J.* **451** (1995), 60–75.

467. A.Gould and J.Villumsen, "Weak lensing by nearby structures," *Astrophys. J.* **428** (1994), L45–L48.

468. M.V.Sazhin and A.M.Cherepashchuk, "Microlensing of double and multiple stars," *Letters to Astronomical J. (Moscow)* **20** (1994), 613–619.

469. M.B.Bogdanov and A.M.Cherepashchuk, "Gravitational microlensing of stars of nonzero angular size," *Astronomical J. (Moscow)* **72** (1995), 873–878.

470. M.V.Sazhin, V.V.Shul'ga and N.P.Gorbatko, "Search for dark objects in the Galaxy," *Astronomical J. (Moscow)* **71** (1994), 837–845.

471. J.Wambsganss and T.Kundić, "Gravitational microlensing by random motion of stars: analysis of light curves," *Astrophys. J.* **450** (1995), 19–26.

472. M.Kiraga and B.Paczyński, "Gravitational microlensing of the Galactic bulge stars," *Astrophys. J.* **430** (1994), L101–L104.

473. A.V.Mandzhos, "Mutual interference and structural properties of images of the source in vicinity of the gravitational lens cusp point. The image structure at the cusp point," *Astronomical J. (Moscow)* **72** (1995), 146–152.

474. A.V.Mandzhos, "Mutual interference and structural properties of images of the source in vicinity of the gravitational lens cusp point. The degree of mutual coherence of images," *Astronomical J. (Moscow)* **72** (1995), 153–160.

475. V.N.Shalyapin, "Calculation of the microlensing effect by the particle-network method," *Astronomical J. (Moscow)* **72** (1995), 668–673.

SUBJECT INDEX

AUTHOR INDEX